Lecture Notes in Computer Science　　10250

Commenced Publication in 1973
Founding and Former Series Editors:
Gerhard Goos, Juris Hartmanis, and Jan van Leeuwen

More information about this series at http://www.springer.com/series/7409

Eva Blomqvist · Diana Maynard
Aldo Gangemi · Rinke Hoekstra
Pascal Hitzler · Olaf Hartig (Eds.)

The Semantic Web

14th International Conference, ESWC 2017
Portorož, Slovenia, May 28 – June 1, 2017
Proceedings, Part II

Springer

Editors
Eva Blomqvist
Linköping University
Linköping
Sweden

Diana Maynard
University of Sheffield
Sheffield
UK

Aldo Gangemi
Paris Nord University
Paris
France

Rinke Hoekstra
Vrije Universiteit Amsterdam
Amsterdam
The Netherlands

Pascal Hitzler
Wright State University
Dayton, OH
USA

Olaf Hartig
Linköping University
Linköping
Sweden

ISSN 0302-9743   ISSN 1611-3349   (electronic)
Lecture Notes in Computer Science
ISBN 978-3-319-58450-8   ISBN 978-3-319-58451-5   (eBook)
DOI 10.1007/978-3-319-58451-5

Library of Congress Control Number: 2017939119

LNCS Sublibrary: SL3 – Information Systems and Applications, incl. Internet/Web, and HCI

Printed on acid-free paper

This Springer imprint is published by Springer Nature
The registered company is Springer International Publishing AG
The registered company address is: Gewerbestrasse 11, 6330 Cham, Switzerland

Preface

This volume contains the main proceedings of the ESWC 2017 conference. The ESWC conference is established as a yearly major venue for discussing the latest scientific results and technology innovations related to the Semantic Web and linked data. At ESWC, international scientists, industry specialists, and practitioners meet to discuss the future of applicable, scalable, user-friendly, as well as potentially game-changing solutions. This 14th edition took place from May 28 to June 1, 2017, in Portorož (Slovenia). Building on its past success, ESWC is also a venue for broadening the focus of the Semantic Web community to span other relevant research areas in which semantics and Web technology plays an important role. Thus, the chairs of ESWC 2017 organized two special tracks putting particular emphasis on usage areas where Semantic Web technologies are facilitating a leap of progress, namely: "Multilinguality" and "Semantic Web and Transparency."

Emerging from its roots in AI and Web technology, the Semantic Web today is mainly a Web of linked data, upon which a plethora of services and applications for all possible domains are being proposed. Some of the core challenges that the Semantic Web aims at addressing are the heterogeneity of content and its volatile and rapidly changing nature, its uncertainty, provenance, and varying quality. This in combination with more traditional disciplines — such as logical modelling and reasoning, natural language processing, databases and data storage and access, machine learning, distributed systems, information retrieval and data mining, social networks, Web science and Web engineering — shows the span of topics covered by this conference. The nine regular tracks, in combination with an in-use and applications track, and the two special tracks, constituted the main technical program of ESWC 2017.

The program also included three exciting invited keynotes. Lora Aroyo (Professor Human Computer Interaction Vrije Universiteit Amsterdam, The Netherlands, and Visiting Professor at Columbia University in the City of New York, USA) focused on the notion of ambiguity, discussing how ambiguity can be captured and even taken advantage of by capturing the diversity of interpretations. In particular, she discussed how to capture this diversity and how to allow machines to deal with it. Kevin Crosbie (Chief Product Officer, RavenPack) discussed the role of semantic intelligence in financial markets. In particular, he focused on the challenges of turning unstructured content into structured data, given that many new kinds of alternative data (social media, satellite imagery, etc.) are being used to complement traditional data for use in predictive modelling for financial trading algorithms. John Sheridan (Digital Director, The National Archives) discussed the use and benefits of Semantic Web technologies for digital archiving, in particular for managing heterogeneous metadata, dealing with uncertainty, and in areas such as provenance and trust.

The main scientific program of the conference comprised 51 papers: 40 research and 11 in-use and application papers, selected out of 183 reviewed submissions, which corresponds to an acceptance rate of 25% for the research papers submitted and 52%

for the in-use papers. A special thanks goes to our process improvement chair, Derek Doran, who helped us establish an improved quality assurance process during the paper selection, ensuring the originality and quality of the research papers that were accepted to the conference. This program was completed by a demonstration and poster session, in which researchers had the chance to present their latest results and advances in the form of live demos. In addition, the PhD Symposium program included ten contributions, selected out of 14 submissions.

This year's edition of ESWC's main scientific program presented a significant number of research papers with a focus on solving typical Semantic Web problems, such as entity linking, discoverability, etc., by using methods and techniques from areas such as machine learning and natural language processing, and reflecting in particular the current interest in deep learning. Work on both the fundamental development and use of Semantic Web technologies in relation to transparency is particularly interesting in view of current governmental and institutional open data initiatives.

The conference program also offered 12 workshops, six tutorials, and an EU Project Networking session. This year, an open call also allowed us to select and support five challenges. These associated events create an even more open, multidisciplinary, and cross-fertilizing environment at the conference, allowing for work-in-progress and practical results to be discussed. Workshops ranged from domain-focused topics, including the biomedical, scientific publishing, e-science, robotics, distributed ledgers, and scholarly fields, to more technology-focused topics ranging from RDF stream processing, query processing, data quality and data evolution, to sentiment analysis and semantic deep learning. Tutorial topics spanned NLP, ontology engineering, and linked data, including specific tutorials on knowledge graphs and rule-based processing of data. Proceedings from these satellite events are available in a separate volume.

The General and Program Committee chairs would like to thank the many people who were involved in making ESWC 2017 a success. First of all, our thanks go to the 24 track chairs and 360 reviewers, including 49 external reviewers, for ensuring a rigorous blind review process that led to an excellent scientific program and an average of four reviews per article. The scientific program was completed by an exciting selection of posters and demos chaired by Katja Hose and Heiko Paulheim.

Special thanks go to the PhD symposium chairs, Rinke Hoekstra and Pascal Hitzler, who managed one of the key events at ESWC, the PhD symposium. The brilliant PhD students will become the future leaders of our field, and deserve both encouragement and mentoring, which Rinke and Pascal made sure we could provide. We also had a great selection of workshops and tutorials, as mentioned earlier, thanks to the commitment of our workshop chairs, Agnieszka Ławrynowicz and Fabio Ciravegna, and tutorial chairs, Anna Lisa Gentile and Sebastian Rudolph.

Thanks to our EU Project Networking session chairs, Lyndon Nixon and Maria Maleshkova, we had the opportunity to facilitate meetings and exciting discussions between leading European research projects. Networking and sharing ideas between projects is a crucial success factor for such large research projects.

We are additionally grateful for the work and commitment of Monika Solanki, Mauro Dragoni, and all the individual challenges chairs, who successfully established a challenge track. The five challenges provided researchers and practitioners with the opportunity to compare their latest solutions in these challenge areas, ranging from topic-focused tasks such as question answering and semantic sentiment analysis, to practical tasks such as storage, semantic publishing, and open knowledge extraction.

We thank STI International for supporting the conference organization, and particularly Alexander Wahler as the conference treasurer. YouVivo GmbH deserve special thanks for the professional support of the conference organization, and for solving all practical matters. Our local chair Marko Grobelnik also deserves a special thanks for selecting the venue and for, together with his local organizers, Marija Kokelj, Monika Kropej, and Spela Sitar, arranging a great on-site experience for our conference attendees.

Further, we are very grateful to Ruben Verborgh, our publicity chair, who kept our community informed throughout the year, and Venislav Georgiev, who administered the website. Of course we also thank our sponsors, listed on the next pages, for their vital support of this edition of ESWC. We would like to stress the great work achieved by the Semantic Technologies coordinators Lionel Medini and Luigi Asprino, who maintained and updated our ESWC mobile app and published our conference dataset. A special thanks also to our proceedings chair, Olaf Hartig, who did an excellent job in preparing this volume with the kind support of Springer.

March 2017

Eva Blomqvist
Diana Maynard
Aldo Gangemi

Organization

Organizing Committee

General Chair

Eva Blomqvist   Linköping University, Sweden

Program Chairs

Diana Maynard   University of Sheffield, UK
Aldo Gangemi   Paris Nord University, France and ISTC-CNR, Italy

Workshops Chairs

Agnieszka Ławrynowicz   Poznan University of Technology, Poland
Fabio Ciravegna   University of Sheffield, UK

Poster and Demo Chairs

Katja Hose   Aalborg University, Denmark
Heiko Paulheim   University of Mannheim, Germany

Tutorials Chairs

Anna Lisa Gentile   University of Mannheim, Germany
Sebastian Rudolph   TU Dresden, Germany

PhD Symposium Chairs

Rinke Hoekstra   Vrije Universiteit Amsterdam, The Netherlands
Pascal Hitzler   Wright State University, USA

Challenge Chairs

Monika Solanki   University of Oxford, UK
Mauro Dragoni   Fondazione Bruno Kessler, Italy

Semantic Technologies Coordinators

Lionel Medini   University of Lyon, France
Luigi Asprino   University of Bologna, Italy

EU Project Networking Session Chairs

Lyndon Nixon   Modul Universität Vienna, Austria
Maria Maleshkova   Karlsruhe Institute of Technology (KIT), Germany

Process Improvement Chair

Derek Doran   Wright State University, USA

Publicity Chair

Ruben Verborgh   Ghent University, Belgium

Web Presence

Venislav Georgiev   STI International, Austria

Proceedings Chair

Olaf Hartig   Linköping University, Sweden

Treasurer

Alexander Wahler   STI, Austria

Local Organization Chair

Marko Grobelnik   Jožef Stefan Institute, Slovenia

Local Organization and Conference Administration

Katharina Vosberg   YouVivo GmbH, Germany
Marija Kokelj   PITEA, Slovenia
Monika Kropej   Jožef Stefan Institute, Slovenia
Spela Sitar   Jožef Stefan Institute, Slovenia

Program Committee

Program Chairs

Diana Maynard   University of Sheffield, UK
Aldo Gangemi   Paris Nord University, France and ISTC-CNR, Italy

Track Chairs

Vocabularies, Schemas, Ontologies

Helena Sofia Pinto   Universidade de Lisboa, Portugal
Silvio Peroni   University of Bologna, Italy

Reasoning

Uli Sattler   University of Manchester, UK
Umberto Straccia   ISTI-CNR, Italy

Linked Data

Jun Zhao   University of Oxford, UK
Axel Ngonga Ngomo   Universität Leipzig, Germany

Social Web and Web Science

Harith Alani   The Open University, UK
Wolfgang Nejdl   Leibniz Universität Hannover, Germany

Semantic Data Management, Big Data, Scalability

Maria Esther Vidal   University of Bonn, Germany and Universidad Simón
                              Bolívar, Venezuela
Jürgen Umbrich   Vienna University of Economics and Business, Austria

Natural Language Processing and Information Retrieval

Claire Gardent   CNRS, France
Udo Kruschwitz   University of Essex, UK

Machine Learning

Claudia d'Amato   University of Bari, Italy
Michael Cochez   Fraunhofer Institute for Applied Information
                              Technology FIT, Germany

Mobile Web, Sensors and Semantic Streams

Emanuele Della Valle   Politecnico di Milano, Italy
Manfred Hauswirth   TU Berlin, Germany

Services, APIs, Processes and Cloud Computing

Peter Haase   metaphacts GmbH, Germany
Barry Norton   Elsevier, UK

Multilinguality

Philipp Cimiano   Universität Bielefeld, Germany
Roberto Navigli   Sapienza University of Rome, Italy

Semantic Web and Transparency

Mathieu d'Aquin   The Open University, UK
Giorgia Lodi   CNR, Italy

In-Use and Industrial Track

Paul Groth   Elsevier Labs, The Netherlands
Paolo Bouquet   Trento University, Italy

Members (All Tracks)

Karl Aberer
Maribel Acosta
Alessandro Adamou
Nitish Aggarwal
Mehwish Alam
Harith Alani
Jose Julio Alferes
Muhammad Intizar Ali
Marjan Alirezaie
Carlo Allocca
Grigoris Antoniou
Alessandro Artale
Sören Auer
Nathalie Aussenac-Gilles
Franz Baader
Michele Barbera
Valerio Basile
Chris Biemann
Antonis Bikakis
Eva Blomqvist
Fernando Bobillo
Piero Bonatti
Kalina Bontcheva
Alex Borgida
Stefan Borgwardt
Paolo Bouquet
Charalampos Bratsas
John Breslin
Christopher Brewster
Carlos Buil Aranda
Paul Buitelaar
Davide Buscaldi
Joseph Busch
Raf Buyle
Elena Cabrio
Jean-Paul Calbimonte
Diego Calvanese
Andrea Calì
Matteo Cannaviccio
Amparo E. Cano
Pompeu Casanovas
Giovanni Casini
Michele Catasta

Irene Celino
Ismail Ilkan Ceylan
Pierre-Antoine Champin
Thierry Charnois
Gong Cheng
Christian Chiarcos
Key-Sun Choi
Philip Cimiano
Michael Cochez
Sergio Consoli
Olivier Corby
Oscar Corcho
Gianluca Correndo
Fabio Cozman
Christophe Cruz
Claudia d'Amato
Enrico Daga
Laura M. Daniele
Jérôme David
Brian Davis
Victor de Boer
Giuseppe De Giacomo
Gerard de Melo
Stefan Decker
Thierry Declerck
Makx Dekkers
Emanuele Della Valle
Tommaso Di Noia
Stefan Dietze
John Domingue
Derek Doran
Mauro Dragoni
Alistair Duke
Michel Dumontier
Mathieu d'Aquin
Maud Ehrmann
Henrik Eriksson
Vadim Ermolayev
Jérôme Euzenat
Nicola Fanizzi
Catherine Faron Zucker
Bettina Fazzinga
Miriam Fernandez

Javier D. Fernández
Sebastien Ferre
Besnik Fetahu
Giorgos Flouris
Antske Fokkens
Achille Fokoue
Muriel Foulonneau
Enrico Francesconi
Irini Fundulaki
Fabien Gandon
Aldo Gangemi
Roberto Garcia
Raúl García-Castro
Claire Gardent
Daniel Garijo
Anna Lisa Gentile
Chiara Ghidini
Alain Giboin
Claudio Giuliano
Birte Glimm
Jose Manuel Gomez-Perez
Julio Gonzalo
Rafael S. Gonçalves
Jorge Gracia
Alasdair Gray
Paul Groth
Tudor Groza
Francesco Guerra
Alessio Gugliotta
Giancarlo Guizzardi
Víctor Gutiérrez Basulto
Christophe Guéret
Asunción Gómez-Pérez
Peter Haase
Siegfried Handschuh
Andreas Harth
Olaf Hartig
Matthias Hartung
Oktie Hassanzadeh
Manfred Hauswirth
Conor Hayes
Johannes Heinecke
Benjamin Heitmann
Sebastian Hellmann
Pascal Hitzler
Rinke Hoekstra

Aidan Hogan
Laura Hollink
Matthew Horridge
Katja Hose
Geert-Jan Houben
Eero Hyvönen
Yazmin Angelica Ibanez-Garcia
Valentina Janev
Krzysztof Janowicz
Mustafa Jarrar
Anja Jentzsch
Clement Jonquet
Md. Rezaul Karim
Tomi Kauppinen
Takahiro Kawamura
Carsten Keßler
Ali Khalili
Sabrina Kirrane
Matthias Klusch
Craig Knoblock
Magnus Knuth
Boris Konev
Jacek Kopecky
Manolis Koubarakis
Adila A. Krisnadhi
Anastasia Krithara
Udo Kruschwitz
Patrick Lambrix
Christoph Lange
Michael Lauruhn
Alberto Lavelli
Agnieszka Lawrynowicz
Danh Le Phuoc
Domenico Lembo
David Lewis
Giorgia Lodi
Steffen Lohmann
Nuno Lopes
Vanessa Lopez
Nikolaos Loutas
Chun Lu
Markus Luczak-Roesch
Ioanna Lytra
Bernardo Magnini
Maria Maleshkova
Pierre Maret

Nicolas Matentzoglu
Andrea Maurino
Diana Maynard
Suvodeep Mazumdar
John P. Mccrae
Fiona McNeill
Nandana Mihindukulasooriya
Daniel Miranker
Riichiro Mizoguchi
Dunja Mladenic
Marie-Francine Moens
Pascal Molli
Elena Montiel-Ponsoda
Gabriela Montoya
Federico Morando
Yassine Mrabet
Claudia Müller-Birn
Hubert Naacke
Ndapandula Nakashole
Amedeo Napoli
Roberto Navigli
Wolfgang Nejdl
Axel Ngonga
Matthias Nickles
Andriy Nikolov
Olaf Noppens
Barry Norton
Andrea Giovanni Nuzzolese
Leo Obrst
Alessandro Oltramari
Magdalena Ortiz
Francesco Osborne
Matteo Palmonari
Jeff Z. Pan
Heiko Paulheim
Terry Payne
Laura Perez-Beltrachini
Silvio Peroni
Rafael Peñaloza
H. Sofia Pinto
Vassilis Plachouras
Axel Polleres
Simone Paolo Ponzetto
María Poveda-Villalón
Valentina Presutti
Laurette Pretorius

Cédric Pruski
Yuzhong Qu
Filip Radulovic
Diego Reforgiato Recupero
Georg Rehm
Achim Rettinger
Chantal Reynaud
Mikko Rinne
Petar Ristoski
Carlos R. Rivero
Giuseppe Rizzo
Riccardo Rosati
Marco Rospocher
Camille Roth
Marie-Christine Rousset
Ana Roxin
Sebastian Rudolph
Harald Sack
Hassan Saif
Sherif Sakr
Muhammad Saleem
Felix Sasaki
Ulrike Sattler
Luigi Sauro
Vadim Savenkov
Monica Scannapieco
Francois Scharffe
Ansgar Scherp
Stefan Schlobach
Jodi Schneider
Pavel Shvaiko
Gerardo Simari
Elena Simperl
Hala Skaf-Molli
Charese Smiley
Monika Solanki
Steffen Staab
Giorgos Stamou
Yannis Stavrakas
Luc Steels
Giorgos Stoilos
Audun Stolpe
Umberto Straccia
York Sure-Vetter
Mari Carmen Suárez-Figueroa
Vojtěch Svátek

Pedro Szekely
Anders Søgaard
Dhavalkumar Thakker
Martin Theobald
Allan Third
Keerthi Thomas
Ilaria Tiddi
Thanassis Tiropanis
Ioan Toma
David Toman
Alessandra Toninelli
Anna Tordai
Yannick Toussaint
Cassia Trojahn
Dmitry Tsarkov
Giovanni Tummarello
Juergen Umbrich
Jacopo Urbani
Ricardo Usbeck
Marieke Van Erp
Frank Van Harmelen

Jacco van Ossenbruggen
Paola Velardi
Chiara Veninata
Ruben Verborgh
Guido Vetere
Maria Esther Vidal
Carlos Viegas Damásio
Serena Villata
Holger Wache
Simon Walk
Shenghui Wang
Nick Webb
Chris Welty
Erik Wilde
Frank Wolter
Feiyu Xu
Peter Yeh
Amrapali Zaveri
Jun Zhao
Antoine Zimmermann

Additional Reviewers

Andrejs Abele
Manuel Atencia
Sotiris Batsakis
Julia Bosque-Gil
Elena Botoeva
Markus Brenner
Benjamin Cogrel
Zlatan Dragisic
Lukas Eberhard
Amosse Edouard
Mezghani Emna
Ronald Ferguson
Johannes Frey
Jinlong Guo
Amit Gupta
Amelie Gyrard
Lavdim Halilaj
Michael Hoffmann
Aidan Hogan
Filip Ilievski
Prateek Jain

Daniel Janke
Amit Joshi
Unmesh Joshi
Naouel Karam
Nazifa Karima
Amit Kirschenbaum
Ruediger Klein
Sarah Kohail
Philipp Koncar
Albert Meroño-Peñuela
Diego Moussallem
Frank Nack
Yaroslav Nechaev
Chifumi Nishioka
Inna Novalija
Jonas Oppenländer
Enrico Palumbo
Alexander Panchenko
George Papadakis
Peter Patel-Schneider
Viviana Patti

Minh Tran Pham
Guangyuan Piao
Andreas Pieris
Danae Pla Karidi
Gustavo Publio
José Luis Redondo-García
Sebastian Ruder
Anisa Rula
Hassan Saif
Ihab Salawdeh
Valerio Santarelli
Marvin Schiller
Lukas Schmelzeisen

Panayiotis Smeros
Tommaso Soru
Ilias Tachmazidis
Steffen Thoma
Trung-Kien Tran
Despoina Trivela
Sahar Vahdati
Massimo Vitiello
Binh Vu
Guohui Xiao
Benjamin Zarrieß
Lu Zhou

PhD Symposium Program Committee

Chairs

Rinke Hoekstra	Vrije Universiteit Amsterdam, The Netherlands
Pascal Hitzler	Wright State University, USA

Members

Chris Biemann	TU Darmstadt, Germany
Michelle Cheatham	Wright State University, USA
Philipp Cimiano	Bielefeld University, Germany
Claudia D'Amato	University of Bari, Italy
Chiara Di Francescomarino	Fondazione Bruno Kessler-IRST, Italy
Anna Lisa Gentile	IBM Research Almaden, USA
Chiara Ghidini	Fondazione Bruno Kessler-IRST, Italy
Pascal Hitzler	Wright State University, USA
Rinke Hoekstra	University of Amsterdam/VU University Amsterdam, The Netherlands
Aidan Hogan	DCC, Universidad de Chile, Chile
Krzysztof Janowicz	University of California, Santa Barbara, USA
Agnieszka Lawrynowicz	Poznan University of Technology, Poland
Matteo Palmonari	University of Milano-Bicocca, Italy
Axel Polleres	Vienna University of Economics and Business, WU Wien, Austria
Stefan Schlobach	Vrije Universiteit Amsterdam, The Netherlands

Jodi Schneider	University of Illinois Urbana Champaign, USA
Juan Sequeda	Capsenta Labs, USA
Monika Solanki	University of Oxford, UK
Vojtěch Svátek	University of Economics, Prague, Czech Republic
Danai Symeonidou	INRA, France
Serena Villata	CNRS, Laboratoire d'Informatique, Signaux et Systèmes de Sophia-Antipolis, France

Steering Committee

Chair

John Domingue	The Open University, UK and STI International, Austria

Members

Claudia d'Amato	Universià degli Studi di Bari, Italy
Mathieu d'Aquin	Knowledge Media Institute KMI, UK
Philipp Cimiano	Bielefeld University, Germany
Oscar Corcho	Universidad Politécnica de Madrid, Spain
Fabien Gandon	Inria, W3C, Ecole Polytechnique de l'Université de Nice Sophia Antipolis, France
Valentina Presutti	CNR, Italy
Marta Sabou	Vienna University of Technology, Austria
Harald Sack	Hasso Plattner Institute (HPI), Germany

Sponsoring Institutions

Gold Sponsors

http://www.iospress.nl/

http://www.sti2.org/

Silver Sponsors

https://www.elsevier.com/

Abstract of Keynotes

Bringing Semantic Intelligence to Financial Markets

Kevin Crosbie

RavenPack, New York, US

Abstract. The most successful hedge-funds in today's financial markets are consuming large amounts of alternative data, including satellite imagery, point-of-sale data, news, social media and publications from the web. This new trend is driven by the fact that traditional factors have become less predictive in recent years, requiring sophisticated investors to explore new data sources. The majority of this new alternative content is unstructured and hence must first be converted into structured analytics data in order to be used systematically. Instead of building such capabilities themselves, financial firms are turning towards companies that specialize in this field. In this talk, Kevin will discuss some of the practical challenges of giving structure to unstructured content, how entities and ontologies may be used to link data and the ways in which semantic intelligence can be derived for use in financial trading algorithms.

Disrupting the Semantic Comfort Zone

Lora Aroyo[1,2]

[1] Vrije Universiteit Amsterdam, Amsterdam, The Netherlands
[2] Columbia University, New York, USA
lora.aroyo@vu.nl

Abstract. Ambiguity in interpreting signs is not a new idea, yet the vast majority of research in machine interpretation of signals such as speech, language, images, video, audio, etc., tend to ignore ambiguity. This is evidenced by the fact that metrics for quality of machine understanding rely on a ground truth, in which each instance (a sentence, a photo, a sound clip, etc) is assigned a discrete label, or set of labels, and the machine's prediction for that instance is compared to the label to determine if it is correct. This determination yields the familiar precision, recall, accuracy, and f-measure metrics, but clearly presupposes that this determination can be made. CrowdTruth is a form of collective intelligence based on a vector representation that accommodates diverse interpretation perspectives and encourages human annotators to disagree with each other, in order to expose latent elements such as ambiguity and worker quality. In other words, CrowdTruth assumes that when annotators disagree on how to label an example, it is because the example is ambiguous, the worker isn't doing the right thing, or the task itself is not clear. In previous work on CrowdTruth, the focus was on how the disagreement signals from low quality workers and from unclear tasks can be isolated. Recently, we observed that disagreement can also signal ambiguity. The basic hypothesis is that, if workers disagree on the correct label for an example, then it will be more difficult for a machine to classify that example. The elaborate data analysis to determine if the source of the disagreement is ambiguity supports our intuition that low clarity signals ambiguity, while high clarity sentences quite obviously express one or more of the target relations. In this talk I will share the experiences and lessons learned on the path to understanding diversity in human interpretation and the ways to capture it as ground truth to enable machines to deal with such diversity.

Keywords: Ambiguity · Crowdsourcing · Disagreement · Diversity · Perspectives · Opinions · Machine-crowd computation · Crowdsourcing ground truth

Semantic Web Technologies
for Digital Archives

John Sheridan

The National Archives, Kew, UK

Abstract. What will people in the future know of today? As the homes for our collective memory archives have a special role to play. Semantic Web technologies address some important needs for digital archives and are being ever more embraced by the archival community.

Archives face a big challenge. The use of digital technologies has profoundly shaped what types of record are created, captured, shared and made available. Digital records are not just documents or email but all sorts of content such as websites, threaded discussions, video, websites, structured datasets and even computer code. Yet, in the digital era, when so much is encoded as 0s and 1s there is no long term solution to the challenge of preservation. All archives can do is make the institutional commitment to continue to invest, through generations of technological change, in the engineering effort required for records to continue to be available.

The National Archives is one of the world's leading digital archives. Our Digital Records Infrastructure, which makes extensive use of RDF and SPARQL, is capable of safely, securely and actively preserving large quantities of data. Our Web Archive provides a comprehensive record of government on the web. We also lead the maintenance of a register of file format signatures that is used relied on by archives and other memory institutions around the world.

As a digital archive we provide value by preserving digital records, keeping them safe for the future. We maintain the context for the records so their evidential value can be understood in the context of their creation and continuing use. We produce records so that they are available for others to access, and we also enable use.

Semantic Web technologies play a key role in each of these areas and are integral to our approach for preserving, contextualising, presenting and enable use of digital records. This presentation will explain why and how we have used semantic web technologies for digital archiving and the benefits we have seen, for managing heterogeneous metadata and also in areas such a provenance and trust. It will explore new opportunities for archives from using Semantic Web technologies in particular around contextual description, with digital records increasingly contextualising each other. This is part of a shift to a more fluid approach where context grows with an archives collection and in relation to other collections. Finally it will also look at the challenges for archives with using Semantic Web technologies in particular around how best to manage uncertainty in our data as we increasingly use probabilistic approaches.

Contents – Part II

PhD Symposium

Contents – Part I

Mobile Web, Sensors, and Semantic Streams Track

Natural Language Processing and Information Retrieval Track

Vocabularies, Schemas, and Ontologies Track

Reasoning Track

In Use and Industrial Track

Applying Semantic Web Technologies to Assess Maintenance Tasks from Operational Interruptions: A Use-Case at Airbus

Ghislain Auguste Atemezing[✉]

MONDECA, 35 Boulevard de Strasbourg, Paris, France
ghislain.atemezing@mondeca.com

Abstract. Airbus, one of the leading Aircraft company in Europe, collects and manages a substantial amount of unstructured data from airlines companies, related to events occurring during the exploitation of an aircraft. Those events are called "Operational Interruptions" (OI) describing observations and the work performed associated by operators in form of short text. At the same time, Airbus maintains a dataset of programmed maintenance task (MPD) for each family of aircraft. Currently, OIs are reported by companies in Excel spreadsheets and experts have to find manually in the OIs the ones that are most likely to match an existing task. In this paper, we describe a semi-automatic approach using semantic technologies to assist the experts of the domain to improve the matching process of OIs with related MPD. Our approach combines text annotation using GATE and a graph matching algorithm. The evaluation of the approach shows the benefits of using semantic technologies to manage unstructured data and future applications for data integration at Airbus.

Keywords: Information retrieval · Tagging system · Graph matching · CA-Manager · GATE · Airbus

1 Introduction

Semantics enable machine-to-machine exchange and automated processing of data to facilitate the integration of business processes and systems [4]. The main goal of having machine readable information is to search, reuse information and develop innovative applications easier. Semantic Web [2] technologies are becoming more and more adopted outside the research communities or academics. The most promising of adopting these technologies is the benefits of using open standards developed by the W3C. Linked (open) data [3] has already shown its application in many domains, such as health, cultural heritage, libraries, etc. What is starting to appear is the use of semantics in aircraft industries where some often tools developed by the semantic web are perceived to be not mature enough by expert domains. This paper describes the challenges of using semantic technologies for matching events occurred during the exploitation of an aircraft

© Springer International Publishing AG 2017
E. Blomqvist et al. (Eds.): ESWC 2017, Part II, LNCS 10250, pp. 3–17, 2017.
DOI: 10.1007/978-3-319-58451-5_1

reported by a company with the official list of maintenance programmed by Airbus during the life-cycle of a given aircraft. The goal is to update the maintenance task (MPD) with new issues if the detection of operational interruptions (OIs) is relevant. The ultimate goal is to anticipate failures reported by the aircraft companies to the manufacturer. The main contributions of this paper are the following: (i) Capture the implicit knowledge of expert domains in RDF for generating gazetteers using SKOS [8] concepts; (ii) propose and implement a semantic annotation workflow to detect relevant OIs concepts and (iii) implement a robust graph matching algorithm to make recommendation to expert for assessment and validation. The results obtained are very promising as we are able to obtain a high level detected OIs (81.52%) with a very limited amount of keywords provided by experts (less than 150 concepts in the gazetteer). The paper is structured as follow: In Sect. 2, an overview of the domain expertise and problem statement are presented, followed by the vocabularies and data model in Sect. 3. Section 4 describes the dataset converted in RDF. The semantic annotation is described in Sect. 5, then the graph matching algorithm is presented in Sect. 6. The experiments and lessons learned are respectively presented in Sects. 7 and 8. Section 9 briefly expose some related work before concluding of the paper.

2   Problem Statement

In this section, we describe the scope of the domain at Aircraft and the main challenge of the provided solution in this paper. We first describe the concepts related to aircraft domain of our study, and then presents the approach using semantic technologies to tackle this real-world use-case.

2.1   Domain Description

Operational interruptions (OIs) are incidents occurring during the use of the aircraft, with an impact on commercial exploitation. All maintenance events are not OIs though: we need a minimal delay time for this to be considered an OI, i.e., if the cause is neutralized very quickly, an OI is not generated). Secondly, the OIs are not only technical incidents due to an aircraft: a crew arriving late because the van had broken down can also be an OI. Also, some elements occurring in the plane does not have a "technical" cause: for example, a collision with a bird (bird strike) will require maintenance, while the plane itself was not involved.

Generally, an OI contains the description of the interruption (what happened: it may be an effect or failure (failure message, vibration, noise, etc.) or a description of what the pilots or ground personnel observed), actions taken to address them (sometimes there are none), and additional metadata as the ATA code[1], the aircraft (type, operator, number, engine, etc.) the "changeability" (e.g., is it due to the plane? to ground maintenance teams? etc.), the operational impact

---

[1] https://en.wikipedia.org/wiki/ATA_100.

(does that re-routed the plane? Did that cancel the flight? etc.), the type of failure, etc. Today, OIs are used primarily for teams studying aircraft reliability to derive statistics and make recommendations.

Scheduled maintenance aims to avoid preemptively that some outages occur, develop into silence and ultimately have a significant impact on the use of the aircraft or its security. Some tasks are performed "before" the effect occurs: inspection of areas, equipment testing, periodic replacement. Not all parts of the aircraft are subject to scheduled maintenance. For example, some devices have sensors, integrated testing means, and thus emit messages in case of failure or when their condition deteriorates - so they perform in these cases, corrective maintenance. Therefore, many OIs concern elements that scheduled maintenance would not have prevented.

On one hand, the scheduled maintenance tasks are described in the Maintenance Planning Document (MPD), with their title, interval, the areas where they occur, the necessary duration, etc. However, the MPD itself does not describe why it is necessary to perform these tasks. On the other hand, the Maintenance Review Board Report (MRBR) falls under analysis by experts of scheduled maintenance, in agreement with the authorities and operators. The objective of these teams is to see if the items to be covered are often detected or not, to adjust the range of preventive actions to more suitable ones. Therefore, they look at reports of scheduled maintenance operations (reports of "Findings"), but need also data from non-scheduled operations to be complete which require the analysis of OIs.

Each MRBR item is a task involving one or more functions, each of which can have multiple causes, which will induce a functional failure in the systems of a plane with effects or consequences on the plane. The effects can be detectable or not. Each row in the MRBR describes a failure. A link exists between MRBR and MPD task by the following rules: (R1)- each MPD task can cover one or more MRBR items (or none) and (R2)- some MPD tasks are derived from further analysis.

This paper solves the following challenge: to identify the indirect link between OIs and MPD task. That is: if the OI describes a cause that would have been avoided by conducting an MPD task, then we associate the corresponding MPD task. Therefore, we seek in OI (description, metadata) elements related to potentially preventable failures of MPD, so as to associate it causes, failures and consequences described in MRBR items. Hence the keywords used by experts to identify potentially interesting OI.

2.2   Our Approach

The main challenge is to assist the expert with a reduced list of OIs candidates that can be matched with some existing tasks in the MPD. The top down approach takes a task, uses the list of keywords and the implicit knowledge of the expert (e.g., experience) and try to match with OI. Currently this is the approach manually used by domain experts by means of formulas in spreadsheets to find and filter sets of OIs based on list of keywords. A second approach consisting a fully automatic process, from OIs annotations and matching without experts in the loop (bottom-up approach) was not acceptable by the domain experts. We define an *hybrid approach* where the expert can assess and validate the results of the approach, consisting of the following (as shown in Fig. 1):

1. OIs are annotated with keywords from the experts.
2. MP tasks already treated are modeled in RDF with associated keywords.
3. Find OIs candidates matching the set of reference tasks by using any relevance information such as ATA references and aircraft metadata.
4. Experts evaluate and validate the suggested OIs candidates.

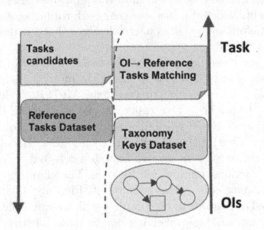

Fig. 1. The hybrid approach used to match the OI descriptions with MPD tasks, using semantic technologies.

3   Vocabularies and Data Model

The legacy data in spreadsheets is converted into RDF for better integration and homogeneity. Two vocabularies is designed to capture the knowledge presented in the diverse data: a vocabulary for operational interruptions (oi-vocab) and a vocabulary for maintenance tasks and analysis experts report (task-vocab). The URI namespaces used for the vocabularies follow the pattern http://data.airbus.com/def/{vocab-prefix}#.

3.1   OI Vocabulary

The OI vocabulary[2] is a lightweight model with one main class and three data properties. An operational interruption is modeled as a subclass of an event, reusing the class event:Event of the event ontology[3]. The data properties oi:ataref, oi:description and oi:workPerform are used to respectively capture the ATA reference, the textual description of an OI and the work performed by an operator. Figure 2 depicts the vocabulary and a sample graph generated with the model.

---

[2] http://data.airbus.com/def/oi.
[3] http://purl.org/NET/c4dm/event.owl.

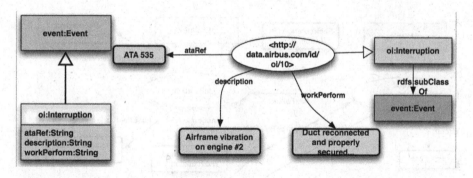

Fig. 2. The left side shows the model and the right side an example of graph based on the OI vocabulary

3.2 Task Vocabulary

A vocabulary[4] for describing the relationships the failures and the maintenance tasks has been created to build a knowledge graph for integrating different silos of documents in Airbus organization. Currently, four main classes are defined: tf:Task, tf:MRBR, tf:Failure and tf:ProgramTask. Figure 3 shows an excerpt of the current usage of the vocabulary.

4   RDF Dataset

One of the main challenge in the generation of RDF data from the different spreadsheets is to create unique URIs to identify instances of the classes from the vocabularies. Each element of the graph should be identified uniquely with a persistent URI. The chosen scheme for the URI for generating RDF dataset is the following: *http://data.airbus.com/id/{class}/{aircraft-family}/{year}/{code}* where {class} is one of (oi, task, mrbr); {aircraft-family} is one of (a330, a320); {Year} a four digit for the year and {code} is generated by concatenating a number of the XLS file name + raw number in the file. For example, this URI <http://data.airbus.com/id/oi/a330/2014/07011231177> represents in the Knowledge Graph the OI for an A330 received in 2014, where the description is within the file "isaim-a330-2014-0701to1231.xls" in line 177. Table 1 shows the generated graphs in RDF for the two aircraft families: A330 and A320 for the period from 2013 to 2015. The conversion process to RDF of all the Excel files is performed by the Datalift [12] platform, using custom CONSTRUCT SPARQL queries to transform columns to specific classes and properties.

5   Semantic Annotation

Since the input of the descriptions and the work performed for the each OI is short text in English, an annotation pipeline based on the GATE architecture

---

[4] http://data.airbus.com/def/tf.

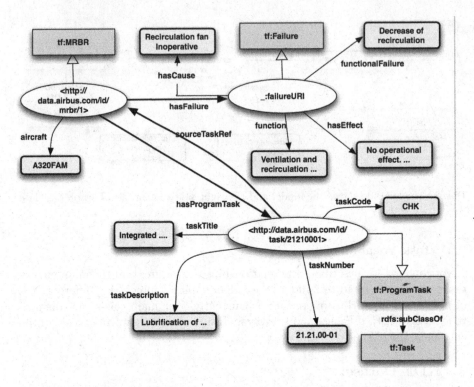

Fig. 3. A sample graph showing the relations between tasks, master documents and failures using the mpd-vocabulary.

Table 1. Overall number of the data converted into RDF by family of Aircraft for the period 2013–2014 (A330) and 2014–2015 (A320).

Class	OI		MRBR- MPD		
Dataset	NbTriples	NbOIs	NbTriples	NbMRBR	NbTask
A320	111,044	27,761	19,481	1,969	880
A330	60,867	21,400	54,457	6,871	4,234

[6,7] is implemented to detect and tag the relevant concepts. CA-Manager is the annotation tool developed at Mondeca to annotate heterogeneous unstructured content with personalized pipelines based on GATE (Fig. 4).

The architecture of the workflow consists of the following modules:

1. skos-ification module, in charge of converting into SKOS concepts the experts keywords.
2. SKOS2Gate module, for creating the gazetteer to be used during the annotation of the OIs.
3. CA-Batch module, for creating multi-thread calls to annotate the text documents.

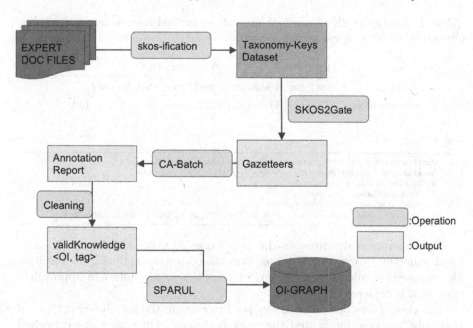

Fig. 4. Workflow of the semantic annotation and knowledge base updates.

4. Cleaning module, in charge of removing non relevant information from the reports generated by the annotator.
5. SPARUL module, which is in charge of updating the dataset in the endpoint using SPARQL updates queries, and enriching the underlying RDF graph.

5.1 SKOS-ification Process

The skos-ification process consists of manually convert the keywords used by the expert of the domain into SKOS concepts that is later used for generating the gazetteers during the annotation. Since the input file are DOCX files, we manually create the RDF file in Turtle. Some rules and hypothesis are made during the process based on different interviews with the experts of the domain (Table 2):

- Ordinal numbers such as "Third", "First", etc. are not taken into account.
- A keyword can not be at the same time in the description AND in the action work performed of the OI. The two sets are DISJOINTS.
- Descriptors SHOULD be short concepts(names), with variants in the case of verbs that are treated as SYNONYMS (see sample below in Turtle format)
- Expressions of type *Not[properly][term]* are modeled as just [term a skos:Concept]. For e.g., in the case of the term "not properly closed", we only model the concept "close" with the variant closed.

Table 2. Number of SKOS concepts manually generated from domain experts file, grouped by description and work performed.

Class	Description		WorkPerformed		Total
Feature	descItem	descAction	workItem	workAction	-
SKOS concept	47	45	22	16	130

```
1  @prefix workAction: <http://data.airbus.com/id/scheme/workAction/> .
2  @prefix skos: <http://www.w3.org/2004/02/skos/core#> .
3  workAction:Change a skos:Concept';
4      skos:prefLabel "Change"@en ;
5      skos:inScheme workAction:keyword ;
6      skos:altLabel "Replace"@en, "Changed"@en, "Replaced"@en , "Replacement"@en .
```

The experiment described in this paper uses 20 tasks with keywords annotated manually by experts. Together with the corresponding task, the resulting file represents a gold standard where the descriptors are linked to appropriate task, as it is depicted in Fig. 5.

The class tf:DescriptionKey is designed to capture the two different types of keywords: the description and the work performed. Once they are converted into SKOS concepts, they are used to identify relevant concepts in OIs text. Furthermore, four properties are used to link to the appropriate namespaces: tf:descItem, tf:descAction, tf:workItem and tf:workAction.

5.2 Content Annotation

The annotation process employed is based on a central component: the Content Augmentation Manager (CA-Manager) [5]. The content augmentation manager (CA-Manager) is in charge of processing any type of content (plain text, XML, HTML, PDF, etc.). This module extracts the concepts and entities detected using text mining techniques with the text input module. The strength of CA-Manager is to combine semantic technologies with a UIMA-based infrastructure[5] which has been enriched and customized to address the specific needs of both semantic annotation and ontology population tasks.

The scenario presented in this paper is built on top of the GATE framework for entity extraction. CA-Manager uses an ontology-based annotation schema to transform heterogeneous content (text, image, video, etc.) into semantically-driven and organized one. The overall architecture of CA-Manager is depicted in Fig. 6. We first create the gazetteer with the SKOS representation of the documents obtained from the experts. We then launch in parallel 10 documents in multi-threads containing OIs. The annotation report contains the valid knowledge section, an RDF/XML document containing the URIs of the concepts detected by the annotator. We then use a python script to map each document with its corresponding URI. Finally, a SPARQL update query is launched to update the dataset containing the OI-Graph.

---

[5] Unstructured Information Management Architecture (http://uima.apache.org).

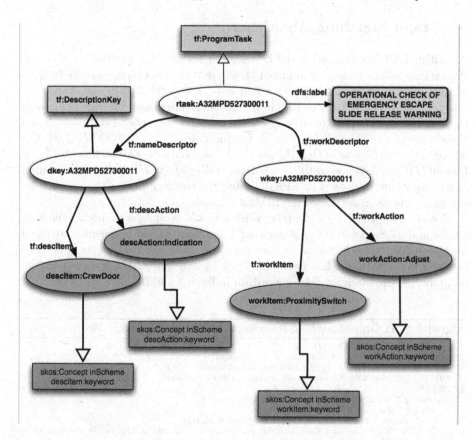

Fig. 5. A sample of SKOS concepts used to model the experts' keywords and linked to a task and maintenance graph

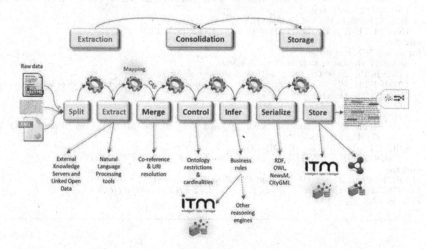

Fig. 6. Pipeline of annotation using CA-Manager

6   Graph Matching Algorithm

Algorithm 1 (Relax version) is the first variant for creating candidates. The initialization sets are presented in lines 1–4, where there are two subsets. We reduce the size of the OIs by taking into account only those with tags in one of the subsets presented (lines 13, 23 and 26) in the initialization phase. The algorithm uses the logical UNION operator to compute the intermediate sets for comparison and the resulting set of candidates in C. In this case, $C = UNION(P, Q, P', Q')$ where $P = UnionOfDescActions$, $Q = UnionOfDescItems$, $P' = UnionOfWorkActions$ and $Q' = UnionOfWorkItems$. All the queries used in the algorithm are based on SPARQL and the computation of the sets involving intersections make use of the LIMES [9] matching tool.

A second approach for a less relax version (LRX-version) is to modify the lines 14, 24 and 27 respectively in Algorithm 1 by doing the adjustments described in Algorithm 2. Depending on how we combine the results makes the difference between the LRX and RLX algorithm (lines 14 and 24), with the corresponding mappings in the modified RLX algorithm in lines 3 and 5.

Algorithm 1. GraphMatching (OI-Graph,MPD-Graph) - RLX Version

---

1: Input: Set of OIs tagged $< OI, tag >$
2: Input: Set of MPD in RDF with tags $< MPD, tag >$.
3: INITIALIZE $Description = \{DescItems, DescActions\}$
4: INITIALIZE $WorkPerformed = \{WorkItems, WorkActions\}$
5: BEGIN
6: **for** each $j \in MPD$ **do**
7:     SELECT the list of tags in $DescActions$.
8:     FIND in the set $< OI, tag >$ with at least one $descItem$
9:     COMPUTE P = card(OIs with at least one of the MPD tag in $DescActions$) = $|UnionOfDescActions|$
10:    SELECT the list of tags in $DescItems$.
11:    FIND in the set $< OI, tag >$ with at least one $descAction$
12:    COMPUTE Q = card(OIs with at least one of the MPD tag in $DescItems$) = $|UnionOfDescItems|$
13:    SELECT OIs in the set (P **OR** Q).
14:    COMPUTE R = UNION(P,Q) = P + Q - INTERSECTION(P, Q)
15: **end for**
16: **for** each $j \in MPD$ **do**
17:    SELECT the list of tags in $WorkActions$.
18:    FIND in the set $< OI, tag >$ with at *least one workItem*
19:    COMPUTE P' = card(OIs with at least one of the MPD tag in $WorkActions$) = $|UnionOfWorkActions|$
20:    SELECT the list of tags in $WorkItems$.
21:    FIND in the set $< OI, tag >$ with at *least one workAction*
22:    COMPUTE Q' = card(OIs with at least one of the MPD tag in $WorkItems$) = $|UnionOfWorkItems|$
23:    SELECT OIs in the set (P' **OR** Q').
24:    COMPUTE $R' = UNION(P', Q') = P' + Q' - INTERSECTION(P', Q')$
25: **end for**
26: SELECT OIs in the set $Description$ **OR** $WorkPerformed$.
27: COMPUTE $C = UNION(R, R') = R + R' - INTERSECTION(R, R')$
28: **return** C
29: END

---

7   Experiments and Results

We perform experiments in both the annotation and the graph matching algorithm. To perform this, we use the datasets for the A330 family of Airbus aircrafts during the year 2013–2014.

7.1   Experiments

We perform the annotation of $21,400$ OIs consisting of the descriptions and the work performed in English. We use a single machine on Windows 10, Core i7; 6 Go RAM. The annotation took 1 h 48 min (6434.464 s), with an error rate by CA-Manager of 2.71%. Moreover, 81.52% of OIs were tagged (17,446), where more than 11K OIs have at least two tags. This is important because the graph matching algorithm uses this set of OIs with at least two tags detected. Table 3 gives more details of the annotation process results.

Algorithm 2. GraphMatching - LRX-Version

1: SAME INIT AS ALGO RLX-version
   .....
2: BEGIN
3: COMPUTE $R = UNION(P, P')$
   ......
4: COMPUTE $R' = UNION(Q, Q')$
   ......
5: COMPUTE $C = INTERSECTION(R, R')$
6: **return** C
7: END

Table 3. Details of the annotation content of the OIs for the A330 family.

Graph	Feature	Number
OI A330-Graph	NbTriples	60,867
	NbOI	21,400
	NbOIs-tagged	17,446
	Max. tags	12
	Min.tag	1
	NbOI-with-one-tag	5,954
	CAM-processed	20,819
	Errors-annotation	581
	Time	6,434.464 s

7.2   Results

We present the results of applying our annotation and matching to assess the possible OIs candidates for three tasks for A330 and the OIs of the same aircraft family during the year 2013–2014 (Table 4).

Table 4. Results of the matching process in terms of size of the candidates using Algorithm 1 with three tasks for A330 and the OIs of the same aircraft family during the year 2013–2014.

Task ID	P	P'	Q	Q'	Inter (P,Q)	Inter (P',Q')	R=Union (P,Q)	R'=Union (P',Q')	Inter (R,R')	Union (R,R')
MPD245000031	545	271	129	0	41	0	633	271	2	**902**
MPD271400031	110	306	0	366	0	250	110	422	7	**525**
MPD235100021	545	271	109	0	19	0	635	271	1	**905**

In a more open criteria, as it is the case with Algorithm 1, we obtain the tasks T1(MPD245000031), T2 (MPD271400031) and T3 (MPD235100021) 902, 525 and 905 potential candidates. That means almost more than 95% of the OIs are not relevant at all to those three tasks. But still many OIs to review by a human expert. So, after some interviews with the domain experts, the LRX-version of the algorithm (see Algorithm 2) is implemented suggesting a more reduced set of candidates, without loosing any good candidate. With this variant, it reduces significantly the number of the candidates (an average of 63%) for reviewing (Table 5).

Table 5. Results of the matching process when using Algorithm 2, the LRX version.

Task ID	P	P'	Q	Q'	Inter (P,P')	Inter (Q,Q')	R=Union (P,P')	R'=Union (Q,Q')	Inter (R,R')
MPD245000031	545	271	129	0	13	0	803	129	**40**
MPD271400031	110	306	0	366	8	0	408	366	**4**
MPD235100021	545	271	109	0	13	0	803	109	**24**

Currently the team in charge of manually detect relevant OIs uses four complex Excel functions to assess the OIs. When presented the solution with the data we received from them, they were satisfied with the results, meaning we were able to use semantics to capture both their expertise and their daily work.

7.3   Evaluation

The experts of this domain are difficult to access. However, we were accompanied by 3 of them during the work to better understand the challenges, to gather relevant datasets. The experts help refining the gazetteers in reviewing some candidates to identify false positives. This sample[6] shows one view of the type of call-for-arms during the duration of the project. Since they are very busy experts

---

[6] https://github.com/gatemezing/eswc2017/blob/master/sampleEvalForTuningGazet teers_eswv2017.pdf.

of the domain, we didn't have access to a full "evaluation" *ala* research fashion (computing precision and recall for a representative set of OIs). Nevertheless, the domain experts gave us those 3 tasks on purpose (see https://goo.gl/CM3KzB for Turtle transcription of the word file received) - as they can easily check the results - for our scenario to see if we were able to miss some relevant OIs. This work also shows an application of semantic web to solve a real business use case, covering all the process of creation the ontologies, the population of the different knowledge bases from heterogeneous data, etc. The output of this work is installed at Airbus-Toulouse in a virtual machine for internal usage.

8 Lessons Learned

One of the barrier in the adoption of Semantic technologies in industry is the difficulty to explain the key concepts underlying RDF model, vocabulary creation and the benefits of changing the paradigm from traditional data management to semantic repositories. This was one of the challenge of this project at Airbus, starting with a real use case to increasingly find the benefits of semantic technology to solve a real world problem. During the time frame of this work, several meetings where held to explain the "why" at every single steps.

The team in charge of collecting the OIs from diverse companies and experts working on maintenance tasks has to progressively make a mental shift with the new paradigms introduced during this project:

- Data model by implementing vocabularies with Protégé [10] from scratch driven by the existing data.
- The notion of unique identifiers to be used across different data silos to represent the same real world object. Hence, a policy for creating persistent URIs was clearly identified as crucial
- The confusion between the RDF model and the different serializations

This work also permits to identify many datasets that could be useful in other services at Airbus to bring more context and integrate them easily. For example, during some interviews, the experts realized that some implicit knowledge can be derived actually from existing datasets, which otherwise are experts' experience. Moreover, the system proposed would not replace the experts but assist them in their work to speed up the process. This is one of the strong requirement of the proposed approach.

Additionally, the conversion process into RDF reveals some errors in the legacy data. For example, sometimes there were not consistent use of the task code across different XLS files. This shows the benefits of our approach compared to existing one as it helps detecting and cleaning errors in the data. As the result of this work, the expert team is looking at a better combination for the output sets to adjust the proposed algorithm based on their evaluation of the proposed candidates.

9   Related Work

This work falls under contributions and research in the intersection of natural-break language processing with GATE, enterprise semantic data management and matching techniques. The authors in [13] examine semantic annotation, identify a number of requirements, and review the generation of semantic annotation systems. Dadzi and et al. [11] developed an integrated methodology to optimize Knowledge reuse and sharing in the aeronautics domain based on ontologies. They proposed an interface for multiple modalities during search of documents based on their approach. While our approach also used domain ontologies to convert legacy data into RDF, it also combines NLP techniques for annotating textual data, and SPARQL queries for matching purposes. Recently, in [1] it is defined a state of the art in semantic annotation models and a scheme that can be used to clarify requirements of end-user use cases. Moreover, the collection of real world applications from industry in [4] tackle some challenges in the earlier adoption of semantic technologies in industry. Although some issues remind challenging today, a diversity of domains applying semantics is visible worldwide.

10   Conclusion and Future Work

This paper presents a semi-automatic approach using semantic technologies to assist the experts of the aircraft domain to improve the matching process of Operational Interruptions with related Maintenance Programming Task. Our approach combines text annotation using GATE alike system for annotation and a graph matching algorithm for suggesting candidates. Annotation of 21,400 OIs show a high amount of tag detected (82%) with a small amount of expert concepts. The graph matching technique shows that by combining suitable combinations, it is possible to reduce to up to 63% the amount of candidates of OIs to be assessed by domain experts. The first results shed the light for future intensive application of semantic technologies at Airbus for many other aspects, such as data integration, data fusion and semantic recommendation tools.

For future work, we plan to improve the semantic annotation by looking at the better management of noises in the text for annotation. Also, we plan to implement fuzzy annotation in the GATE pipeline to find a trade-off between performance and recall. Currently we are using a small gazetteer, we plan to integrate external sources to have a much bigger scope for the terms to detect.

Acknowledgments. We would like to thank the Airbus team in Toulouse and ATOS colleagues for their valuable input and partnership.

References

1. Andrews, P., Zaihrayeu, I., Pane, J.: A classification of semantic annotation systems. Semant. Web **3**(3), 223–248 (2012)
2. Berners-Lee, T., Hendler, J., Lassila, O., et al.: The semantic web. Sci. Am. **284**(5), 28–37 (2001)

3. Bizer, C., Heath, T., Berners-Lee, T.: Linked data - the story so far. Int. J. Semant. Web Inf. Syst. **5**, 1–22 (2009)

4. Cardoso, J., Hepp, M., Lytras, M.D.: The Semantic Web: Real-World Applications from Industry, vol. 6. Springer Science & Business Media, Heidelberg (2007)

5. Cherfi, H., Coste, M., Amardeilh, F.: Ca-manager: a middleware for mutual enrichment between information extraction systems and knowledge repositories. In: 4th Workshop SOS-DLWD Des Sources Ouvertes au Web de Données, pp. 15–28 (2013)

6. Cunningham, H.: Gate, a general architecture for text engineering. Comput. Humanit. **36**(2), 223–254 (1996)

7. Kenter, T., Maynard, D.: Using gate as an annotation tool. University of Sheffield, Natural language processing group (2005)

8. Miles, A., Bechhofer, S.: SKOS simple knowledge organization system reference. W3C (2009). https://www.w3.org/TR/skos-reference/

9. Ngomo, A.-C.N., Auer, S.: Limes - a time-efficient approach for large-scale link discovery on the web of data. In: Proceedings of IJCAI (2011)

10. Noy, N.F., Sintek, M., Decker, S., Crubézy, M., Fergerson, R.W., Musen, M.A.: Creating semantic web contents with protege-2000. IEEE Intell. Syst. **2**, 60–71 (2001)

11. Dadzie, A.-S., Bhagdev, R., Chakravarthy, A., Chapman, S., Iria, J., Lanfranchi, V., Magalhães, J., Petrelli, D., Ciravegna, F.: Applying semantic web technologies to knowledge sharing in aerospace engineering. J. Ind. Manuf. **20**, 611–623 (2009)

12. Scharffe, F., Atemezing, G., Troncy, R., Gandon, F., Villata, S., Bucher, B., Hamdi, F., Bihanic, L., Képéklian, G., Cotton, F., et al.: Enabling linked-data publication with the datalift platform. In: Proceedings of AAAI Workshop on Semantic Cities (2012)

13. Uren, V., Cimiano, P., Iria, J., Handschuh, S., Vargas-Vera, M., Motta, E., Ciravegna, F.: Semantic annotation for knowledge management: requirements and a survey of the state of the art. Web Semant. Sci. Serv. Agents World Wide Web **4**(1), 14–28 (2006)

Modeling Company Risk and Importance in Supply Graphs

Lucas Carstens[1]([✉]), Jochen L. Leidner[1], Krzysztof Szymanski[2], and Blake Howald[3]

[1] Thomson Reuters, Corporate Research and Development,
30 South Colonnade, England E14 5EP, UK
lucas.carstens@thomsonreuters.com
[2] Thomson Reuters, Platform Group, Slaska 23/25, 81-001 Gdynia, Poland
[3] Thomson Reuters, Platform Group, 610 Opperman Drive, Eagan, MN 55123, USA

Abstract. Managing one's supply chain is a key task in the operational risk management for any business. Human procurement officers can manage only a limited number of key suppliers directly, yet global companies often have thousands of suppliers part of a wider ecosystem, which makes overall risk exposure hard to track. To this end, we present an industrial graph database application to account for direct and indirect (transitive) supplier risk and importance, based on a weighted set of measures: criticality, replaceability, centrality and distance. We describe an implementation of our graph-based model as an interactive and visual supply chain risk and importance explorer. Using a supply network (comprised of approximately 98,000 companies and 220,000 relations) induced from textual data by applying text mining techniques to news stories, we investigate whether our scores may function as a proxy for actual supplier importance, which is generally not known, as supply chain relationships are typically closely guarded trade secrets. To our knowledge, this is the largest-scale graph database and analysis on real supply relations reported to date.

Keywords: Supply chain analysis · Graph analysis · Risk analysis · Vulnerability analysis · Linked data · Procurement

1 Introduction

A supply chain is a complex network of interconnected actors that continually exchange goods, with the goal of producing value for all actors in the supply chain. Though supply chains are growing ever more involved, and remain as vital as ever to companies' success, many companies operate with little insight beyond their first tier suppliers and customers. This means that any disruption occurring removed from a company's immediate view risks to be met with little preparedness, and without mitigation strategies in place. To alleviate such risks of being unprepared it is in the interest of companies to increase visibility in supply chains, identifying not only actors they directly interface and exchange

© Springer International Publishing AG 2017
E. Blomqvist et al. (Eds.): ESWC 2017, Part II, LNCS 10250, pp. 18–32, 2017.
DOI: 10.1007/978-3-319-58451-5_2

goods with, but also those residing in subsequent tiers. In addition much of the management of supply chains within companies are founded upon ad hoc methods, relying heavily on human expert knowledge and intuition.

With our work we present a novel approach to investigating the structure of a company's supply chain, based on insights extracted from free text. We represent relations between companies as a graph, where companies are represented as nodes and supply relations as directed edges, pointing from a supplier to a customer (or *consignee*). Not only does this allow us to interpret relations between companies in a formally defined manner, but it additionally provides the opportunity to investigate links between companies beyond their first tier suppliers and customers. More specifically, we use this graph to identify *peers* of a company within its supply chain that are not only particularly relevant, but that are also exposed to certain risks and thus increase the potential for supply chain disruptions. Our graph-based model captures the connectedness of the supplier-consignee supply chain ecosystem in conjunction with the strength of the relationships and the risk exposure of each company entity, which transitively affects potentially large parts of the graph. Specifically, we have developed a solution that is comprised of two APIs, which together provide an aggregate view of peers that are important suppliers to a company, while also being exposed to certain risks. Peers of a company are extracted from a graph database. A pre-specified number of neighbors, from within a pre-specified distance from the node, are extracted and subsequently scored for their importance to the company and their risk. Such a graph model has the potential to serve as the basis for numerous subsequent experiments, including exploring the resilience (see literature review below).

The remainder of this paper is organized as follows. In Sect. 2 we discuss related work at the intersection of two or more of the fields of interest to our work. In Sect. 3 we describe our data sets as well as the construction of a graph database to store and access it. The method of extracting suppliers from the graph database and subsequently scoring their importance and risk is described in Sect. 4. We describe the APIs through which the scoring methods are invoked, as well as the system with which we represent the API output, in Sect. 5. We present an empirical evaluation of the quality of the importance scores in Sect. 6, before concluding the paper in Sect. 7.

2   Related Work

Our work spans a number of fields, touching upon risk and graph analysis, as well as the more nascent area of scientific supply chain analysis, all of which we base on content extracted from the web. Little research has considered the application of both risk and graph analysis to supply chains and modeling. The work of Wagner and Neshat [22], who investigate supply chain risk quantification and mitigation based on graph theory, presents a notable exception. In the remainder of this section we thus review work that resides at the intersection of at least two of the areas we are concerned with. For more specific surveys of the individual fields refer, for example, to [1] for an overview of graph analysis, to [4] for research on risk and to [20] for a summary of recent advances in supply chain management and procurement.

Supply Graphs. Recent trends in the analysis of supply chains have highlighted the value of representing supply chains as graphs, or networks, rather than as flat structures and relational databases. To this end, Borgatti and colleagues [6] provide an overview of social network analysis, geared towards supply chain research. In the same vein, Kim and colleagues [12] interpret supply chains as networks and apply social network analysis metrics, such as *closeness* or *betweenness centrality*, to evaluate the flow of materials through a supply chains, as well as contractual relationships. Interpreting supply chains as graphs produces wholly new opportunities to investigate structural characteristics and transitive links of complex relations. This is not limited to risk or importance, the focus of our work, but extends well beyond these metrics. For example, Tan and colleagues [19] propose the use graphs to identify innovation potential throughout a network of interlinked companies. Further exploiting graph capabilities, Xu and colleagues [23] describe an evolutionary mechanism that dynamically grows and alters supply networks, reflecting the dynamic nature of supply relations.

Risk and Importance in Supply Chains. Much of the development in assessing risk in supply chains is based on qualitative studies, using expert opinion and case studies. For example, Blome colleagues [5] investigate whether the 2008 financial crisis has had an impact on how risk is managed and, more specifically, whether any of the stages of risk analysis, risk mitigation and risk monitoring have changed. Similarly, Hallikas and colleagues [8] conduct case studies on eleven companies, operating in either the electronics or metal industry, to illustrate challenges that network co-operation brings to risk management. Aqlan and colleagues [3] describe a risk assessment framework that produces risk scores for *suppliers, customers, manufacturers, transportation* and *commodities*. For each stakeholder, experts are consulted to identify the main risk factors. This produces a quantification similar to what we describe in our work, joining impact potential with the risk of this impact actually materializing. Ghadge and colleagues [7] describe a framework comprised of an iterative process to identify, assess and mitigate supply chain risks. They focus on risk assessment, which is comprised of risk modeling and sensitivity analysis, using both a *risk register* and data collected through interviews, company reports, etc. Harland and colleagues [9] describe a *network risk tool* to address the same challenges. The authors focus on risks arising from product and service complexity, outsourcing, globalization and e-business. Based on a set of surveys and focus groups, Juettner [11] seeks to identify and understand business requirements for SCRM from the perspective of professionals working in the field. To structure overarching issues encountered in her analysis into these levels, Juettner first identifies the extent to which organizations already manage risks in their supply chain and then determines critical issues that arise as part of the implementation of risk management. Simchi-Levi, Schmidt and Wei [17] present a dynamic graph model, which includes recovery time. Unlike our model, their data is obtained from human questionnaires, not automatic text mining.

Risk and Importance in Graphs. The use of *attack graphs* represents one of the more popular approaches to interpreting risk and adverse events in graphs. Attack graphs, as well as *attack trees* are used to model all possible attacks, or *exploits*, on a network. In an early proposal for the application of attack-graphs to the identification of risks in physical networks Phillips and Swiler [15] coin *network-vulnerability analysis*. In a more recent development, Alhomidi and Reed [2] use a *genetic algorithm (GA)* [13] to model a large number of possible *paths* in attack graphs, where each path connects the source of an attack on a network to the target of the attack. In each path, nodes are assigned with a probability that represents the likelihood of the node being exploited by an attacker, as well as an expected loss, accrued when a node is indeed attacked. In an adoption of attack graphs, Poolsappasit and colleagues [16] propose a framework for dynamically managing security risks called *Bayesian attack graphs (BAG)*. The overall risk of each possible path in an attack graph is calculated as a product of the attack success likelihoods and the value of the expected loss incurred. Based on data for 371 banks that failed during the 2008 financial crisis, Huang and colleagues [10] study the systemic risk of financial systems. To do so they propose a cascading failure model to describe the risk propagation process during crises. A bi-partite banking network model is proposed, where one type of node represents banks and another represents assets held by banks. The resulting graph is *shocked* by decreasing the *total market value* of an asset, leading to a decrease in value for every bank that holds the affected asset. Stergiopoulos and colleagues [18] extend the notion of cascading failure models to include *graph centrality* measures to help identify the nodes most critical in identifying and mitigating failures. Graph centrality here is used as a proxy for identifying the *most important* nodes within a graph so that any risk mitigation strategy can be based on both the importance of nodes and their susceptibility to failure, in general. None of the work surveyed above provides a larger-scale supply chain graph model, which can be used for the analysis of and experimentation with supply relation scoring methods. Below we present such a model, as well as its implementation.

3   Building a Supply Graph

Our analysis of supply relations between companies is based on a graph database, where the nodes represent companies and edges signify directed supply relations, pointing from a supplier to a customer. Supplier/customer pairs are extracted automatically from specifically news articles. Each node in the graph is assigned a set of attributes, namely (i) their business sector, (ii) a credit risk score, (iii) the company name and (iv) a closeness centrality score. We describe the data extraction process, as well as the node attributes, in detail below, for two separate supply graphs. On the one hand we conduct experiments, as described in Sect. 6, on the full graph (SPR^+), with all its attributes. On the other hand,

Table 1. Dataset: summary statistics.

Number of nodes	98,402
Number of vertices	217,188
Average path length	6.614
Average degree	4.414
Average closeness centrality	0.225

necessitated by the proprietary nature of this data, we make available a second dataset (SPR^-) for research purposes[1]; in it, company names and business sectors are anonymized. Table 1 summarizes the main characteristics of the data.

3.1 Supply Relations Data

Both datasets, $SPR^{+/-}$ are comprised of supply relations between two companies, where each individual relation and the companies involved are automatically extracted form text snippets. While we describe a static snapshot of the dataset for our experiments, an underlying *RDF triple store* of supply relations is continually updated, both to add new relations and to remove those considered out of date. A snippet corresponds to a sentence, extracted either from a news article or a *Security and Exchange Commission (SEC)* filing. A logistic regression model was trained on a set of 45,000 snippets, while the test set was comprised of 20,000 snippets. The training and test data were aggregated using the following procedure:

1. Identify companies in a document, using *Calais*[2];
2. split documents into sentences;
3. choose candidate sentences that contain two companies, as well as one or more of a pre-specified set of patterns; and
4. using *Mechanical Turk*, label companies in the candidate sentences as suppliers, customers, or neither.

Patterns are based on a set of indicative *n-grams*, as well as variations of these n-grams to catch terms such as *powered by*, *contracts with*, etc. Each candidate sentence has been labeled by two separate *Turkers* and any disagreement was addressed by presenting the instance to a third annotator. The regression model has been tuned to yield high precision, focusing on the extraction of high quality evidence sentences, while relying on the fact that eventually, highly indicative sentences will be introduced into the dataset. The classifier produced an F_1-score $F_1 = 0.57$ (with *precision* $= 0.76$ and *recall* $= 0.46$) on the test set (note that a random baseline classifier would achieve an accuracy of 0.5). Data is stored in an RDF triple store from which we can extract a subset or, to

---

[1] see http://bit.ly/TRSupplyChainRisk.
[2] http://www.opencalais.com/.

populate our graph database, the entire set using *Sparql* queries. This triple store is continually updated to add additional relations found in unseen text; multiple patterns producing the same supplier-consignee pairs are aggregated to a single triple. Each triple has a confidence score assigned, based on the classifier output as well as the amount of examples found for a specific relation.

3.2 Company Attributes

To score a company according to its importance as a supplier to a customer, as well as the risk it is exposed to, we assign a set of attributes to each company, in addition to its name for identification purposes. The importance of a company is then determined based on how a supplier's attributes compares to those of the customer, as well as both their position in the overall graph.

Business Sector. Each company in the supply graph is labeled with the *business sector* it operates in. We use the *Thomson Reuters Business Classification (TRBC)*[3] scheme for this purpose, a widely used industry standard. The TRBC scheme offers classification of companies at various levels of abstraction, i.e. economic sectors, the most abstract level, business sectors, industry groups, industries, and activities. To strike a compromise between informativeness and the ability to group various companies we label companies with their business sector, meaning that we distinguish between 28 different labels, such as *Renewable Energy*, *Industrial goods*, etc.

Credit Risk. To identify the risk companies are exposed to we score them according to a credit risk measure. This score broadly signifies the likelihood of a company defaulting on one or more of their debt obligations within a year. A score between zero and 100 is used to signify the likelihood, where a lower score represents a higher likelihood of default.

Closeness Centrality. One of the aims of our importance scoring is to incorporate both attributes of individual companies and those formalizing a company's role within a larger graph of companies. To this end we have chosen to score each node in the graph according to its *closeness centrality*. Closeness centrality measures a node's centrality in a graph as the sum of the length of the shortest paths between the node and all other nodes in the graph. This sum is usually normalized by division with the total node count N (minus one so as not to count the node itself) to represent the average length of the shortest paths, or distance $d(y, x)$, giving

$$C(x) = \frac{N - 1}{\sum_y d(y, x)}, \tag{1}$$

3.3 Database

The data described above is initially extracted from the RDF store and represented as separate node and edge tables, which are, in turn, used to populate a

---

[3] http://financial.thomsonreuters.com/en/products/data-analytics/market-data/indices/trbc-indices.html.

graph database, implemented using Neo4j. While the original dataset described here contains proprietary data and can hence not be published, we make available an anonymized version of the node and edge tables. To interact with the database we use *Cypher*, the graph query language developed as part of Neo4j. To load the node table we call the following command

```
1  USING PERIODIC COMMIT
2  LOAD CSV WITH HEADERS FROM 'file:///file_path/nodes.csv' as line
3  WITH line MERGE (ID:permID {name: TOINT(line.permID)})
4  SET ID.trbc = line.TRBC, ID.centrality = line.centrality,
5  ID.company_name = line.company_name,
6  ID.ccgr = line.CCGR;
```

and, to load the edge table, we call

```
1  USING PERIODIC COMMIT
2  LOAD CSV WITH HEADERS FROM 'file:///file_path/rels.csv' as line_b
3  MATCH (sup:permID {name: TOINT(line_b.supplier)})
4  WITH sup, line_b MATCH (cus:permID {name: TOINT(line_b.customer)})
5  MERGE (sup)-[:supplies]->(cus)
```

This populates a previously initialized Neo4j instance, which can then be queried. In our case we want to evaluate the importance and risk of a pre-defined number of suppliers to a specific customer. To do so, we need to identify a single node within the database, i.e. the customer, and query for neighbours whose directionality points towards that node, i.e. the company's suppliers. Depending on the setting of the query we may do this recursively to not only retrieve direct suppliers, but suppliers of suppliers, also. We generally refer to direct suppliers as *first-tier* suppliers, to suppliers of suppliers as *second-tier* suppliers, and so forth. The below retrieves up to 1,000 first- and second-tier suppliers of the node *0123456789*.

```
1  MATCH (n:permID {name: 0123456789}),
2  p=shortestPath((x)-[:supplies*1..2]->(n))
3  WITH LENGTH(p) AS lp, x LIMIT 1001
4  RETURN
5  x.name, x.trbc, x.ccgr, x.centrality, x.company_name, lp;
```

Note that we set the limit to 1,001 because the node we are searching for is included in the limit, as well. While we have used Neo4j as the database of choice we opted to run graph analyses using *Gephi*[4]. On the one hand we have used Gephi to calculate closeness centrality scores for nodes in the graph, as described above. On the other hand Gephi provided a natural interface to run initial analyses on the graph to determine its overall structure. This includes calculating the measures reported in Table 1.

---

[4] https://gephi.org.

3.4   World Input-Output Database (WIOD)

The *World Input-Output Database (WIOD)* [21] provides data on the distribution of supply activities between business sectors.[5] We use this data as part of our importance calculation, where we compare the business sector the supplier operates in with the business sector the customer resides in. The WIOD allows us to deduce whether these two industries have a strong relation, in terms of relative volume exchanged between the business sectors, compared to other business sector combinations. The WIOD is comprised of supply data between a total of 43 countries and compares business sectors based on the *International Standard Industrial Classification (ISIC)*. Data is collected for the period between 2000 and 2014. Since for our work we have assigned TRBC codes to the companies that comprise the supply graph we use a mapping between TRBC and ISIC codes that has been created internally to align the WIOD with our data.

4   Scoring Method

The graph database described in the previous section facilitates the analysis of supply relations between companies within the context of a larger network. In this section we describe how we use the graph database to identify relevant suppliers of a customer through multiple tiers of the supply graph and score them according to two metrics, (1) importance and (2) risk. Importance, described in detail in Sect. 4.1, scores suppliers of a company based on a combination of metrics, incorporating both the structure of the graph and the supplier's position in it, and attributes of the supplier itself. With it we aim to quantify the adverse impact that a disruption to the supply from a specific supplier would have on a specific customer, where a high importance, i.e. a score close to 1, reflects a high potential adverse impact. Risk, described in detail in Sect. 4.2, is scored according the *credit risk scores* assigned to each company in the graph.

4.1   Scoring Supply Chain Importance

We calculate importance scores $I = (i_0, ..., i_n)$ for suppliers $S = (s_0, ..., s_n)$, each represented by a node in a graph, retrieved from the graph database. The nodes are retrieved in relation to node c, representing a customer, representing the n companies closest to c. Each importance score i_m is an aggregate of four measures:

a. *Criticality*: The proportion of goods the business sector of q receives from the business sector of i_m (based on WIOD data, see Sect. 3.4);

$$a = \frac{Criticality}{m};$$ (2)

where m is a normalization constant $m = 34.27$. The constant represents the strongest tie between any two industries in the WIOD dataset.

---

[5] http://www.wiod.org/home.

b. *Replaceability*: The sum of how many s operate in the same business sector as s_m (based on TRBC codes):

$$b = 1 - \left(\frac{Replaceability}{n - 1} \right) ; \tag{3}$$

c. *Centrality*: a metric of the importance of s_m to the overall graph (we can use *closeness centrality*, for instance);

$$c = Centrality \tag{4}$$

d. *Distance*: the (step-)distance between s_m and c.

$$d = \frac{n - Distance}{n - 1} \tag{5}$$

We then aggregate i_m as follows:

$$i_m = \frac{\left(\frac{a+b+c+d}{4} \right)}{max \in I} \tag{6}$$

The above operations normalize all individual scores to a value in the range $[0; 1]$. The scores are also normalized so that a value closer to one reflects a higher importance. We also normalize i_m by dividing its result by the maximum score of all i, so that the *most important* node always has a score of one and all other nodes are scored in relation to it. We catch the fringe case that yields division by zero programmatically (where $n = 1$), in which case we can simply set $i_0 \leftarrow 1$.

4.2   Scoring Supply Chain Risk

The second metric according to which we score the suppliers is credit risk. Akin to importance we calculate risk scores $R = (r_0, ..., r_n)$ for suppliers $S = (s_0, ..., s_n)$ of customer c. Each risk score r_m is based on a single attribute of a node, namely one of two scores, (1) *Credit Combined Global Rank* or (2) *Private Company SmartRatios Global Rank*. Score (1) is assigned to public companies, while score (2) is used for private companies. Both scores are extracted from proprietary Thomson Reuters solutions. The coverage of risk scores for companies that comprise the supply chain agreement dataset is roughly 26%. To cover the gaps we heuristically determined risk scores for companies without a risk score. In a first step we grouped companies based on their business sector, using TRBC codes, and calculated the average risk for each business sector, using the available scores. Companies with missing risk scores were then assigned the average risk score according to their TRBC code. Once each node in the graph had a risk score assigned we normalized the score so that its range is between zero and one, and a higher score represents a higher risk.

5   System Description

The implementation of our scoring methodology is comprised of two components. On the one hand we have implemented two APIs to expose scoring algorithms, one each to execute the importance and risk scoring for the suppliers retrieved form the graph database and returning the results as *json* files. On the other hand we set up a profile for an existing interface to dynamically visualize the results.

5.1   Application Programming Interfaces

The scoring algorithms described in the previous section are accessed through separate Application Programming Interfaces (APIs), each of which accept as arguments the following three parameters; *company ID*, *node count* and *depth count*. In the original dataset we use *permIDs*[6] as company IDs, which have been replaced by random ten-digit IDs in the public dataset.

The node count determines how many neighbours of the node representing the company ID are retrieved from the graph, while the depth count determines from how many tiers we retrieve neighbours. The system accepts two API calls, one to score supply chain importance and another to score risk. Each of the two APIs returns a *json* file with the following format:

```
1   {
2        "dimensionName": "Supply chain importance",
3        "peers":[
4        {
5               "eid": "0022446688",
6               "name": "c",
7               "score": 1,
8               "baseEntity": true
9        },{
10              "eid": "8800224466",
11              "name": "s_0",
12              "score": i_0,
13              "baseEntity": false
14       },{
15              "eid": "6688002244",
16              "name": "s_1",
17              "score": i_1,
18              "baseEntity": false
19       },...{
20              "eid": "4466880022",
21              "name": "s_n",
22              "score": i_n,
23              "baseEntity": false
24       }]
25  }
```

---

[6] https://permid.org/.

The output JSON file is comprised of the header and two types of blocks. The header identifies which dimension the scores in the JSON file represent. In the example this is the importance score. The first block following the header represents the input entity, i.e. the customer passed to the API. The *baseEntity* label is set to true to represent this and the score is set to a placeholder value of one. Each subsequent block represents a supplier of the *baseEntity*, which may be a supplier at any tier, depending on the parameter settings. Each block is comprised of the ID (*eid*), uniquely identifying the company, the company's name, its importance score and the *baseEntity* flag set to *false*. The output of the risk scoring API produces the same structure, the only difference being the dimension name in the header.

5.2 Interface

The importance and risk calculations, exposed through the APIs described above, are queried, and results visualized, using an internal application called *Jersey*. Jersey provides an interface with functionalities to search for an entity, here companies, and retrieve and visualize the entity's peers. Once a user submits a query, Jersey requests data through our APIs and renders the returned json files as shown in Fig. 1 or 2. Figure 1 shows importance as a single slider, aggregated as described in Sect. 4.1. Figure 2, on the other hand, returns the components of our importance score individually. In both cases the user can use the sliders to adjust the weighting of the individual scores, depending on individual preferences, with the second view allowing more granular weighting. Additionally Jersey offers options to use different weighting mechanisms and render more or fewer results.

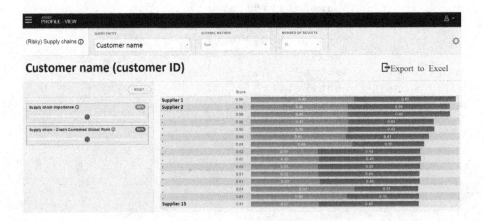

Fig. 1. Our application's web interface.

Fig. 2. Importance scores can be broken down into components.

6   Experiments

The risk score we have used is an industry standard measure of credit risk and we have thus focused our evaluation on determining whether the importance score, which we have developed as part of our work, produces a representative measure of the actual importance of a supplier to its customer. As described in Sect. 3.1, the supply chain agreements between two companies that comprise the graph are extracted from natural language text. To evaluate the veracity of how we score these relations on the importance we use the snippets from which the relations have been extracted. To build an evaluation dataset we have done the following:

1. Select 200 snippets that contain correctly classified supply relations.
2. Identify the corresponding supply agreement and score their importance using our application.
3. Select the 20 highest scoring, as well as the 20 lowest scoring snippets
4. Randomly match a high scoring snippet with a low scoring snippet to create a set of 200 pairs of snippets

Two annotators were then asked to the snippet in a pair that described a more *important* supply relation. To determine the veracity of the generated importance scores we then measured correlation between the scores and human judgments. The annotations produced rather low inter-annotator agreement, with a *Cohen's kappa* of $\kappa = 0.2499$. As expected, then, a *Pearson correlation coefficient* produces a similarly low score of $p = 0.1451$.

Discussion. Considering the low kappa score it appears that deciding the importance based solely on comparing snippets may not be a viable approach to judging the veracity of our importance score. The random matching of snippets without pre-selection may have also hurt our evaluation. Note, however, that we do not calculate importance based on judging the snippets and thus low correlation need not mean that our importance scores do not reflect reality. What we can conclude, based on the low kappa value, is that human judgement of importance based on short snippets does not seem viable and, based on the low p value, that we cannot confirm the veracity of our importance scores. A further limitation of our approach is that, because no fine-grained data at the level of each Other challenges that may have hurt performance are the data gaps described in Sect. 4.2 and the simple, equal weighting that produces the aggregate importance score, as described in Sect. 4.1 individual transaction between a supplier and consignee was available for this study, we had to estimate the connections, somewhat crudely, taking sector-to-sector flows as proxies. This will no doubt have had adversarial effects, and it is hard to quantify them. Nevertheless, we believe that having a quantitative model is valuable, and if and when more granular data becomes available, more refined models can be compared to our model to demonstrate their merit.

7   Summary, Conclusions and Future Work

With this paper we have presented a novel approach to evaluating suppliers of companies according to their importance and risk. We have developed a dataset comprised of roughly 98,000 companies and 220,000 supply relations, which we represent as a directed graph. Using a combination of a company's attributes and their position within the graph we determine their importance as suppliers to a specific company. This has allowed us not only to develop a principled representation of a supply graph, it also allows us to investigate supply beyond the first tier suppliers of a company.

One of the main shortcomings of our work is the limited insight we can gain from the evaluation. In future work we will need to identify new ways of evaluating the importance scores, based on which we can develop new weighting algorithms. Beyond this immediate concern the following three main avenues for further research present themselves. First, extending the scope of risks we measure will add to the informativeness of the overall scoring. For example, we may include evidence from text-mining based risk analysis approaches as described in [14]. Secondly, an extension of our approach to languages other than English will vastly expand the solution's usefulness from a practical application point of view. Whether such an extension ought to be based on machine translation or purpose built models for each language in scope remains to be seen. Finally, we believe that developing a mechanism by which we can either learn the weights of the individual importance scores or determine them through a grid search, rather than simply weighting them equally, may further enhance the quality of the importance scores assigned to companies.

Acknowledgments. We would like to thank Khalid Al-Kofahi and the CTO office for supporting this work and thank Giuseppe Saltini, Shai Hertz, Yoni Mataraso and Geoffrey Horrell for discussions and data.

References

1. Aggarwal, C.C.: An introduction to social network data analytics. In: Aggarwal, C.C. (ed.) Social Network Data Analytics, pp. 1–15. Springer, Heidelberg (2011)
2. Alhomidi, M., Reed, M.: Attack graph-based risk assessment and optimisation approach. Int. J. Netw. Secur. Appl. **6**(3), 31 (2014)
3. Aqlan, F., Lam, S.S.: A fuzzy-based integrated framework for supply chain risk assessment. Int. J. Prod. Econ. **161**, 54–63 (2015)
4. Bisias, D., Flood, M.D., Lo, A.W., Valavanis, S.: A survey of systemic risk analytics. US Department of Treasury, Office of Financial Research 0001 (2012)
5. Blome, C., Schoenherr, T.: Supply chain risk management in financial crises - a multiple case-study approach. Int. J. Prod. Econ. **134**(1), 43–57 (2011)
6. Borgatti, S.P., Li, X.: On social network analysis in a supply chain context. J. Supply Chain Manage. **45**(2), 5–22 (2009)
7. Ghadge, A., Dani, S., Chester, M., Kalawsky, R.: A systems approach for modelling supply chain risks. Supply Chain Manage. Int. J. **18**(5), 523–538 (2013)
8. Hallikas, J., Karvonen, I., Pulkkinen, U., Virolainen, V.-M., Tuominen, M.: Risk management processes in supplier networks. Int. J. Prod. Econ. **90**(1), 47–58 (2004)
9. Harland, C., Brenchley, R., Walker, H.: Risk in supply networks. J. Purch. Supply Manage. **9**(2), 51–62 (2003)
10. Huang, X., Vodenska, I., Havlin, S., Stanley, H.E.: Cascading failures in bi-partite graphs: model for systemic risk propagation. Sci. Rep. 3, Article no: 1219 (2013). doi:10.1038/srep01219
11. Jüttner, U.: Supply chain risk management: understanding the business requirements from a practitioner perspective. Int. J. Logist. Manage. **16**(1), 120–141 (2005)
12. Kim, Y., Choi, T.Y., Yan, T., Dooley, K.: Structural investigation of supply networks: a social network analysis approach. J. Oper. Manage. **29**(3), 194–211 (2011)
13. Mitchell, M.: An Introduction to Genetic Algorithms. MIT Press, Cambridge (1998)
14. Nugent, T., Leidner, J.L.: Risk mining: company-risk identification from unstructured sources. In: IEEE International Conference on Data Mining, ICDM, pp. 1308–1311 (2016)
15. Phillips, C.A., Swiler, L.P.: A graph-based system for network-vulnerability analysis. In: Proceedings of the 1998 Workshop on New Security Paradigms, Charlottsville, VA, USA, September 22–25, 1998, pp. 71–79 (1998)
16. Poolsappasit, N., Dewri, R., Ray, I.: Dynamic security risk management using Bayesian attack graphs. IEEE Trans. Dependable Sec. Comp. **9**(1), 61–74 (2012)
17. Simchi-Levi, D., Schmidt, W., Wei, Y.: From superstroms to factory fires: managing unpredictable supply chain disruptions. Harv. Bus. Rev. **92**(1), 96–100 (2014)
18. Stergiopoulos, G., Kotzanikolaou, P., Theocharidou, M., Gritzalis, D.: Risk mitigation strategies for critical infrastructures based on graph centrality analysis. IJCIP **10**, 34–44 (2015)
19. Tan, K.H., Zhan, Y., Ji, G., Ye, F., Chang, C.: Harvesting big data to enhance supply chain innovation capabilities: an analytic infrastructure based on deduction graph. Int. J. Prod. Econ. **165**, 223–233 (2015)
20. Tayur, S., Ganeshan, R., Magazine, M.: Quantitative Models for Supply Chain Management, vol. 17. Springer, Heidelberg (2012)
21. Timmer, M.P., Dietzenbacher, E., Los, B., Stehrer, R., Vries, G.J.: An illustrated user guide to the world input-output database: the case of global automotive production. Rev. Int. Econ. **23**(3), 575–605 (2015)

22. Wagner, S.M., Neshat, N.: Assessing the vulnerability of supply chains using graph theory. Int. J. Prod. Econ. **126**(1), 121–129 (2010)
23. Xu, N.-R., Liu, J.-B., Li, D.-X., and Wang, J.: Research on evolutionary mechanism of agile supply chain network via complex network theory. In: Mathematical Problems in Engineering 2016 (2016)

Declarative Data Transformations for Linked Data Generation: The Case of DBpedia

Ben De Meester[(⊠)], Wouter Maroy, Anastasia Dimou, Ruben Verborgh, and Erik Mannens

Ghent University - imec - IDLab, Ghent, Belgium
{ben.demeester,wouter.maroy,anastasia.dimou,
ruben.verborgh,erik.mannens}@ugent.be

Abstract. Mapping languages allow us to define how Linked Data is generated from raw data, but only if the raw data values can be used *as is* to form the desired Linked Data. Since complex data transformations remain out of scope for mapping languages, these steps are often implemented as custom solutions, or with systems separate from the mapping process. The former data transformations remain case-specific, often coupled with the mapping, whereas the latter are not reusable across systems. In this paper, we propose an approach where data transformations (i) are defined declaratively and (ii) are aligned with the mapping languages. We employ an alignment of data transformations described using the Function Ontology (FnO) and mapping of data to Linked Data described using the RDF Mapping Language (RML). We validate that our approach can map and transform DBpedia in a declaratively defined and aligned way. Our approach is not case-specific: data transformations are independent of their implementation and thus interoperable, while the functions are decoupled and reusable. This allows developers to improve the generation framework, whilst contributors can focus on the actual Linked Data, as there are no more dependencies, neither between the transformations and the generation framework nor their implementations.

Keywords: Data transformations · FnO · Linked Data generation · RML

1 Introduction

Workflows that generate Linked Data from (semi-)structured data encompass both *schema* and *data* transformations [22]. *Schema* transformations involve (re-)modeling the original data, describing how objects are related, and deciding which vocabularies and ontologies to use [18]. *Data* transformations are needed

---

The described research activities were funded by Ghent University, imec, Flanders Innovation & Entrepreneurship (AIO), the Fund for Scientific Research Flanders (FWO Flanders), and the European Union.

E. Blomqvist et al. (Eds.): ESWC 2017, Part II, LNCS 10250, pp. 33–48, 2017.
DOI: 10.1007/978-3-319-58451-5_3

to support any changes in the structure, representation or content of data [22], for instance performing string transformations or computations.

Schema transformations – also called *mappings* – are defined as a collection of rules that specify correspondences between data in different schemas [13]. Lately, schema transformations are declaratively defined using mapping languages such as the W3C-recommended R2RML [7] or its extension RML [12]. Mapping languages detach rule definitions from the implementation that executes them. This renders the rules *interoperable* between implementations, whilst the systems that process those rules are *use-case independent*.

However, Linked Data generation systems usually assume data transformations are done beforehand. For instance, the R2RML specification explicitly mentions that data transformations or computations should be performed before generating Linked Data by generating a virtual table based on the result-set of an SQL statement (i.e., an SQL view) [7]. Other relevant W3C recommendations and working drafts do not take data transformations into account. More precisely, when discussing the "Convert Data to Linked Data" step, the Linked Data Best Practices [18] recommends using mapping languages – which only implies *schema* transformations – and does not distinguish data transformations elsewhere. Similarly, CSVW [27] specifies how to generate Linked Data from CSV by directly mapping the raw data values *as is*.

Systems that do include data transformations exist, but show one or more of the following limitations: the schema and data transformations are *uncombinable*, the allowed data transformations are *restricted*, the system is *case specific*, or the data transformations are *coupled* with the implementation.

For instance, the DBpedia Extraction Framework [2] (DBpedia EF) that generates Linked Data for one of the most widely known datasets, requires very specific data transformations, which are not available in existing systems. Thus, a case-specific hard-coded framework that depends on an internal set of parsing functions to generate the data values in the correct format was created. These parsing functions are *coupled* with the DBpedia EF, the schema and data transformations are *uncombinable*, and the overall system is *case specific*. Specifically for the DBpedia EF, these parsing functions are of high value. Indeed, they were created to parse manually entered (i.e., ambiguous and error-prone) data and are used for (and thus evaluated on) the entire Wikipedia corpus.

In this paper, we propose an approach that enables (i) *declarative and machine-processable data transformation descriptions* and (ii) the *alignment* of schema and data transformation descriptions. To validate this approach, we employ transformations described using the Function Ontology (FnO) [10], and align them with the RDF Mapping Language (RML) [12].

We apply our approach to the DBpedia EF. In the resulting system:

Schema and data transformations are *combinable*:
No separate systems need to be integrated.
Data transformations are *independent* of the mapping processor:
They are not restricted by the processor's capabilities.

Declarative transformations are *interoperable*:
The implementation can be case-independent.
Data transformations are *reusable*:
Their implementation is no longer dependent on the generation system.

We built and used the extended RMLProcessor with mapping documents to create the same DBpedia dataset, allowing more types of schema transformations, and enabling developers to work separate from contributors. The built Function Processor allows for an easier integration of data transformation libraries with other frameworks, and the DBpedia data parsing functions are made available independently, so other use cases can benefit from these data parsing tasks without needing to re-implement them.

The paper is organized as follows: after investigating the state of the art in Sect. 2, we detail why aligning declarative schema and data transformations is needed in Sect. 3. In Sect. 4, we introduce our approach, which we apply to RML and FnO, and provide a corresponding implementation. In Sect. 5, we explain how the DBpedia EF currently works, and prove how applying the proposed approach enables a fully declarative system with the same functionality as the existing DBpedia EF. Finally, we summarize our conclusions in Sect. 6.

2   State of the Art

Linked Data generation workflows require both *schema* and *data* transformations to generate the desired Linked Data [22]. Nevertheless, even though data transformations are often required [16], recommendations or best practices were not established so far, leading to a broad range of diverse approaches.

The simplest approaches rely on *custom solutions* which try to address both schema and data transformations in a coupled and hard-coded way, such as the DBpedia EF [2]. However, those approaches require new development cycles every time a modification or extension is desired. For instance, any change on the data transformations performed to generate the DBpedia dataset requires extending the DBpedia EF. There are cases where such approaches do allow certain configurations, yet those configurations are limited and, at least for the DBpedia EF, they focus on schema transformations rather than data transformations.

Similarly, *case-specific solutions* were established, which also couple the schema and data transformations. For instance, XSLT- or XPath-based approaches were established for generating Linked Data from data originally in XML format, such as by Lange [20]. In these cases, the range of possible transformations is limited by the respective language or syntax potential, while they can be performed prior or while the mapping is performed. Similarly, even mapping languages, such as HIL [14], D2RQ [6], or R2RML [7] can be considered, as their range of possible data transformations is determined by the range of transformations that can be defined when the data is retrieved from the data source, e.g., data transformations supported by SQL, performed before the mapping.

Other solutions first perform a direct mapping [1] to Linked Data, and then perform schema and data transformations on that generated Linked Data.

The range of possible data transformations then often depends on SPARQL, as is the case of Datalift [25]. More customization is enabled by solutions that allow embedded scripts inside mapping documents such as R2RML-F [11]. However, they require existing libraries (and their dependencies) to be embedded (or possibly rewritten) within the mapping document, and are inherently limited to the standard libraries provided by the runtime environment (e.g., runtime environments often – for safety reasons – disallow file Input/Output operations).

There are also *query-oriented* languages that combine SPARQL with other languages or custom alignments to the underlying data structure. For instance, XSPARQL [5] maps data in XML format, R3M [15] data in relational databases and Tarql[1] data in CSV. Query-oriented languages are restricted to data transformations which can be translated when the query translation is performed, such as R3M that requires bidirectional transformations to retain read-write access [16].

Besides the aforementioned custom solutions, there are Linked Data generation workflows which rely on distinct systems to perform the schema and data transformations. These types of transformations cannot always be distinguished, as data transformations may affect the original schema. Such *data transformation tools* typically couple the transformation rules with their implementation, being either format specific (e.g., XSLT for data in XML format), or generic (e.g., Open Refine[2]). As the latter are targeted to contributors, they are often interactive. Thus, most data transformation systems can be configured and this happens often using a User Interface (UI), of which one of the most widely known is OpenRefine. Other systems – specifically for generating Linked Data – include the Linked Data Integration Framework (LDIF) [26], Linked Pipes [19], and DataGraft [24]. Their support for data transformations range from a fixed predefined set of transformations (e.g., LDIF and Linked Pipes) to an embedded scripting environment (e.g., OpenRefine and DataGraft).

Lately, different approaches emerged that define data transformations declaratively, such as Hydra [21] for Web Services, VOLT [23] for SPARQL, or FnO [10] as technology-independent abstraction. Hydra or VOLT depend on the underlying system (Web Services and SPARQL, respectively), thus their use is inherently limited to that system. On the one hand, using Hydra descriptions for executing transformations only works online, and requires all data to be transferred over HTTP, which is not always possible due to size or privacy concerns. On the other hand, VOLT only works for data already existing in a SPARQL endpoint. Describing the transformations using FnO does not include this dependency, thus allows for reuse in other use cases and technologies.

3   Limitations and Requirements

In this section, we discuss current schema and data transformation systems limitations (Sect. 3.1) and requirements (Sect. 3.2).

---

[1] https://tarql.github.io/.
[2] http://openrefine.org/.

3.1   Limitations of Current Systems

Considering the current Linked Data generation systems discussed above, we come across data transformations which are (i) *uncombinable*, (ii) *restricted*, (iii) part of a *case specific* system, or (iv) *coupled*, as we discuss below.

Uncombinable. When schema and data transformations are executed in successive steps (e.g., in the DBpedia EF, R2RML, or Datalift), additional integration is needed between them. However, schema and data transformations often depend on each other. Data transformations could influence the attributes of objects and vice versa. For instance, the calculated population of a settlement decides whether it is called a "town" or "city". This integration thus becomes complex, hurts interoperability, or limits the allowed transformations.

Restricted. Data transformations are embedded, defined and coupled within the system that executes them. Both in dedicated data transformation systems as when data transformations are embedded in mapping languages, the range and type of transformations used is limited to the ones implemented by the underlying system. Either a fixed set of data transformations is provided (e.g., LDIF, Unified Views), thus no other transformations can be defined, or a restricted scripting environment allows the definition of data transformations (e.g., OpenRefine, R2RML-F). In both cases limitations exist, e.g., using additional libraries, file manipulations, or external services are often disallowed. As such, existing tools cannot be applied for every use case. Supporting specialized use cases then usually requires providing separate systems (e.g., GeoSPARQL [4] for the geospatial domain). For instance, Blake et al. [23] unveiled quality issues in DBpedia as the current extraction framework does not support basic geographic calculations, such as calculating the population density.

Case Specific. Hard-coded systems couple the reference to a certain transformation with its implementation, and also mapping languages and dedicated systems support an opinionated set of transformations. As such, they can only be used for certain cases, and they require changes to the source code to apply any modifications or extensions, i.e., new developments cycles.

Coupled. So far, data transformations definitions are coupled with the implementation that executes them. For instance, data transformations specified by OpenRefine cannot be reused in other systems, and data transformations implemented in hard-coded systems are only available for that system and not reusable by others. Similarly, the coverage of data transformations differs across data sources, e.g., it is different between different SQL dialects for relational databases, XQuery for XML documents, and JSONPath for JSON documents. Chances of discrepancies between different systems (and the Linked Data they produce) are thus very high.

3.2 Requirements for Future Systems

In this paper, we argue that data transformations should be (i) *declaratively* specified, and (ii) *aligned* with declarative mapping languages. By specifying data transformations declaratively, just as for mapping languages, we decouple the transformations from the implementation that executes them. By aligning them with mapping languages instead of embedding them within the mapping languages, we remove the burden of the mapping processor to provide all needed functionality, allowing the implementations of the data transformations to exist separately from the generation system. This way, we achieve data transformations which are *reusable*, *interoperable*, *independent*, and *combinable*, as detailed below.

Reusable. Data transformations implementations should be reusable across use cases and systems, not necessarily only for Linked Data generation.

Interoperable. The declarations for data transformations should remain independent of the underlying implementation, i.e., be interoperable. This strictly separates the concerns of developers with those of contributors: developers can implement and improve the tools without being required to obtain domain knowledge, whilst contributors can focus on data modeling without being needed to get acquainted with the systems' source code. The generation of these declarations can be facilitated using a (graphical) editor [17].

Independent. Schema and data transformation declarations should be independent from each other. As such, their corresponding implementations also remain independent of each other, without enforcing mutual limitations. As such, (custom) data transformations can be integrated in the mapping process, but it is not required, i.e., they can still be executed in advance, and the mapping languages can still be used without data transformations.

Combinable. Data transformations should be usable not only in separate steps, but be combinable, e.g., with schema transformations. This enables, e.g., joining and meanwhile transforming multiple input values, or conditionally change the schema depending on the data transformations and vice versa.

4 Declarative Data Transformations

We provide a solution that implements the aforementioned declarative, machine processable data transformations which are aligned with schema transformations to Linked Data. Its main components are (i) the FnO ontology (Sect. 4.1), which enables describing functions in a declarative and machine processable way without making assumptions of their implementation; and (ii) the RML language (Sect. 4.2) that allows defining schema transformations (i.e., mappings) for generating Linked Data, independent of the input format. The Function Map is introduced, as an extension of RML, to facilitate the alignment of the two as explained in Sect. 4.3. Details regarding our proof-of-concept implementations are summarized in Sect. 4.4. For the remainder of this paper, we will use the following prefixes:

```
PREFIX fno:   <http://w3id.org/function/ontology#>
PREFIX grel:  <http://semweb.datasciencelab.be/ns/grel#>
PREFIX rr:    <http://www.w3.org/ns/r2rml#>
PREFIX rml:   <http://semweb.datasciencelab.be/ns/rml#>
PREFIX fnml:  <http://semweb.datasciencelab.be/ns/fnml#>
```

4.1   The Function Ontology (FnO)

The Function Ontology (FnO) [8,10] allows agents to declare and describe functions uniformly, unambiguously, and independently of the technology that implements them. As mentioned in Sect. 2, we choose FnO over other declarative languages as it does not depend on the underlying system or implementation. A *function* (`fno:Function`) is an activity which has input parameters, output, and implements certain algorithm(s). A *parameter* (`fno:Parameter`) is the description of a function's input value. An *output* (`fno:Output`) is the description of the function's output value. An *execution* (`fno:Execution`) assigns values to the parameters of a function for a certain execution.

The actual implementation of the function can be retrieved separately from its description. Depending on the system, different implementations can be retrieved/used, e.g., a system implemented in Java can retrieve the implementation as a Java archive (JAR), whilst a browser-based system might rely on external APIs. Via content negotiation, different systems can request and discover different implementations of the same described function [9], given that these implementations exist. This allows a mapping processor to parse any function description, and retrieve and trigger the corresponding implementation for executing it.

For instance, `grel:toTitleCase`[3] (Listing 1, line 1) is a function that renders a given string into its corresponding title cased value. It expects a string, indicated by the `grel:stringInput` property (line 4) as input. An Execution (line 6) can be instantiated to bind a value to the parameter. The result is then bound to that Execution via the `grel:stringOutput` property (line 9).

4.2   The RDF Mapping Language (RML)

R2RML [7] is the W3C-recommended mapping language for defining mappings of data in relational databases to the RDF data model. Its extension RML [12] broadens its scope and covers also schema transformations from sources in different (semi-)structured formats, such as CSV, XML, and JSON. RML documents [12] contain rules defining how the input data will be represented in RDF. The main

---

[3] Specified from the description as provided by OpenRefine on https://github.com/OpenRefine/OpenRefine/wiki/GREL-String-Functions#totitlecasestring-s.

```
1   grel:toTitleCase a fno:Function ;
2     fno:name    "title case" ;
3     dcterms:description "return the input string in title case" ;
4     fno:expects ( [ fno:predicate grel:stringInput  ] ) ;
5     fno:output  ( [ fno:predicate grel:stringOutput ] ) .
6   :exe a fno:Execution ;
7     fno:executes grel:toTitleCase ;
8     grel:stringInput  "This is an input STRING." ;
9     grel:stringOutput "This Is An Input String." .
```

Listing 1. Function descriptions and Executions using FnO

```
1   <#Mapping> rml:logicalSource <#InputX> ;
2     rr:subjectMap [ rr:template "http://ex.com/{ID}"; rr:class foaf:Person ];
3     rr:predicateObjectMap [ rr:predicate foaf:knows;
4       rr:objectMap [ rr:parentTriplesMap <#Acquaintance> ]].
5   <#Acquaintance> rml:logicalSource <#InputY> ;
6     rr:subjectMap [ rml:reference "acquaintance"; rr:termType rr:IRI; rr:class ex:Person]].
```

Listing 2. RML mapping definitions

building blocks of RML documents are Triples Maps (Listing 2: line 1). A Triples Map defines how triples of the form (subject, predicate, object) will be generated.

A Triples Map consists of three main parts: the Logical Source, the Subject Map and zero or more Predicate-Object Maps. The Subject Map (line 2, 6) defines how unique identifiers (URIs) are generated for the mapped resources and is used as the subject of all RDF triples generated from this Triples Map. A Predicate-Object Map (line 3) consists of Predicate Maps, which define the rule that generates the triple's predicate (line 3) and Object Maps or Referencing Object Maps (line 4), which define how the triple's object is generated. The Subject Map, the Predicate Map and the Object Map are Term Maps, namely rules that generate an RDF term (an IRI, a blank node or a literal). A Term Map can be a *constant-valued term map* (line 3) that always generates the same RDF term, or a *reference-valued term map* (line 6) that uses the data value of a referenced data fragment in a given Logical Source, or a *template-valued term map* (line 2) that uses a valid string template that can contain referenced data fragments of a given Logical Source.

Other languages used for mapping (such as CSVW, XPath, and SPARQL) are dependent on the input format (CSV, XML, and SPARQL, respectively). RML abstracts the input source format, making it applicable in more use cases. Moreover, as the schema transformations are declared in RDF, the integration with external vocabularies or data sources is inherently available.

4.3   Model Integration

Typically, mapping languages refer to raw data values. Therefore, aligning them with declarative data transformations requires a way to refer to terms which are derived from raw data, but after applying certain transformations, i.e., functions.

In the case of [R2]RML, Term Maps determine how to generate an RDF term relying on references to raw data. Therefore, a new type of Term Map was

```
1   <#Person_Mapping>
2     rml:logicalSource      <#LogicalSource> ; # Specify the data source
3     rr:subjectMap           <#SubjectMap>   ; # Specify the subject
4     rr:predicateObjectMap <#NameMapping>    . # Specify the predicate-object-map
5
6   <#NameMapping>
7     rr:predicate dbo:title                  ; # Specify the predicate
8     rr:objectMap <#FunctionMap>             . # Specify the object-map
9
10  <#FunctionMap>
11    fnml:functionValue [                     # The object is the result of the function
12      rml:logicalSource <#LogicalSource>   ; # Use the same data source for input
13      rr:predicateObjectMap [
14        rr:predicate fno:executes          ; # Execute `grel:titleCase`
15        rr:objectMap [ rr:constant  grel:titleCase ] ] ;
16      rr:predicateObjectMap [
17        rr:predicate grel:inputString       ;
18        rr:objectMap [ rr:reference "name" ] ] # Use as input the "name" reference
19    ] .
```

Listing 3. Alignment RML and FnO

introduced, the Function Map (`fnml:FunctionMap`, Listing 3: line 10). A Function Map is a Term Map generated by executing a function, instead of using a constant or a reference to the raw data values. In contrast to an RDF Term Map that uses values referenced from a Logical Source to generate an RDF term, a Function Map uses values referenced from a Logical Source to execute a function (line 12). Once the function is executed, its output value is the term generated by this Function Map. To this end, the `fnml:functionValue` property was introduced to indicate which instance of a function needs to be executed to generate an output and considering which values (line 11). Such a function is described using FnO.

This extension of one class and one property allows us to align RML and FnO, without creating additional dependencies between the two. This is possible as they are both declarative and described in RDF.

4.4   Implementation

As a proof of concept, we extended the RMLProcessor to support the Function Map, available at github.com/RMLio/RML-Mapper/tree/extension-fno. In addition, we implemented a generic Function Processor in Java which can be found at github.com/FnOio/function-processor-java that uses the function declarations described in FnO to retrieve and execute their relevant implementations. When the RMLProcessor encounters a Function Map[4], it extracts the function identifier (i.e., its URI) and the parameter values as described in the mapping document or from the data sources[5], and sends those to the Function Processor. When receiving an unknown function identifier, the Function Processor discovers the relevant implementations online [9], and obtains an implementation to be executed locally

---

[4] https://git.io/vSPDg.
[5] https://git.io/vSPD6.

if available[6]. Based on the function description using FnO, the Function Processor automatically detects how to execute the needed function and returns the resulting value back to the RMLProcessor.

We extracted GREL functions and the DBpedia parsing Functions (see Sect. 5) as independent libraries at github.com/FnOio/grel-functions-java and github.com/FnOio/dbpedia-parsing-functions-scala, respectively. Their descriptions using FnO are available at semweb.datasciencelab.be/ns/grel and semweb.datasciencelab.be/ns/dbpedia-functions respectively. Thus, we can (re-)use these functions separately from their original systems (i.e., OpenRefine and the DBpedia EF), but we can also – when using their descriptions in mapping documents – require them as data transformations within the RMLProcessor.

Our resulting extension of the RMLProcessor overcomes the limitations as stated in Sect. 3. It is capable of combining schema and data transformations. It could already process [R2]RML statements, and now, it can also extract the Function Map and allows the Function Processor to perform the data transformations. Next, the Function Processor is independent of the RMLProcessor, thus no limitations are enforced between them, and the system does not depend on the use case, as all schema and data transformations are specified in the mapping document and the implementations of the needed data transformations are obtained on the fly. Finally, all data transformations are available as stand-alone libraries, independent of the use case, the Function Processor, or the RMLProcessor.

We also extended the RMLEditor [17] to support the definition of Function Maps so users can easily edit mapping documents with declarative data transformations, without needing prior knowledge about RML or FnO. The default version of the RMLEditor considers the GREL functions, but any other function may be available. A screencast showcasing how the RMLEditor was extended can be found at www.youtube.com/watch?v=-38pkkTxQ1s. In total, users from 16 companies and research institutes profit from this RMLEditor extension in addition to the DBpedia community.

5   Application to DBpedia

In this section, we show the current DBpedia generation workflow (Sect. 5.1), the changes we implemented (Sect. 5.2), and validate our approach (Sect. 5.3).

5.1   Current Generation Workflow with the DBpedia EF

DBpedia is a crowd-sourced community effort to extract structured information from Wikipedia and make this information available on the Web [2]. Data from DBpedia is generated in two parts: The first maps data from the relationships already stored in the underlying relational database tables and the second directly extracts data from the article texts and infobox templates within the

---

[6] Currently, Java snippets and JARs are supported, as the latter allows using additional dependencies in the implemented functions.

articles [3]. Figure 1 shows the current DBpedia EF, specifically focused on the RDF generation from infobox templates (i.e., the second part). The grey area denotes the DBpedia-specific implementation, and the cogs denote the successive processing steps.

Infobox templates are text fragments inside the article text with specific syntax to denote certain visualizations (e.g., '{{' and '}}' denote the beginning and ending of an infobox table, respectively). The DBpedia EF consists of the following steps: step a, all Wikipedia pages containing infobox templates for the relational database are selected. Then, step b, only the significant templates which are contained in these pages are selected and extracted. Step c, each template is then parsed to generate the desired triples (i.e., the subject and predicate-object pairs). Afterwards, step d, object values are further post-processed, i.e., (i) when these object values contain Wiki links, suitable URI references are generated (the bottom arrow of step d in Fig. 1), otherwise, (ii) uniform typed literals are generated by parsing the strings and numeric values (the top arrow of step d). The data of DBpedia is structured using the dedicated DBpedia Ontology[7]: a cross-domain ontology, which has been manually created based on the most commonly used infoboxes within Wikipedia.

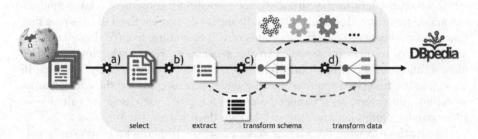

select   extract   transform schema   transform data

Fig. 1. The current generation workflow: successive hard-coded processes (a) select pages, (b) extract infoboxes, (c) transform the schema, and (d) transform the data, either by generating URIs (bottom arrow) or by using hard-coded parsing functions (top arrow).

The extracted infobox contains a textual representation of a list of key-value pairs, e.g., the item 'established = 4 October 1830'. After assigning per key a fixed predicate from the DBpedia Ontology and a fixed data type to the value [3], each value is processed individually according to that datatype. Wiki links are converted to meaningful URIs, but other values need to be parsed. However, since there are not many restrictions on the design of Wikipedia templates, the format of these manually entered values can be very diverse. For instance, when revisiting the previous example, the same date can be written down as '04-10-1830', '1830, 4 10', 'October 4th 1830', etc. Many other types of discrepancies occur, for example, using different numbering formats

---

[7] http://dbpedia.org/ontology.

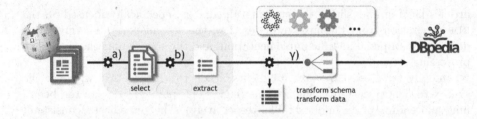

Fig. 2. The new generation workflow: after (a) selecting pages and (b) extracting the infoboxes using the original framework, (γ) both schema and data transformations are combined using an interoperable mapping processor and reusable parsing functions, specified by a fully declarative mapping document.

(e.g., '1 000 000' vs '1E6' vs '1 million'), or using different units than specified in the template (e.g., 'area_km2 = 11,787 sqmi'). This situation is aggravated because information in Wikipedia is crowd-sourced, thus these differences in cultures and countries – coming from different contributors of Wikipedia – can occur within one page, together with already existing inaccuracies inherent to manual entries, such as typos and misspellings.

To accommodate to this situation, the DBpedia EF consists of a large amount of parsing functions that fruitfully handle most edge cases. Each of these parsing functions were tested against thousands of values coming from Wikipedia. They are thus very robust and essential to the generation framework. However, they form an internal set of functions, hard-coded in the framework. Each change in these parsing functions requires another development cycle for the entire framework, but, moreover, they cannot be reused for other use cases. As valuable as these parsing functions are, they are hidden deep within the DBpedia EF.

Hence, the following limitations arise. First, the DBpedia EF successively performs the schema and then the data transformations, which limits its capabilities, e.g., it is currently not possible to join multiple values from the infobox templates to form one output value, nor is it possible to connect with external data sources. Second, all transformations are hard-coded. Changes require knowledge of the source code and involve new development cycles. Third, all parsing functions are embedded in the framework, making them non-reusable and use-case specific.

5.2   New Generation Workflow with RML and FnO

We apply our system that enables declarative data transformations which can be aligned with schema transformations to the the DBpedia EF as can be seen in Fig. 2. In step *a*, Wikipedia pages containing infobox templates are selected. Then, in step *b*, the significant templates are selected and extracted from those pages. Finally, in step γ, on these templates, schema and data transformations are performed together to achieve the resulting RDF.

Steps *a* and *b* provide the input data and have not changed. Step γ however is performed using the RMLProcessor, the transformations are declared using a DBpedia mapping document, and the DBpedia parsing functions are used as stand-alone library. The generation of the DBpedia mapping document in RML based on the existing mappings in the DBpedia EF has been done in previous work[8]. This work has been extended to include the data transformation descriptions.

fnoio.github.io/dbpedia-demo/ allows users to try out the possible customizations of the new DBpedia EF. Changes can be made to a mapping document – used for the country-infobox template – both for schema and data transformations. Both the DBpedia parsing functions as the GREL functions are loaded. It is thus possible to, e.g., change string values using GREL functions, or use a different parsing function, whilst also changing the schema transformations, without needing prior knowledge of the DBpedia EF.

5.3   Validation

By applying our approach to DBpedia, we have not only created a fully declarative system that is capable of extracting the same RDF data from the Wikipedia infoboxes as the current DBpedia EF, we also achieve the following:

Combinable **schema and data transformations.** *Before*, schema and data transformations were executed in successive steps in the DBpedia EF. Consequently, the data transformations were executed based on the data type as assigned by the schema transformations, and transformations applying to both schema and data were not supported. *Now*, data transformations can be specified within the structure, not just the data type, and joining multiple input values, or conditionally assigning types based on the data values becomes possible.

Independent **schema and data transformations.** *Before*, all data parsing functions needed to be hard-coded inside the DBpedia EF, as existing tools did not provide the required data transformation capabilities. *Now*, all data parsing functions are separate libraries, and no dependencies exist between these data parsing functions and the DBpedia EF.

An *interoperable* **system.** *Before*, the DBpedia EF was a hard-coded system depending on a custom mapping document that mapped keys to predicates of the DBpedia Ontology, after which hard-coded data transformations were performed. Every change in the generation process required a new development cycle. This explains why the DBpedia EF has been developed by only forty-two developers[9]. *Now*, no dependencies exist between the implementation and the specification of the generated Linked Data, as schema transformations, data transformations, and their alignment are all specified declaratively. The adjusted RMLProcessor remains a use-case independent system, and the

---

[8] www.mail-archive.com/dbpedia-discussion@lists.sourceforge.net/msg07837.html.

[9] See github.com/dbpedia/extraction-framework.

declarations do not depend on any implementation, separating the concerns of the contributors with those of the developers.

Reusable **data transformations.** *Before*, all data parsing functions were embedded in the DBpedia EF, making it even harder for developers to improve its code. The core team that improved the DBpedia EF parsing functions consisted of barely six out of the forty-two people. *Now*, all parsing functions exist as a stand-alone library, without dependencies to the original DBpedia EF, RML or FnO. They can be used and improved or extended by anyone, for any use case. The common problem of parsing manually entered data has just become more easy as this set of functions can now freely be used: it has been tested on the Wikipedia corpus, is capable of resolving many typos and ambiguities, and now no longer depends on the use case or data source type. Its usage has been made user-friendly by including data transformations in the RMLEditor.

6   Conclusion and Future Work

Linked Data generation encompasses both schema and data transformations. However, in this paper, we identified that data transformations in current Linked Data generation processes are uncombinable with the schema transformations, restricted by the mapping language, part of a case-specific system, or non-reusable.

Our proposed approach specifies data transformations declaratively and aligns them with declarative schema transformations. We employed this approach by aligning FnO with RML and provided an implementation by extending the RMLProcessor and building the Function Processor. As validated on the DBpedia EF, schema and data transformations remain independent but are combinable. The created system is interoperable and data transformations are reusable across systems and data sources. The DBpedia EF now supports more schema and data transformations, separates the concerns between contributors and developers, and the DBpedia parsing functions are available as independent libraries.

In the future, we aim to reuse well-tested descriptive data transformations, such as the DBpedia parsing functions to facilitate different use cases.

References

1. Arenas, M., Bertails, A., Prudhommeaux, E., Sequeda, J.: A direct mapping of relational data to RDF. W3C Recommendation (2012). http://www.w3.org/TR/rdb-direct-mapping/
2. Auer, S., Bizer, C., Kobilarov, G., Lehmann, J., Cyganiak, R., Ives, Z.: DBpedia: a nucleus for a web of open data. In: Aberer, K., et al. (eds.) ASWC/ISWC - 2007. LNCS, vol. 4825, pp. 722–735. Springer, Heidelberg (2007). doi:10.1007/978-3-540-76298-0_52

3. Auer, S., Lehmann, J.: What have innsbruck and leipzig in common? Extracting semantics from wiki content. In: Franconi, E., Kifer, M., May, W. (eds.) ESWC 2007. LNCS, vol. 4519, pp. 503–517. Springer, Heidelberg (2007). doi:10.1007/978-3-540-72667-8_36
4. Battle, R., Kolas, D.: GeoSPARQL: enabling a geospatial Semantic Web. Semant. Web J. **3**(4), 355–370 (2011)
5. Bischof, S., Decker, S., Krennwallner, T., Lopes, N., Polleres, A.: Mapping between RDF and XML with XSPARQL. J. Data Semant. **1**(3), 147–185 (2012)
6. Cyganiak, R., Bizer, C., Garbers, J., Maresch, O., Becker, C.: The D2RQ Mapping Language. Technical report (2012). http://d2rq.org/d2rq-language
7. Das, S., Sundara, S., Cyganiak, R.: R2RML: RDB to RDF Mapping Language. Working group recommendation, W3C, September 2012. http://www.w3.org/TR/r2rml/
8. De Meester, B., Dimou, A.: The Function Ontology. Unofficial Draft (2016). https://w3id.org/function/spec
9. De Meester, B., Dimou, A., Verborgh, R., Mannens, E.: Discovering and using functions via content negotiation. In: 15th International Semantic Web Conference: Posters & Demonstrations Track. CEUR Workshop Proceedings, vol. 1690 (2016)
10. De Meester, B., Dimou, A., Verborgh, R., Mannens, E.: An ontology to semantically declare and describe functions. In: Sack, H., Rizzo, G., Steinmetz, N., Mladenić, D., Auer, S., Lange, C. (eds.) ESWC 2016. LNCS, vol. 9989, pp. 46–49. Springer, Cham (2016). doi:10.1007/978-3-319-47602-5_10
11. Debruyne, C., O'Sullivan, D.: R2RML-F: towards sharing and executing domain logic in R2RML mappings. In: Workshop on Linked Data on the Web. CEUR Workshop Proceedings, vol. 1593 (2016)
12. Dimou, A., Vander Sande, M., Colpaert, P., Verborgh, R., Mannens, E., Van de Walle, R.: RML: a generic language for integrated RDF mappings of heterogeneous data. In: Proceedings of the 7th Workshop on Linked Data on the Web. CEUR Workshop Proceedings, vol. 1184 (2014)
13. Euzenat, J., Shvaiko, P.: Ontology Matching. Springer, New York (2013)
14. Hernández, M., Koutrika, G., Krishnamurthy, R., Popa, L., Wisnesky, R.: HIL a high-level scripting language for entity integration. In: Proceedings of the 16th International Conference on Extending Database Technology. ACM (2013)
15. Hert, M., Reif, G., Gall, H.C.: 'Semantic Web 2.0' - write-enabling the Web of Data. In: 6th Workshop on Semantic Web Applications and Perspectives (2010)
16. Hert, M., Reif, G., Gall, H.C.: A comparison of RDB-to-RDF mapping languages. In: Proceedings of the 7th International Conference on Semantic Systems. ACM (2011)
17. Heyvaert, P., Dimou, A., Herregodts, A.-L., Verborgh, R., Schuurman, D., Mannens, E., Walle, R.: RMLEditor: a graph-based mapping editor for linked data mappings. In: Sack, H., Blomqvist, E., d'Aquin, M., Ghidini, C., Ponzetto, S.P., Lange, C. (eds.) ESWC 2016. LNCS, vol. 9678, pp. 709–723. Springer, Cham (2016). doi:10.1007/978-3-319-34129-3_43
18. Hyland, B., Atemezing, G., Villazón-Terrazas, B.: Best Practices for Publishing Linked Data. WG Note, W3C, January 2014. http://www.w3.org/TR/ld-bp/
19. Klímek, J., Škoda, P., Nečaský, M.: LinkedPipes ETL: evolved linked data preparation. In: Sack, H., Rizzo, G., Steinmetz, N., Mladenić, D., Auer, S., Lange, C. (eds.) ESWC 2016. LNCS, vol. 9989, pp. 95–100. Springer, Cham (2016). doi:10.1007/978-3-319-47602-5_20

20. Lange, C.: Krextor - an extensible framework for contributing content math to the web of data. In: Davenport, J.H., Farmer, W.M., Urban, J., Rabe, F. (eds.) CICM 2011. LNCS (LNAI), vol. 6824, pp. 304–306. Springer, Heidelberg (2011). doi:10. 1007/978-3-642-22673-1_29

21. Lanthaler, M.: Hydra Core Vocabulary. Unofficial Draft, June 2014. http://www. hydra-cg.com/spec/latest/core/

22. Rahm, E., Do, H.H.: Data cleaning: problems and current approaches. IEEE Data Eng. Bull. **23**(4), 3–13 (2000)

23. Regalia, B., Janowicz, K., Gao, S.: VOLT: a provenance-producing, transparent SPARQL proxy for the on-demand computation of linked data and its application to spatiotemporally dependent data. In: Sack, H., Blomqvist, E., d'Aquin, M., Ghidini, C., Ponzetto, S.P., Lange, C. (eds.) ESWC 2016. LNCS, vol. 9678, pp. 523–538. Springer, Cham (2016). doi:10.1007/978-3-319-34129-3_32

24. Roman, D., Nikolov, N., Putlier, A., Sukhobok, D., Elvesaeter, B., Berre, A., Ye, Xi., Dimitrov, M., Simov, A., Zarev, M., Moynihan, R., Roberts, B., Berlocher, I., Kin, K.S., Lee, T., Smith, A., Heath, T.: DataGraft: one-stop-shop for open data management. Semant. Web J. (2016)

25. Scharffe, F., Atemezing, G., Troncy, R., Gandon, F., Villata, S., Bucher, B., Hamdi, F., Bihanic, L., Képéklian, G., Cotton, F., Euzenat, J., Fan, Z., Vandenbussche, P.Y., Vatant, B.: Enabling linked data publication with the datalift platform. In: Proceedings AAAI Workshop on Semantic Cities (2012)

26. Schultz, A., Matteini, A., Isele, R., Bizer, C., Becker, C.: LDIF - linked data integration framework. In: Proceedings of the Second International Conference on Consuming Linked Data. CEUR Workshop Proceedings, vol. 782, pp. 125–130 (2011)

27. Tennison, J., Kellogg, G., Herman, I.: Generating RDF from Tabular Data on the Web. W3C Recommendation, December 2015. https://www.w3.org/TR/csv2rdf/

BalOnSe: Temporal Aspects of Dance Movement and Its Ontological Representation

Katerina El Raheb[1,2($\boxtimes$)], Theofilos Mailis[1,2], Vladislav Ryzhikov[3],
Nicolas Papapetrou[1], and Yannis Ioannidis[1,2]

[1] University of Athens, Athens, Greece
kelraheb@di.uoa.gr
[2] Athena Research Center, Athens, Greece
[3] Free University of Bozen-Bolzano, Bolzano, Italy

Abstract. In this paper, we propose an approach to describe the temporal aspects of ontological representation of dance movement. By nature, human movement consists of complex combinations of spatiotemporal events, a fact that creates a big challenge for representing, searching, and reasoning about movement-related content, such as movement annotations on video dances. We have defined MoveOnto, a movement ontology whose expressive power captures movements that range from body states and transitions based on the semantics of Labanotation, to generic actions or specialized vocabularies of specific dance genres, e.g., ballet or folk. We combine the ontology description with temporal reasoning in Datalog-MTL, based on temporal rules of the movement events. Finally, we present the specifications and requirements for dance exploration from a user's perspective and describe the architecture of BalOnSe, a specific system that is currently under implementation on top of MoveOnto according to them. BalOnSe consists of a web-based application with semantic annotation, search, and browsing on the movements, as well as a backend with archival and query processing functionality based on temporal rules.

1 Introduction

Dance multimedia data such as videos can be found in large volumes in multimedia channels, but also in dedicated archives and collections. Various recent efforts are aiming at developing and bringing state-of the art information technologies to the area of digitisation, archiving and preservation of intangible cultural heritage and performing-art content [1–3,7], as well as investigating bodily knowledge and widening the access to such content through enhanced experiences in various contexts, e.g., education. The question is how to facilitate *findability* and provide usable ways to access, search and browse multimedia content, based on the information related to the movement itself through concepts that are interesting for the users. This information could vary from generic concepts regarding actions e.g., step, turn, jump to specific terminology of dance syllabi e.g., pirouette, arabesque, as well as any verbal descriptions that convey something about the form or the quality of movements.

© Springer International Publishing AG 2017
E. Blomqvist et al. (Eds.): ESWC 2017, Part II, LNCS 10250, pp. 49–64, 2017.
DOI: 10.1007/978-3-319-58451-5_4

Semantic information about movements in a video can be added in two ways: automated extractions of patterns and movement recognition and manual annotation of experts. In addition to the above, explicit information can be inferred by reasoning if we add to the system rules to represent common and expert knowledge about movement.

The semantic representation of movement includes many challenges. We describe these challenges and how we address them in our work.

Complexity of the human body and its movement: The human body can create endless different combinations depending on the context. These can vary from functional everyday movements and the execution of specific actions, to various dance techniques, expressive gestures, and non-verbal communication. To address this challenge, we have adopted a modular approach based on theoretical basis on movement analysis such as Labanotation and chorological methodologies. In particular, we have developed a core ontology based on MoveOnto, and we develop specific ontologies to describe the terminology of syllabi of different dance genres. We focus on dance genres that allow this modular approach, since they have specific well-defined vocabularies, and structured rules. The entities of the different ontologies (core and specific) are linked through rules which reflect basic and common knowledge about the movement according to technique and the dedicated literature [16,20].

Temporal aspects: Movement descriptions imply a temporal description. Motion, as well as stillness, in dance is always connected with a duration, and depending on the orchestration and coordination of the different body parts, the temporal aspects might vary in complexity. This complexity is again related to the dance style and technique and the level of detail which is needed depending on the context. We address this challenge by using DatalogMTL [9], a language for representing temporal ontologies. The language allows both temporal representation and reasoning.

Segmentation and Discretization: Semantic descriptions, the construction of concepts and entities, require clear time and space bounds to identify movement entities. Nevertheless, the human motoric usually includes complex synchronizations and coordination of different body parts that occur in a continuous manner. For example, at the semantic level walking is considered to be a continuous sequence of altering steps from one step to another. On the other hand, if we need to analyze and observe movement there is not always an absolute measure to tell us when exactly the first step ends and when the second step ends. We address this challenge by providing a usable interface where expert users can add annotations by choosing terms from the ontology according to their knowledge and perspective. The terms in the ontology vary from more complex recognized sequences to very simple movements. These annotations are stored in the system for further analysis. We base the segmentation, on the notions of *kinemes, morphokinemes* and *motifs* as structural, recognisable components of a dance genre, as have been introduced by Kaeppler [17] in an analogous way that phonemes, morphemes and words are the morphological components of language.

To put our ideas into practice we have developed *BalOnSe*, a system that allows to add semantically rich temporal annotations about movement on multimedia content, *store* these annotations, and *search* the content based on the annotated metadata. The *BalOnSe* platform is build upon the *MoveOnto* ontology, a movement ontology whose expressive power captures movements that range from body states and transitions based on the semantics of Labanotation, to generic actions or specialized vocabularies of specific dance genres, e.g., ballet or folk. We combine the ontology description with temporal reasoning in DatalogMTL, based on temporal rules of the movement events. In this paper we describe the theoretical basis and the architecture of our implementation, focusing on the representation and management of the temporal aspects, a feature that was not included in the previous version of the system [15].

2   Requirements and Specifications

The architecture (Sect. 4) of the BalOnSe platform is designed to allow modularity and extensibility, i.e., the integration of different rules and ontologies, including more details about the movement, terminology of other dance genres, context of the performance or the recording. To sum up, the current version of BalOnSe platform provides the following characteristics: *(i)* Annotation with archival functionalities and user moderation; *(ii)* Semantic-domain specific search (search the content based on movement and dance terminology); *(iii)* Rich predefined vocabularies for key-word search; *(iv)* Usable user interface including video streaming, and preview of statistics of annotations; *(v)* Temporal Reasoning; *(vi)* Modularity and extensibility.

Through the BalOnSe interface, the user may *(i)* employ ontological assertions to annotate a dance video recording using a rich ontology that contains both temporal and non-temporal predicates; *(ii)* browse through dance-multimedia content by querying the ontological video annotations; *(iii)* preview the rich multimedia content that resides in the BalOnSe deposit. A key characteristic of our system is that it allows further development and integration of other dance-specific ontologies in order to describe different dance genres. By extending our core ontology, that is designed in order to describe sequences of movements, we have built dance-specific vocabularies for different types of genres, i.e. ballet. Thus, a user may query a ballet repository using movement sequences that are defined in the ballet syllabus.

To what follows, we outline the different technical functional and non-functional specifications resulting from the aforementioned scenarios, based also on similar approaches [18].

Integrated Data Access: One of the features of the BalOnSe platform, is its ability to incorporate annotations that have been already made on dance-performance content. These annotations may be stored in a relational database endpoint and can be easily accessed. This is due to the datalogMTL translator that follows an Ontology Based Data Access (OBDA) [21] approach as a means to enhance end-user direct data access to relational databases.

Utilising Implicit Information: In databases it is typically assumed that only explicit data matters, i.e. the explicit data that are stored in the system. Nevertheless, our system aims at inferring additional information by utilising the initial annotations and some domain-specific knowledge related to dance and movement analysis. This could be reflected by logical formalism, that allows to derive implicit information from the data stated explicitly, typically using some forms of background knowledge, via a reasoning procedure. The use of implicit information can greatly increase the practical benefit of the BalOnSe system.

Temporal Data Processing: Our applications requires for the existence of a temporal component that will allow to correlate annotations representing movement sequences of a dance-recording video. Existing languages, such as the OWL 2 language are not designed to represent temporal information. Thus we should adopt a language that allows for an expressive temporal component. We identify the specific refined characteristics for the underlying temporal language: *(i)* It should provide for a natural way to annotate movements within a dance video recording of a performance. The annotations of video sequence should refer to an absolute time component, i.e. its event should be mapped to a specific time interval within the video that describes its starting and ending point in seconds or subdivisions of seconds. The user should be able to annotate in absolute time intervals of events within a video sequence. *(ii)* It should provide for a natural way to describe movement patterns and sequences in terms of recognised movement forms for the dance genre. In contrast to the absolute time that applies when annotating a dance video recording, a movement pattern description, even if expressed in notation, does not depend on absolute time-intervals. Instead, similar to music scores, dance sequence descriptions are expressed on beats, i.e. temporal ratios instead of specific time intervals. This means that a mapping is needed between the time expressed in beats in patters and the absolute time intervals coming from the annotations. *(iii)* The corresponding language should allow to have a temporal component when describing the background knowledge of our domain. The later will allow to effectively encode complex dance sequences via simple predicates. *(iv)* Finally the corresponding formalism should allow for an expressive query component. The latter will allow to query our metadata for specific events and the time intervals that these events occurred.

Modular Ontological Representation: Ontology modularisation can be interpreted as decomposing potentially large and monolithic ontologies into (a set of) smaller and interlinked components (modules) [4]. An ontology module can be considered as a loosely coupled and self-contained component of an ontology maintaining relationships to other ontology modules. In our domain of interest, we have the core dance ontology that is used to describe basic human body motion (based on Labanotation) and domain specific ontologies that describe specific dance genres, e.g., ballet. The latter provides users of ontologies with the knowledge they require, reducing the scope as much as possible to what is strictly necessary.

3   Movement Description

The BalOnSe platform allows users to enrich multimedia content such as videos with annotations by choosing specific terms from the MoveOnto ontology. Move-Onto includes terminology of movement on different levels of abstraction. To represent choreographies in an expressive searchable way, we need a strong theoretical basis that allows to describe the required elements of dance (grammar and syntax). The theoretical basis for the MoveOnto representation is based on the Labanotation system for describing movement [16] and Kaeppler's choreological approaches that analyse the structural parts of dance in motifs, morphokinemes and kinemes. Under this perspective, the notions of kineme and morphokineme [17], as dance segments are analogous to phoneme and morpheme in morphology linguistics. Kinemes are actions or positions, which have no meaning as units alone, but constitute the building blocks of a dance genre. A morphokine is defined as the smallest kinetic unit that has meaning, where meaning here does not reflect any pictorial or narrative notion, but it is used to indicate movements that are recognized as units from the people practicing a specific dance genre. The third level of dance structure is the motif level. A motif is a frequently occurring combination of morphokinemes that themselves form a short entity.

3.1   MoveOnto

In order to provide a more human readable and computer understandable format, MoveOnto was developed to capture the semantics of Labanotation [12], a notation system for recording and analyzing human movement. It is a symbolic language which allow to create dance notations on paper. The symbols are put in specific columns on the staff which is read from bottom to the top. The vertical axis represents time, while different columns of the staff are dedicated to different body parts. The reader interprets the different symbols based on their shape (movement type, directions, and other symbols), their color (level of movement) and their size (duration). A Labanotation score, as in Fig. 1(a), can be seen as a complex timeline with parallel slots.

The main objective of MoveOnto is the interpretation of symbols into entities that are both human- and machine-understandable. Moreover, in our work, the use of the ontology provides the ability to express complex relationships, restrictions, and rules about the concepts, creating hierarchies and graphs of movement entities and properties, and as a result provides a rich vocabulary for describing dance movements. In fact, in a Labanotation score, since the different movements are represented as events, related to time intervals, all Allen's interval relationships [5] are possible to be found on a dance score, to represent the relative synchronisation of states and transitions of movement.

MoveOnto considers three different levels of describing movement [13,14]: *(i)* a *Labanotation Concepts* level based on Labanotation; *(ii)* a *Generic Concepts Movement* level that is used to describe terms such as turn, step, slide, and *(iii)* *Specific Vocabulary Concepts* level that is used to represent the terminology related to a specific dance genre, such as the Ballet syllabus.

In this paper we provide some examples of the semantic and temporal relations between the syllabus (specific movements that are meaningful for the genre), and their representation in Labanotation and MoveOnto. In our approach, we create a hierarchy/taxonomy of movements, by classifying them into abstract categories. This hierarchy supports the scalability of the system by giving the opportunity to search movements in different levels of detail. As MoveOnto is based on DatalogMTL, it can express complex inference rules and relationships with a temporal component. Reasoning capabilities support reuse of entities, and allow the system itself to infer explicit knowledge from the stored dance knowledge, e.g., "a Gesture is an Action".

3.2   Time Representation and DatalogMTL

In our previous approach in order to represent time and synchronisation in the core ontology, we adopted a reification strategy. Time intervals were represented as OWL2 individuals called Temporal Entities [12] and in order to represent the sequential order between time points, the functional property hasNext was adopted. Each time interval corresponded to a set of facts representing the dance movements, while the duration was expressed by the datatype property hasDuration, relating each time-interval individual to a specific duration.

In this work we propose the use of DatalogMTL to represent our knowledge. DatalogMTL is a language for representing temporal ontologies. The advantages that the DatalogMTL language provides, compared to our previous approach, are the following: *(i)* It provides for a natural way to represent time in dance movements, since it does not demand the introduction of individuals that corresponded to time intervals and a datatype property to represent duration. *(ii)* DatalogMTL allows having a temporal component in complex rules. Thus complicated dance combinations can be effectively encoded via simple predicates. For example, a jeté is a complex ballet jump (Fig. 1(f)) that involves a set of different body movements that take part on time. *(iii)* DatalogMTL provides for an expressive query component, which allows querying the dance ontology for specific events and the time points that these events occurred. This provides for very interesting capabilities since it allows to search for specific movement sequences and find similarities between different types of dance.

3.3   Movement Representations and Rules

In the following section we describe some examples of simple movements, expressed in verbal descriptions, Labanotation, and DatalogMTL and we show the different levels of complexity that might occur in temporal representation. A DatalogMTL *program*, Π, is a set of rules about our domain, and a *data instance*, $\mathcal{D}$, is a finite set of facts, i.e., in our case video annotations.

The MoveOnto ontology makes the following assumptions: *(i)* Simple and more complex movements are represented as unary predicates that are used to characterise video segments. For example an assertion in $\mathcal{D}$ such as

$$Left\_Leg\_Gesture\_Middle\_Back(: Video1)@[12\,s, 13\,s]$$

means that a left-leg-gesture-middle-back movement occurred in Video1 during the $[12\,s, 13\,s]$ time period. *(ii)* The time in Π (rules) is represented in beats, in a similar way time is represented in dance notations, and expressed in practice when a dance sequence is taught and analysed. *(iii)* The time in $\mathcal{D}$ (facts) is represented in seconds of each video file. *(iv)* In the system, the durational value of the beat is mapped into seconds or subdivisions of seconds for each video file according to its tempo, in a similar way the tempo is given in beats per minute (bpm).

Example 1. In Fig. 1(b) we give an example of an Arabesque Allongé pose, a morphokineme that makes sense for a specific dance genre, in this case ballet. The Arabesque morphokineme, either as a pose or action Fig. 1(d), is a case of describing a single pose or movement with no complex temporal relationship of its integral parts. It consists of different simple kinemes that *(i)* are happening at the same time period, simultaneously; *(ii)* share the exact same duration. The DatalogMTL definition of the Arabesque Allongé pose is the following:

$$\boxplus_{[0,1]} Arabesque\_Allonge(x) \leftrightarrow \boxplus_{[0,1]} Right\_Support\_Midde\_Place(x) \wedge$$
$$\boxplus_{[0,1]} Left\_Leg\_Gesture\_Middle\_Back(x) \wedge \boxplus_{[0,1]} Right\_Arm\_Gesture\_Middle\_Front(x) \wedge$$
$$\boxplus_{[0,1]} Left\_Arm\_Gesture\_Middle\_Back(x) \wedge \boxplus_{[0,1]} Right\_Palm\_Facing\_Place\_Low(x) \wedge$$
$$\boxplus_{[0,1]} Left\_Palm\_Facing\_Place\_Low(x) \tag{1}$$

Based on the previous rule, the occurrence of the separate simultaneous movements (involving the legs, arms, and palms) of an $1\,s$ duration implies an Arabesque Allongé movement of the same duration and vice versa.

Within our knowledge we may have different levels of details.

In Fig. 1(c), a more generic Arabesque is represented where only information about the legs is specified. The more generic movement is represented into MoveOnto by the following DatalogMTL rule:

$$\boxplus_{[0,1]} Right\_Arabesque(x) \leftrightarrow \boxplus_{[0,1]} Right\_Support\_Midde\_Place(x) \wedge$$
$$\boxplus_{[0,1]} Left\_Leg\_Gesture\_Back(x) \tag{2}$$

while we may also add rules for basic movement details:

$$Left\_Leg\_Gesture\_Back(x) \leftarrow Left\_Leg\_Gesture\_Middle\_Back(x) \tag{3}$$

DatalogMTL allows to query our knowledge using the different levels of detail. Thus an assertion $Arabesque\_Allonge(: Video123)@[23, 24]$ in $\mathcal{D}$, along with the rules in Eqs. 1, 2, and 3 imply that

$Right\_Support\_Midde\_Place(: Video123)@[23, 24]$ and $Right\_Arabesque(: Video123)@[23, 24]$.

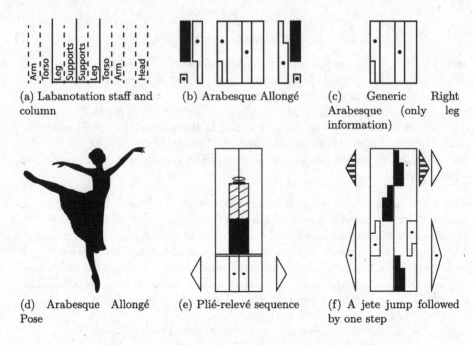

(a) Labanotation staff and column

(b) Arabesque Allongé

(c) Generic Right Arabesque (only leg information)

(d) Arabesque Allongé Pose

(e) Plié-relevé sequence

(f) A jete jump followed by one step

Fig. 1. Labanotation examples

Example 2. In Fig. 1(e) we give an example of a Plié-relevé movement, a more complex morphokineme that includes sequences of simple kinemes: *(i)* the complex movement is described as a set of sequential movements that are all happening during specific intervals; *(ii)* all movements overlapping in time, share the exact same duration and are happening simultaneously. In other words, all parallel movements are happening in sequential intervals. These sequential intervals might be equal in time or not.

$$\boxplus_{[0,3]} Plie\_Releve(x) \leftrightarrow \boxplus_{[0,1]} Right\_Support\_Middle\_Place(x)$$
$$\boxplus_{[0,1]} Left\_Support\_Middle\_Place(x) \boxplus_{[1,2]} Right\_Support\_Low\_Place(x)$$
$$\boxplus_{[1,2]} Left\_Support\_Low\_Place(x) \boxplus_{[2,3]} Right\_Support\_High\_Place(x)$$
$$\boxplus_{[2,3]} Left\_Support\_High\_Place(x) \tag{4}$$

DatalogMTL allows to infer interesting facts that are happening within a Plié-Relevé-movement interval (similarly with the previous example):

$$\boxplus_{[0,1]} First\_Position\_Middle(x) \leftrightarrow \boxplus_{[0,1]} Right\_Support\_Middle\_Place(x) \wedge$$
$$\boxplus_{[0,1]} Left\_Support\_Middle\_Place(x) \tag{5}$$

$$\boxplus_{[0,1]} Plie\_in\_First\_position(x) \leftrightarrow \boxplus_{[0,1]} Right\_Support\_Low\_Place(x) \wedge$$
$$\boxplus_{[0,1]} Left\_Support\_Low\_Place \tag{6}$$

as well as expressing information that involves sequential movements:

$$\boxplus_{[1,2]} \, Support\_Bending(x) \leftrightarrow \boxplus_{[0,1]} Right\_Support\_Middle\_Place \wedge$$
$$\boxplus_{[0,1]} \, Left\_Support\_Middle\_Place(x) \wedge \boxplus_{[1,2]} Right\_Support\_Low\_Place(x) \wedge$$
$$\boxplus_{[1,2]} \, Left\_Support\_Low\_Place(x) \tag{7}$$

What the above rule says is that if the description starts with an interval where the dancer is in normal position, followed by an interval which defines a low level as a destination, this should allow us to subsume that the dancer has bend his/her knees, so there is an action of bending there. Thus by combining the rules in Eqs. 4 and 7 we infer that the Plié-Relevé movement implies a Support-Bending movement as well.

Example 3. In Fig. 1(f) we give an example of a Jeté jump where parallel and sequential movements occur in combination, thus more complex synchronisation needs to be represented: *(i)* the complex movement is described as a set of sequential movements that are all happening during specific intervals of the same or different durations; *(ii)* movements might occur in equal intervals, overlap, start with or end with, so all combinations of intervals relations are possible.

$$\boxplus_{[0,4]} \, Jete\_Jump(x) \boxplus_{[0,1]} \leftrightarrow Right\_Step\_Low\_Place(x) \wedge$$
$$\boxplus_{[1,2]} \, Right\_LegGesture\_Middle\_Back(x) \wedge \boxplus_{[1,2]} Left\_LegGesture\_Middle\_Forward(x) \wedge$$
$$\boxplus_{[2,3]} \, Left\_Step\_Middle\_Low(x) \wedge \boxplus_{[3,4]} Right\_Step\_Middle\_Low(x) \wedge$$
$$\boxplus_{[0,2]} \, Right\_Arm\_Gesture\_Middle\_Right(x) \wedge \boxplus_{[0,2]} Left\_Arm\_Gesture\_Middle\_Left(x) \wedge$$
$$\boxplus_{[3,4]} \, Right\_Arm\_Gesture\_Middle\_Right(x) \wedge \boxplus_{[3,4]} Left\_Arm\_Gesture\_High\_Left(x) \wedge$$
$$\boxplus_{[3,4]} \, Torso\_Gesture\_High\_Right(x) \tag{8}$$

4   BalOnSe Platform

BalOnSe is an integrated system that consists of multiple components to support end-to-end annotation and provide access on ontologies about dance. From a user perspective, the BalOnSe system allows to: *(i)* create temporal ontological annotations of multimedia content; *(ii)* search by movement predicates; *(iii)* preview the existing multimedia content that satisfies the queries. Query evaluation is done via the systems query enrichment, unfolding, and execution modules that allow executing complex temporal queries on relational databases. In this section we give some details of the BalOnSe components that address the challenges and specifications described in Sect. 2.

4.1   BalOnSe Components

The application provides for the following components in order to perform the intended actions:

1. *Multimedia Content Annotation:* the application provides for an archival system for the videos, along with the corresponding user interfaces (UIs) that allow users to annotate video content and store the corresponding annotations in a relational database system.

The query formulation, transformation, execution, and video visualisation procedures are performed in a sequence of stages presented in Fig. 2:

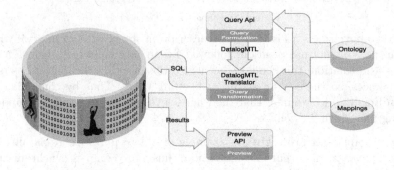

Fig. 2. BalOnSe architecture

2. *Query Formulation:* Data sources can be explored using a query formulation tool that allows the user to navigate through the ontology and the corresponding temporal component. The current version of our query formulation tool is based on DatalogMTL queries. Future work involves the adoption of faceted browsing techniques that allow end-users to explore a collection of information by applying multiple filters.
3. *Query Transformation:* The aforementioned expressions are sent to the corresponding query transformation engine for processing. The processing includes rewriting against the ontology and further unfolding into relational SQL queries based on the corresponding mappings [9].
4. *Query Execution:* SQL queries are executed by a relational database management system.
5. Preview: The results of query execution are video fragments corresponding to the complex movements that were queried. The user is able to preview the corresponding video fragments.

4.2   User Interface

A key characteristic of the BalOnSe platform is its simplicity for annotating video content and search. The platform's web-based interfaced is illustrated in Fig. 3 and can be logically divided to 5 basic interfaces. The user is able to perform annotations using labels of predefined vocabularies on the core ontology and morphokinemes on domain-specific ontologies (Fig. 3D). Each annotation has a starting and an ending time point. For formulating queries, we have adopted a transparent approach where the user asks only for specific labels corresponding to the terminology of (morpho)kinemes (Fig. 3B). Finally, the user can directly search for multimedia objects based on their titles (Fig. 3A), and furthermore preview multimedia content (Fig. 3C).

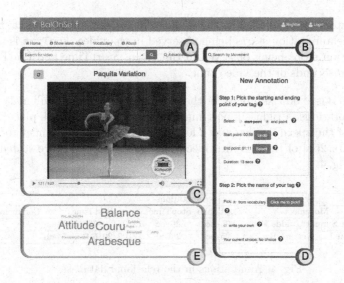

Fig. 3. BalOnSe user interface

4.3   Annotations Storage and Management

The application provides an archival system for the videos while both video metadata and user annotations are stored in a relational database. By taking advantage of the modular characteristics of our ontological representation, the Multimedia Content Annotation Interface allows the user to annotate a dance video either by adopting simple movements (kinemes) on the core ontology, or by using predefined combinations of simple movements (morphokinemes) on the domain-specific ontology (e.g. the ballet ontology).

Example 4. The dance expert may annotate by using the core ontology, that in Video_123 a Right_Support_Midde_Place movement took place during the period [25 s, 26 s] and a Left_LegGesture_Back movement took place during [25 s, 26 s]. Alternatively, the dance expert may assert that a Right_Arabesque movement was performed between [25 s, 26 s]. Based on the definition of Right Arabesque in Eq. 2, the two annotations are equivalent.

Database Schema: We will briefly present the database schema, which implements a basic entity relationship model with three main entities: (User) **Accounts**, **Videos**, and **Annotations**. We focus on the **Annotations** table, which stores our temporal information:

Annotations is the table that stores the experts annotations. A new record is saved for each annotation that is chosen by the user for a specific video segment. It has the following attributes: The **VideoId** contains an identifier that is unique for each video file. The **Movement** attribute represents either a kineme in the core ontology, or a morphokineme in some of its extensions. **Comment** is a user defined characterisation of the annotation the user just created. **PostTime** is the time the user made the annotation, this attribute provides the opportunity to have log history of tags added by the same users in different times. StartTime is the time the annotation starts being observed in the video. EndTime is the time the annotation stops being observed in the video. StartTime and EndTime are expressed in seconds of the video time.

Example 5. Based on Example 4 for the core ontology, the user will make the following annotations: a right-support-middle-place movement took place between [25 s, 26 s] of the specific video and a left-leg-gesture-back movement took place between [25 s, 26 s] of the specific video. The two annotations are stored as displayed in Fig. 4.

Annotations						
VideoId	Movement	StartTime	StopTime	PostTime	Comment	UserId
123	Right_Support_Midde_Place	25 sec	26 sec	12/09/2016 @ 10:54am	-	24
123	Left_Leg_Gesture_Back	25 sec	26 sec	12/09/2016 @ 10:54am	-.	24

Fig. 4. Annotations in the relational database

Therefore, for each video all the annotations provided by different users are stored, as well as different annotations of the same user. The database stores the multiple tags along with the information of the user that provided them. In the current version, the annotations of users are stored and each time the video appears, a tag cloud shows all existing annotations.

4.4 Query Transformation and Execution

This section is dedicated to the enrichment and unfolding stages that occur during a query execution cycle. The DatalogMTL translator is responsible for transforming DatalogMTL conjunctive queries and their temporal component, to relational SQL queries.

Ontology Mappings: The relationship between the ontological vocabulary and the schema of the data is maintained by a set of mappings. Intuitively, a mapping assertion consists of a source (an SQL query retrieving values from the database) and a target (defining RDF triples with values from the source). The DatalogMTL translator exposes relational databases as virtual datalog assertions by linking the terms (predicates) in the ontology to the data sources through mappings. The W3C standard for expressing customised mappings from relational databases to RDF datasets is R2RML (R2RML: RDB to RDF Mapping Language) [11].

DatalogMTL translator extends R2RML mappings, so that they incorporate both predicates of higher arity, as well as temporal information. For the technical details of these extensions the reader may refer to [9]. We will try to explain with a simple example how R2RML mappings work for our use cases.

Example 6. The ontology instance assertions in Fig. 4 can be populated from a database by means of the following mappings:

```
Right_Support_Midde_Place(:Video_{VideoId})@[StartTime,StopTime]
   ←── SELECT VideoId, StartTime,StopTime FROM Annotations
       WHERE Movement='Right_Support_Midde_Place'
Left_Leg_Gesture_Back(:Video_{VideoId})@[StartTime,StopTime]
   ←── SELECT VideoId,StartTime, StopTime FROM Annotations
       WHERE Movement='Left_Leg_Gesture_Back'.
```

By combining the mappings and the data in the `Annotations` table (Fig. 4) we will get the facts: $Right\_Support\_Midde\_Place(: Video123)@[25\,s, 26\,s]$ and $Left\_Leg\_Gesture\_Back(: Video123)@[25\,s, 26\,s]$. This data can then be queried using DatalogMTL, by translating the DatalogMTL queries into SQL queries over the relational databases. This translation process is transparent to the user and performed in the rewriting and unfolding steps.

Rewriting, Unfolding, & Execution: DatalogMTL a uses two-step approach to answer the user queries described in Sect. 3: *(i)* First, an initial DatalogMTL query Q_1 together with the ontology is rewritten into a query Q_2 over the virtual relations that represent temporal data in conceptualised form. Example 6 demonstrated a query that populates one such relation, namely, `Right_Support_Midde_Place`, which has columns `VideoId`, `StartTime`, `StopTime` to represent temporal information. *(ii)* Q_2 is then combined with all the relevant mappings to obtain the final query Q_3 that can be evaluated over the raw data.

The process to obtain the query Q_3 involves the computation of *temporal joins* and *coalescings*. To give an idea, for two tables T and S with columns `from,to` both containing intervals, temporal join is a table with the same columns that contains all the non-empty intervals i, such that $i = j \cap k$, for some j from T and k from S. On the other hand, coalescing, for a table T as above, is a table T' which contains only one "covering" interval $[i_1, j_n]$ for a sequence of (right-)overlapping intervals $[i_1, j_1], \ldots, [i_n, j_n]$.

Example 7. A simple DatalogMTL query asking for the videos that contain a Right-Arabesque movement and the time that this movement was performed has the form $Q_1(x, \delta) = Right\_Arabesque(x)@\delta$ where x corresponds to the video file and δ corresponds to the time interval that the movement was performed. By taking into account the definition of the right-arabesque movement in Eq. 2 and the mappings in Example 6, the rewriting and unfolding procedure will execute a complicated query Q_3 over the table `Annotations` of the relational database. For space limitations, we won't provide the full rewriting and unfolding of the initial query, instead, the interested reader may refer to [9] for more information on the corresponding procedures. The query answer will be the tuple $(: Video\_123, 25, 26)$ meaning that $Video\_123$ contains a $Right\_Arabesque$ during the $[25\,s, 26\,s]$ time interval.

5   Novelty and Impact

Related Work: We describe previous advancement in the field of multimedia annotation for performing arts. Bertini et al. [8] have presented an overview of approaches and algorithms that exploit ontologies to perform semantic video annotation and an adaptation of the First Order Inductive Learner technique to the Semantic Web Rule Language. Ramadoss and Rakummar [22] have presented the system architecture of a manual annotation tool, a semiautomatic authoring tool, and a search engine for the choreographers, dancers, and students of pop Indian dance. Singh et al. [24] presented the Choreographic Notebook, a multimodal annotation tool supporting the use of text and digital ink to be used during the production process of contemporary dance. A similar multimodal annotation approach has been presented by Cabral et al. [10] in the Creation-Tool, while Ribeiro et al. [23] have extended this methodology through addressing the issues of transforming the 2D annotations into 3D visualizations. Another relevant work, in the field of non-verbal communication, is the Anvil interface and the corresponding schema of manual annotation for conversational gestures, which eventually supports the recreation of 2D animation based on time and special descriptions of the gestures on videos [19].

Novelty: In comparison to the existing approaches, BalOnSe is the first platform that allows to represent time within terminological knowledge, allowing to define complex morphokinemes where parallel and sequential movements occur in combination. Moreover, it is the only such platform that uses an OBDA approach, separating the assertional from the terminological knowledge and exploiting database technologies to store and retrieve information.

Impact: The BalOnSe platform is currently in a beta version and its functionality has been presented to dance experts and got positive feedback. In particular, the experts were asked to assess the platform with respect to the specifications introduced in Sect. 2. The experts gave a positive feedback on how the proposed solution addresses the *Integrated Data Access* requirement: the OBDA approach allows to integrate to the ontology video annotations that have already been collected and stored in external data sources such as relational databases. The experts gave positive comments on the platform's ability to derive implicit information using logical reasoning on temporal information, thereby satisfying requirements for *Utilization of Implicit Information* and *Temporal Data Processing*. Finally, the *Modular Ontological Representation* approach that has been adopted by our system, was highly welcomed since it allows to represent information in different levels of abstraction by having the core dance ontology to describe kinemes and domain specific ontologies that introduce morphokinemes for specific dance genres.

6   Conclusions and Future Work

We have proposed an approach to describe the temporal aspects of ontological representation of dance movement. Our approach is based on the MoveOnto ontology that adopts the DatalogMTL language to represent movement in time. To put our ideas into practice we have developed the BalOnSe platform, the first platform that allows to store and query performing-art content containing complex temporal information.

Future work involves extending our platform with faceted browsing techniques, a suitable paradigm for querying ontology repositories [6]. Moreover, we intend to examine DatalogMTL extensions that will allow us to infer conclusions when there is no perfect synchronisation between sequences of movements. Therefore, we want to extend the DatalogMTL semantics with fuzzy values to express the degree of similarity between movements.

Acknowledgements. This paper includes part of the research which is conducted in the framework of the WhoLoDancE H2020 EU funded project.

References

1. The eclap e-library of performing arts. http://www.eclap.eu/
2. Itreasures, capturing the intangible. http://i-treasures.eu
3. Whole body interaction learning for dance education. www.wholodance.eu
4. Abbes, S.B., Scheuermann, A., Meilender, T., d'Aquin, M.: Characterizing modular ontologies. In: 7th International Conference on Formal Ontologies in Information Systems (2012)
5. Allen, J.F., Hayes, P.J.: A common-sense theory of time. In: IJCAI, vol. 85 (1985)
6. Arenas, M., Grau, B.C., Kharlamov, E., Marciuska, S., Zheleznyakov, D., Jimenez-Ruiz, E.: Semfacet: semantic faceted search over yago. In: 23rd International Conference on World Wide Web (2014)
7. Bellini, P., Nesi, P.: Modeling performing arts metadata and relationships in content service for institutions. Multimedia Syst. **21**, 427–449 (2015)
8. Bertini, M., Del Bimbo, A., Serra, G.: Learning ontology rules for semantic video annotation. In: Proceedings of the 2nd ACM Workshop on Multimedia Semantics, pp. 1–8. ACM (2008)
9. Brandt, S., Kalayci, E.G., Kontchakov, R., Ryzhikov, V., Xiao, G., Zakharyaschev, M.: Ontology-based data access with a horn fragment of metric temporal logic. In: Proceedings of the Thirty-First AAAI Conference on Artificial Intelligence, February 4–9, 2017, San Francisco, California, USA, pp. 1070–1076 (2017)
10. Cabral, D., Valente, J.G., Aragão, U., Fernandes, C., Correia, N.: Evaluation of a multimodal video annotator for contemporary dance. In: Proceedings of the International Working Conference on Advanced Visual Interfaces, pp. 572–579. ACM (2012)
11. Das, S., Sundara, S., Cyganiak, R.: R2RML: RDB to RDF Mapping Language (2012)
12. El Raheb, K., Ioannidis, Y.: A labanotation based ontology for representing dance movement. In: International Gesture Workshop (2011)
13. El Raheb, K., Ioannidis, Y.: From dance notation to conceptual models: a multi-layer approach. In: International Workshop on Movement and Computing (2014)
14. El Raheb, K., Ioannidis, Y.: Modeling abstractions for dance digital libraries. In: Joint Conference on Digital Libraries (2014)
15. El Raheb, K., Papapetrou, N., Katifori, V., Ioannidis, Y.: Balonse: ballet ontology for annotating and searching video performances. In: 3rd International Symposium on Movement and Computing (2016)
16. Guest, A.H.: Labanotation: The System of Analyzing and Recording Movement. Routledge, New York (2014)

17. Kaeppler, A.L.: Dance structures: perspectives on the analysis of human movement. Akad. Kiadó (2007)
18. Kharlamov, E., et al.: How semantic technologies can enhance data access at siemens energy. In: Mika, P., et al. (eds.) ISWC 2014. LNCS, vol. 8796, pp. 601–619. Springer, Cham (2014). doi:10.1007/978-3-319-11964-9_38
19. Kipp, M., Neff, M., Albrecht, I.: An annotation scheme for conversational gestures: How to economically capture timing and form. Lang. Res. Eval. **41**(3–4), 325–339 (2007)
20. Miles, A., Grant, G.: The Gail Grant Dictionary of Classical ballet in Labanotation. Dance Notation Bureau, New York (1976)
21. Poggi, A., Lembo, D., Calvanese, D., De Giacomo, G., Lenzerini, M., Rosati, R.: Linking data to ontologies. J. Data Semant. **10**, 133–173 (2008)
22. Ramadoss, B., Rajkumar, K.: Semi-automated annotation and retrieval of dance media objects. Cyber. Syst. Int. J. **38**(4), 349–379 (2007)
23. Ribeiro, C., Kuffner, R., Fernandes, C., Pereira, J.: 3d annotation in contemporary dance: enhancing the creation-tool video annotator. In: Proceedings of the 3rd International Symposium on Movement and Computing, p. 41. ACM (2016)
24. Singh, V., Latulipe, C., Carroll, E., Lottridge, D.: The choreographer's notebook: a video annotation system for dancers and choreographers. In: Proceedings of the 8th ACM Conference on Creativity and Cognition, pp. 197–206. ACM (2011)

Reasoning on Engineering Knowledge: Applications and Desired Features

Constantin Hildebrandt[1(✉)], Matthias Glawe[1], Andreas W. Müller[2], and Alexander Fay[1]

[1] Institute of Automation Technology, Helmut-Schmidt-University, Hamburg, Germany
{c.hildebrandt,matthias.glawe,alexander.fay}@hsu-hh.de
[2] Data Architecture and Frameworks, Schaeffler Technologies AG & Co. KG, Herzogenaurach, Germany
andreas_w.mueller@schaeffler.com

Abstract. The development and operation of highly flexible automated systems for discrete manufacturing, which can quickly adapt to changing products, has become a major research field in industrial automation. Adapting a manufacturing system to a new product for instance requires comparing the systems functionality against the requirements imposed by the changed product. With an increasing frequency of product changes, this comparison should be automated. Unfortunately, there is no standard way to model the functionality of a manufacturing system, which is an obstacle to automation. The engineer still has to analyze all documents provided by engineering tools like 3D-CAD data, electrical CAD data or controller code. In order to support this time consuming process, it is necessary to model the so-called skills of a manufacturing system. A skill represents certain features an engineer has to check during the adaption of a manufacturing system, e.g. the kinematic of an assembly or the maximum load for a gripper. Semantic Web Technologies (SWT) provide a feasible solution for modeling and reasoning on the knowledge of these features. This paper provides the results of a project that focused on modeling the kinematic skills of assemblies. The overall approach as well as further requirements are shown. Since not all expectations on reasoning functionality could be met by available reasoners, the paper focuses on desired reasoning features that would support the further use of SWT in the engineering domain.

1 Introduction

Decreasing life-cycles and lot sizes [1], increasing numbers of product variants [2] as well as more decentralized manufacturing systems [3] have become a business standard in the manufacturing industry. These circumstances create a dynamic environment, which state-of-the-art manufacturing systems struggle with, since current manufacturing systems are not flexible enough yet to handle this dynamic environment [4]. Applying Semantic Web technologies (SWT) in the engineering domain has become a promising approach in order to gain flexibility in the industrial manufacturing domain. Especially for repetitive tasks, the use of knowledge-based systems provides advantages in terms

E. Blomqvist et al. (Eds.): ESWC 2017, Part II, LNCS 10250, pp. 65–78, 2017.
DOI: 10.1007/978-3-319-58451-5_5

of effort savings, time savings and quality assurance [5–7]. Concerning the operations planning of a manufacturing system, ontologies may be used for modeling the knowledge about the planning domain [8–10]. Once the domain knowledge has been modeled, it can be used for resource allocation, planning and scheduling. As shown in these and other contributions, modeling of manufacturing domain knowledge can be accomplished successfully by using ontologies. On top of such ontologies, rules or queries can be defined to infer additional knowledge for the manufacturing system, e.g. for maintenance [11, 12], security [13], validation [14] or process planning [15]. Typical results of reasoning could be a possible process plan, a security threat or a necessary maintenance task. Especially for these applications, reasoning support is crucial [16] and determines the success of using SWT in the engineering domain. Unfortunately, the use of SWT for reasoning on engineering knowledge seems to lack in support of some features which are necessary in the engineering domain and are addressed by this paper. Therefore, an application from the engineering domain is described in this paper which requires certain reasoning features. The availability of these reasoning features in SWT would significantly support further applicability and, thus, industrial acceptance of SWT in the engineering domain. The structure of this paper is as follows. Section 2 introduces the goal and concept of the underlying research project. Section 3 shows how the knowledge base was built in OWL, while Sect. 4 describes the requirements and desired features on reasoning that have been identified and which aspects are not fulfilled up to now. Section 5 provides a summary and an outlook.

2 Concept of Extracting Knowledge About Skills

According to [17] a skill is the ability of a resource to implement a certain type of manufacturing, logistic or other production related process. In order to represent the skill of a manufacturing system with SWT, a concept for the automatic generation of a skill description based on engineering data has been created by the authors. In this concept, depicted in Fig. 1, an engineer performs design work as usual with a 3D-CAD tool. In the CAD tool, design results are stored in a native file that contains information usable for inferring the mechanical skills of the system. Information stored by the CAD tool are e.g. the modeled kinematic and the material of the mechanical components. From the modeled kinematic, the potentially reachable positions can be inferred, and the chosen material is important to determine the maximum load of a gripping unit. This CAD file, which is usually in a proprietary tool format, is imported by a so-called "Mapping Component" (MAP). Since CAD files may contain lots of information, the purpose of the MAP is to obtain the necessary information for inferring a skill from the CAD file in a semantically proper way and inserting it into a target ontology, which contains all needed concepts and properties. This mapping is based on a concept that was initially introduced by [13] for mapping Computer Aided Engineering Exchange (CAEX) data into an ontological structure, and works as follows. The MAP imports a set of SWRL [21] rules that describe which entities or relations in the CAD file should be mapped to the corresponding structures in the target ontology. The rule's antecedent describes the entities and relation that need to be found in the CAD file, the consequent describes the individuals, object properties and data properties that need to be created in the ontology, whenever the entities in the antecedent are found. Hence, the mapping rules provide a possibility to define the mapping procedure on an abstract semantic level and are, thus, independent of the CAD data format used.

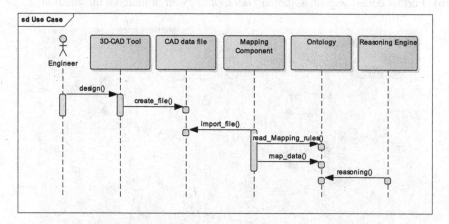

Fig. 1. Sequence diagram of the mapping process

An example of a SWRL rule for the MAP is shown in Listing 1. As the example illustrates, if the MAP finds an entity ?x in the CAD file, that is part of an entity ?y, the object property consistsOf should link the two respective individuals in the ontology. While the antecedent does not have any inner meaning to the ontology, it triggers certain methods in the MAP, which represents a more or less complex analysis on the CAD file.

Listing 1. example of a mapping rule

```
is_CADPartOf(?x, ?y) -> consistsOf(?y, ?x)
```

After finalizing the mapping procedure, the ontology contains all relevant information modeled in the CAD file, which is needed for further inference. The representation of this information is shown in Sect. 3.

3 Representing Engineering Data in an OWL Ontology

This section shows the modeling of engineering data in the underlying project. As an excerpt, the use-case on the left side of Fig. 2 is introduced as an example for a kinematic skill. The modeled object represents an assembly that contains two linear actors and a gripper. The following information is explicitly included in the exemplary 3D CAD file:

(I1) Information about the existence of objects, i.e. components and assemblies
(I2) Information about the hierarchy of objects, i.e. the parts an assembly consists of

(I3) Geometric information about the shape of objects
(I4) Geometric information about the position of objects
(I5) Information about movement restrictions of parts, i.e. how a part can or cannot move
(I6) Further details, e.g. an assumed *Tool Center Point* or material information

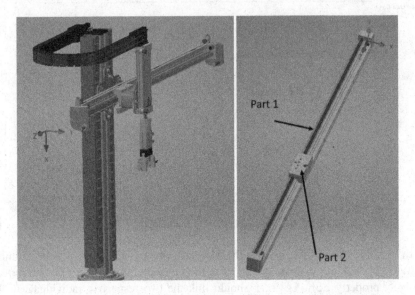

Fig. 2. 3D-CAD complete assembly (left) and linear actor assembly (right)

The ontology used as the target for the mapping was modeled in Protégé. Three main concepts were used in order to model the domain information (see left side of Fig. 3). The concept Description subsumes concepts that represent mathematical and physical descriptions of real world entities, e.g. vector descriptions. The concept ProductionSystem classifies the hierarchical structure of a production system, while the concept FunctionalSkill is used for classifying skills of a production system. The object properties that were used for modeling relations between objects are shown on the right side of Fig. 3, while the data properties are shown in Fig. 4.

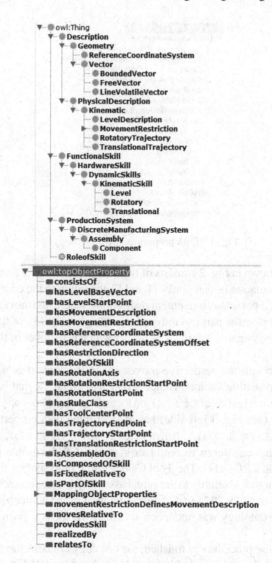

owl:Thing
  Description
    Geometry
      ReferenceCoordinateSystem
      Vector
        BoundedVector
        FreeVector
        LineVolatileVector
    PhysicalDescription
      Kinematic
        LevelDescription
        MovementRestriction
        RotatoryTrajectory
        TranslationalTrajectory
  FunctionalSkill
    HardwareSkill
      DynamicSkills
        KinematicSkill
          Level
          Rotatory
          Translational
  ProductionSystem
    DiscreteManufacturingSystem
      Assembly
        Component
  RoleofSkill

owl:topObjectProperty
  consistsOf
  hasLevelBaseVector
  hasLevelStartPoint
  hasMovementDescription
  hasMovementRestriction
  hasReferenceCoordinateSystem
  hasReferenceCoordinateSystemOffset
  hasRestrictionDirection
  hasRoleOfSkill
  hasRotationAxis
  hasRotationRestrictionStartPoint
  hasRotationStartPoint
  hasRuleClass
  hasToolCenterPoint
  hasTrajectoryEndPoint
  hasTrajectoryStartPoint
  hasTranslationRestrictionStartPoint
  isAssembledOn
  isComposedOfSkill
  isFixedRelativeTo
  isPartOfSkill
  MappingObjectProperties
  movementRestrictionDefinesMovementDescription
  movesRelativeTo
  providesSkill
  realizedBy
  relatesTo

Fig. 3. Classes of the ontology (left), object properties of the ontology (right)

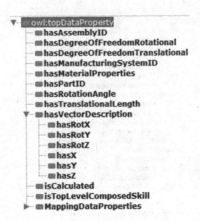

Fig. 4. Data properties of the ontology

The assembly shown in Fig. 2 consists of two components `Part1` and `Part2`. This corresponds to information of categories I1 and I2 according to the enumeration above. Furthermore, `Part2` possesses movement restrictions, where one movement restriction describes explicitly how the part can move along a single axis (I5). In the depicted case it is the individual `MovementRest_Trans_1`, which is member of the class `MovementRestriction`.

In order to represent the respective movement restrictions, it is necessary to use individuals for representing vectors. Hence, every individual `?x` that is a member of a vector class (i.e. `BoundedVector(?x)` or `FreeVector(?x)` or `LineVolatileVector(?x)`) (see Fig. 3 left side) has at least three data properties. These data properties are `hasX(?x,float)`, `hasY(?x,float)`, `hasZ(?x,float)`. Since every vector has to have a reference coordinate system (RCS) for its interpretation, every part or assembly has a RCS (I4). The Tool Center Point, representing the desired point to attach a tool or another assembly to the actual assembly, is also represented by a vector description (`Vector_TCP_P2`) (I6). A representation of the geometrical shape of the CAD objects in the ontology was not necessary, since the focus was on modeling kinematics.

After the mapping procedure is finished, the ontology contains the necessary information for the reasoning process, which was beforehand modeled in the CAD file. A small example of mapped individuals is shown in Fig. 5.

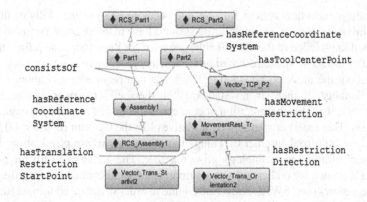

Fig. 5. Example of mapped individuals

In order to support an engineer in checking the functionality of the system, e.g. checking the kinematic, it is now necessary to infer information about the combined movement descriptions of the contained parts. This is essential, since there is no direct information in the CAD file about the combination of single parts movements. The knowledge on how to derive combined kinematics from basic descriptions has been modeled with SWRL. Query-based inference like SPARQL Protocol and RDF Query Language combined with SPARQL Inferencing Notation (as was done by [12] for instance) were also considered. However, the goal was to maintain continuous logical expressivity throughout the entire process of deriving skill descriptions. Thus, SWRL was chosen as it allows for handling and processing knowledge at the OWL level.

For the purpose of mapping and inferencing, a total of 17 rules has been defined. Nine rules are used for mapping and eight rules are used for inferring knowledge about the kinematic skill. Due to the paper's extend, the next section will only introduce a few rule examples, while the next section focuses on desired reasoning features in this context.

4   Reasoning on Engineering Information

As it is pointed out in [16] the reasoning support is a very important requirement when it comes to choosing the modeling language. This section describes desired features in reasoning from an engineering point of view, as found in the project described in Sects. 2 and 3. Beforehand, a basic requirement is introduced as R0 that is based on recent initiatives in the engineering domain.

4.1   R0: Standardized Rule Functionality

Manufacturing systems are becoming more flexible by using data for smart concepts like diagnosis systems [11] or even context-aware maintenance systems [12]. However, as soon as systems need to collaborate with other systems, it is essential for the systems to share a common understanding of concepts, i.e. a common semantic [18]. If a manufacturing execution system for instance needs to analyze the skills of different manufacturing systems, all manufacturing systems have to use the same semantic about the modeled knowledge of the skill. If rules are used for knowledge modeling, the rule functionality has to be standard-based as well. Otherwise, there is no common understanding about the modeled knowledge. Thus, the first requirement is about standard rule functionality. If there is no standard "off-the-shelf" reasoner available for a reasoning task, then the interpretation of concepts is becoming an issue for engineering applications. This receives attention by initiatives like the German Industry 4.0 [18] or the American Industrial Internet of Things initiative [19], where physical production systems and information technology grow together. The authors of [20] for instance present an approach for orchestrating the resources of small companies according to an order. The authors used SWRL and a rule engine in order to define additional functionality. Sharing the knowledge about orchestration is becoming an issue with ongoing decentralization of systems without a common understanding about rule functionalities.

Since SWRL based on [21] is a de-facto standard for rules in Semantic Web, it was chosen to state the rules for extracting kinematic skills. The target skill to be inferred by the reasoning process, is visualized as a plane in Fig. 6 (left side). Since there are two linear actors that do not move in the same direction, the resulting kinematic skill is represented by a plane that has a certain height and width. This plane description relates to the RCS of the highest assembly in the object hierarchy. This is the desired result, which has to be inferred through rule-based inference from the information input. The information input is shown on the right side of Fig. 6. To ensure comparable results in different applications, a common reasoner should be used to infer the information needed to represent the desired skill.

Fig. 6. Kinematic skill of the assembly (left side), geometric description of a linear actor (right side)

4.2 D1: Creating New Individuals Representing Generated Knowledge

The first desired feature (D1) relates to creating individuals in order to represent inferred knowledge. Since the information about a skill is something that was not explicitly created by an engineer, a possibility is needed to create individuals that represent this inferred skill information. For instance (see Listing 2), if there is an assembly ?a that has a RCS ?RCSa and consists of a part ?c that has at least one degree of freedom, the movement restrictions of this component ?c define a kinematic ?Kin that was not known to the model before. Therefore, an individual to represent this kinematic has to be created and further detailed. For creating individuals, the SWRL syntax provides an extension through the SWRL Extensions built-in library[1] called swrlx:makeOWL-Thing(?a,?b), where ?a defines the variable that binds the individual and ?b defines the number of individuals to be created.

Listing 2. Creating new individuals with SWRL extension

```
Assembly(?a), hasReferenceCoordinateSystem(?a,?RCSa),
consistsOf(?a,?c), Component(?c),
hasDegreeOfFreedomTranslational(?c,?DoFT),
hasDegreeOfFreedomRotational(?c,?DoFR),
swrlb:add(?DoFSum,?DoFT,?DoFR),
swrlb:greaterThan(?DoFSum,0),
swrlx:makeOWLThing(?Kin,1) ->
movesRelativeTo(?c,?RCSa),
movementRestrictionDefinesMovementDescription(?c,?Kin),
Kinematic(?Kin)
```

As stated in [22] this is not supported by DL Reasoners of Protégé, although the Protégé SWRL tab supports the syntax. For this reason, it became necessary to use an external rule engine in order to define the functionality for this extension. Unfortunately, this violates requirement R0. In general, many engineering applications require the generation of new knowledge and therefore the creation of new individuals when using an ontology. Therefore, creating individuals with a defined functionality through the use of a maintained reasoner as analyzed in [23] is a desired feature in engineering.

4.3 D2: Calculations

Knowledge in the engineering domain is often based on mathematical descriptions like vectors for instance. In cases where this knowledge is processed or evaluated in a new context, mathematical operations need to be performed. For the extraction of kinematic skills numerous operations are necessary, e.g. calculation of RCS offsets, combination of translational/rotational kinematics or combination of planes or even solid body descriptions. For SWRL only basic mathematical operations are defined by the Built-In

---

[1] http://swrl.stanford.edu/ontologies/built-ins/3.3/swrlx.owl.

library [21]. Due to the fact that even increasing a vector length by one leads to the use of three swrlb:add operators, the rules are becoming very large. This in turn leads to problems in creating or maintaining the rules. For instance, creating a translational kinematic skill out of a part's movement restrictions requires the following calculation (see also right side of Fig. 6):

$$\vec{sv} = \vec{T} + \left(\frac{\vec{O} \cdot \left(\vec{RS} - \vec{T} \right)}{\vec{O} \cdot \vec{O}} \right) \vec{O}; \vec{ev} = \vec{S} + \alpha \vec{O}$$

In this calculation, $\vec{sv}$ would point to the start and $\vec{ev}$ to the end of the translation. In order to define this as a rule, it requires 16 basic math operations and the plane calculation in Fig. 6 requires even more. This number of course does not include the rest of the antecedent and the consequent that have to be defined. In order to solve the problem, a rule engine for defining easier-to-handle math operators had to be used. A desirable solution would be a Protégé plug-in, where complex math operations, e.g. vector algebra, can be created through combination of basic math operations so that R0 is still satisfied and using math operations in SWRL is becoming more flexible. To the best of the authors' knowledge, there are no ways to create such plugins in an easy-to-use way and to implement them in maintained reasoners.

4.4   D3: Defeasible Reasoning

Most engineering methods are based on an iterative procedure, where knowledge about objects is created, updated and sometimes even fully retracted afterwards. In [24] for instance, the whole engineering process from the first idea to the detailed model may be in a feedback loop. The reason for this is that the engineering process is sliced into phases where sometimes assumptions have to be made in order to proceed. These assumptions about an object may be retracted as soon as new knowledge is accessible or processed. Revisiting the extraction of the kinematic skill it is possible to define a rule for the combination of two translational skills of the two linear actors in Fig. 6 (left side). This is shown in Listing 3. If an assembly ?a3 exists, which contains two different assemblies ?a1 and ?a2 that provide a calculated skill, then these two assemblies ?a1 and ?a2 provide a new kinematic skill ?s1. This new kinematic skill is shown in Fig. 6 (left side). In order to represent this skill as the top-level skill, so that it can be found by querying for instance, the data property isTopLevelComposedSkill(?s1,true) should be assigned to the skill ?s1. As soon as another overlying assembly (e.g. the gripper) is found during the process, the value of the data property needs to be changed to false since ?s1 does not represent the top-level skill anymore. Furthermore, retracting knowledge is also necessary for calculation. If a skill should be calculated by combining two skills, it is necessary to indicate whether it has already been calculated before (isCalculated(?s1,true)). Otherwise, stepwise calculations under open world assumption would not be possible, because the actual calculation step could not be determined.

Listing 3. Declaring the state of knowledge after applying a rule

```
Assembly(?a1), Assembly(?a2), Assembly(?a3),
consistsOf(?a1,?a2), consistsOf(?a1,?a3),
providesSkill(?a2,?s2), providesSkill(?a3,?s3),
swrlb:notEqual(?a2,?a3), isCalculated(?s2,true),
isCalculated(?s3,true), consistsOf(?a2,?c2),
Component(?c2), consistsOf(?a3,?c3), Component(?c3),
hasToolCenterPoint(?c3,?TCPc3),
isAssembledOn(?c2,?TCPc3), swrlx:makeOWLThing(?s1,1),
swrlx:makeOWLThing(?Kina1,1) -> KinematicSkill(?s1),
providesSkill(?a1,?s1), hasRoleOfSkill(?s1,Combined),
isComposedOfSkill(?s1,?s2), isComposedOfSkill(?s1,?s3),
isCalculated(?s1,false),
hasMovementDescription(?s1,?Kina1), Kinematic(?Kina1),
isTopLevelComposedSkill(?s1,true)
```

Even though this example is simple, retracting knowledge occurs at various points during the engineering process. Therefore, reasoning on engineering knowledge requires defeasible reasoning features, so that facts or assumptions can be changed through inference whenever downstream facts make this necessary.

It was expected that declaring the data property that needs to be changed (e.g. isCalculated) as a functional data property would help to solve this problem. However, no reasoner supports checking the characteristics of data properties while processing the rules. In order to solve this issue, an own rule engine had to be used, which checks whether an existing data property is functional and in this case changes the value of the data property. Again, this leads to a violation of R0. Even though there has been recent research interest on defeasible reasoning [25–27], none of the reasoners that are analyzed by [23] provides this feature.

4.5 Implementation

A rule engine for executing the mapping rules was created based on the OWL API.NET (https://owlapinet.codeplex.com/). This rule engine was designed to infer the mapping rules and to create the required OWL individuals. Due to the limitations described before, the authors decided to extend this rule engine to infer the non-mapping rules as well. The rule engine's implementation was focused on solving the actual problem. To do so, methods to create new individuals (using swrlx:makeOWLThing operator) and to execute calculations (using self defined SWRL math operators) were implemented. To support the needed aspect of defeasible reasoning the rule engine interpretes data properties with functional characteristics like flags. If an inferred data property exists, the data property's value is changed to the actual inferred value (for an example see Listing 3).

Although the rule engine was able to solve the given problem, it is very proprietary and implements only those parts that were needed to solve the given use case.

Furthermore, constant maintenance and enhancement cannot be ensured. For transferring such problems to a wider scale the implementation of the described desired features within an open and maintained reasoner is be needed.

5 Summary, Conclusion and Outlook

In this paper, an application of Semantic Web technologies in the engineering domain was introduced. The use-case of the application is a system that should support an engineer in checking the functionality (e.g. the kinematics) of a manufacturing system in order to adapt it to new environments. Therefore, a solution for modeling 3D-CAD information in an OWL ontology was shown, while the CAD data was automatically mapped into the OWL ontology using a mapping component based on standard technology. Since the needed information about the kinematic of a manufacturing system is not contained explicitly in the CAD data, it had to be made explicit through rule-based inference. Due to the fact, that a standardized semantic is a basic requirement for Industry 4.0 or similar initiatives, the utilization of both an "off-the-shelf" reasoner and a standardized rule language were defined as a project goal. Unfortunately, some compromises about this basic requirement had to be made, since available reasoners did not support all necessary features for this application. In order to support the use of Semantic Web technologies in the engineering domain, three desired features were derived in this paper. Including these features in an actively maintained reasoner would support the use of Semantic Web technologies in the engineering domain, since the use of project-specific rule functionality defined with rule engines would not be necessary anymore.

References

1. Wiendahl, H.-P., ElMaraghy, H.A., Nyhuis, P., Zäh, M.F., Wiendahl, H.-H., Duffie, N., Brieke, M.: Changeable manufacturing - classification, design and operation. CIRP Ann. Manufact. Technol. **56**(2), 783–809 (2007). doi:10.1016/j.cirp.2007.10.003
2. Hu, S.J., Zhu, X., Wang, H., Koren, Y.: Product variety and manufacturing complexity in assembly systems and supply chains. CIRP Ann. Manufact. Technol. **57**(1), 45–48 (2008). doi:10.1016/j.cirp.2008.03.138
3. Mourtzis, D., Doukas, M.: Decentralized manufacturing systems review challenges and outlook. Logist. Res. **5**(3–4), 113–121 (2012). doi:10.1007/s12159-012-0085-x
4. Vogel-Heuser, B., Diedrich, C., Fay, A., Jeschke, S., Kowalewski, S., Wollschlaeger, M., Göhner, P.: Challenges for software engineering in automation. JSEA **07**(05), 440–451 (2014). doi:10.4236/jsea.2014.75041
5. Strube, M., Runde, S., Figalist, H., Fay, A.: Risk minimization in modernization projects of plant automation — a knowledge-based approach by means of semantic web technologies. In: Factory Automation (ETFA 2011), Toulouse, France, pp. 1–8 (2011)
6. Runde, S., Fay, A.: Software support for building automation requirements engineering—an application of semantic web technologies in automation. IEEE Trans. Ind. Inf. **7**(4), 723–730 (2011). doi:10.1109/TII.2011.2166784
7. Legat, C.: Knowledge-based technologies for future factory engineering and control. IFAC Proc. Volumes **45**(6), 44–48 (2012). doi:10.3182/20120523-3-RO-2023.00447

8. Puttonen, J., Lobov, A., Lastra, M.: Semantics-based composition of factory automation processes encapsulated by web services. IEEE Trans. Ind. Inf. **9**(4), 2349–2359 (2013). doi: 10.1109/TII.2012.2220554

9. Legat, C., Schütz, D., Vogel-Heuser, B.: Automatic generation of field control strategies for supporting (re-)engineering of manufacturing systems. J. Intell. Manuf. **25**(5), 1101–1111 (2014). doi:10.1007/s10845-013-0744-z

10. Harcuba, O., Vrba, P.: Ontologies for flexible production systems. In: 20th Conference on Emerging Technologies and Factory Automation (ETFA). International Conference on Emerging Technologies & Factory Automation,. Luxembourg, Luxembourg. IEEE, Institute of Electrical and Electronics Engineers, Piscataway (2015)

11. Bunte, A., Diedrich, A., Niggemann, O.: Integrating semantics for diagnosis of manufacturing systems. In: 21st IEEE International Conference on Emerging Technologies and Factory Automation, ETFA, Berlin, Germany, 6–9 September 2016. IEEE, Institute of Electrical and Electronics Engineers (2016)

12. Aarnio, P., Vyatkin, V., Hästbacka, D.: Context modeling with situation rules for industrial maintenance. In: 21st IEEE International Conference on Emerging Technologies and Factory Automation, ETFA. Berlin, Germany, 6–9 September 2016. IEEE, Institute of Electrical and Electronics Engineers (2016)

13. Glawe, M., Tebbe, C., Fay, A., Niemann, K.-H.: Knowledge-based engineering of automation systems using ontologies and engineering data. In: 7th International Conference on Knowledge Engineering and Ontology Development, Lisbon, Portugal, 12–14 November 2015

14. Abele, L., Legat, C., Grimm, S., Muller, A.W.: Ontology-based validation of plant models. In: IEEE 11th International Conference on Industrial Informatics (INDIN), Bochum, Germany, pp. 236–241 (2015)

15. Ramis Ferrer, B., Ahmad, B., Vera, D., Lobov, A., Harrison, R., Martínez Lastra, J.L.: Product, process and resource model coupling for knowledge-driven assembly automation. Automatisierungstechnik **64**(3), 231–243 (2016). doi:10.1515/auto-2015-0073

16. Negri, E., Fumagalli, L., Garetti, M., Tanca, L.: Requirements and languages for the semantic representation of manufacturing systems. Comput. Ind. **81**, 55–66 (2016). doi:10.1016/j.compind.2015.10.009

17. Pfrommer, J., Stogl, D., Aleksandrov, K., Escaida Navarro, S., Hein, B., Beyerer, J.: Plug & produce by modelling skills and service-oriented orchestration of reconfigurable manufacturing systems. Automatisierungstechnik **63**(10), 790–800 (2015). doi:10.1515/auto-2014-1157

18. Usländer, T., Epple, U.: Reference model of industrie 4.0 service architectures. Automatisierungstechnik **63**(10), 858–866 (2015). doi:10.1515/auto-2015-0017

19. Industrial Internet Consortium: Industrial Internet Reference Architecture (2015). https://www.iiconsortium.org/IIRA-1-7-ajs.pdf

20. Jules, G.D., Saadat, M., Li, N.: On designing a unified ontology for holonic manufacturing networks. In: Fathi, M. (ed.) Integration of Practice-Oriented Knowledge Technology: Trends and Prospectives, pp. 207–220. Springer, Heidelberg (2013)

21. Horrocks, I., Patel-Schneider, P.F., Boley, H., Tabet, S., Grosof, B., Dean, M.: SWRL: a semantic web rule language combining OWL and ruleML. In: W3C-World Wide Web Consortium. https://www.w3.org/Submission/SWRL/#8, zuletzt geprüft am 16 September 2016

22. Roda, F., Zanni-Merk, C.: An intelligent data analysis framework for supporting perception of geospatial phenomena. In: Ferrario, R., Kuhn, W. (eds.) Formal ontology in information systems. In: Proceedings of the 9th International Conference (FOIS 2016). Frontiers in artificial intelligence and applications, vol. 283. IOS Press, Amsterdam (2016)
23. Matentzoglu, N., Leo, J., Hudhra, V., Parsia, B., Sattler, U.: A survey of current, stand-alone OWL Reasoners. In: Informal Proceedings of the 4th International Workshop on OWL Reasoner Evaluation (2015)
24. Pahl, G., Beitz, W., Blessing, L., Feldhusen, J., Grote, K.-H., Wallace, K. (eds.): Engineering Design. A Systematic Approach, 3rd edn. Springer, London (2007). http://site.ebrary.com/lib/alltitles/docDetail.action?docID=10230457
25. Basseda, R., Gao, T., Kifer, M., Greenspan, S., Chell, C.: Representing flexible role-based access control policies using objects and defeasible reasoning. In: Bassiliades, N., Gottlob, G., Sadri, F., Paschke, A., Roman, D. (eds.) RuleML 2015. LNCS, vol. 9202, pp. 376–387. Springer, Cham (2015). doi:10.1007/978-3-319-21542-6_24
26. Rakib, A., Haque, H.M.U.: Modeling and verifying context-aware non-monotonic reasoning agents. In: ACM/IEEE International Conference on Formal Methods and Models for Codesign (MEMOCODE), Austin, TX, USA, pp. 61–69 (2015)
27. Casini, G., Meyer, T., Moodley, K., Sattler, U., Varzinczak, I.: Introducing defeasibility into OWL ontologies. In: Arenas, M., et al. (eds.) ISWC 2015. LNCS, vol. 9367, pp. 409–426. Springer, Cham (2015). doi:10.1007/978-3-319-25010-6_27

A Compressed, Inference-Enabled Encoding Scheme for RDF Stream Processing

Jérémy Lhez[1]([✉]), Xiangnan Ren[1,2,3], Badre Belabbess[1,2], and Olivier Curé[1]

[1] LIGM (UMR 8049), CNRS, ENPC, ESIEE, UPEM, 77454 Marne-la-vallée, France
{jeremy.lhez,xiangnan.ren,badre.belabbess,olivier.cure}@u-pem.fr
[2] Atos, Bezons, France
{xiangnan.ren,badre.belabbess}@atos.fr
[3] ISEP - LISITE, 75006 Paris, France

Abstract. The number of sensors producing data streams at a high velocity keeps increasing. This paper describes an attempt to design an inference-enabled, distributed, fault-tolerant framework targeting RDF streams in the context of an industrial project. Our solution gives a special attention to the latency issue, an important feature in the context of providing reasoning services. Low latency is attained by compressing the scheme and data of processed streams with a dedicated semantic-aware encoding solution. After providing an overview of our architecture, we detail our encoding approach which supports a trade-off between two common inference methods, *i.e.*, materialization and query reformulation. The analysis of results of our prototype emphasize the relevance of our design choices.

1 Introduction

Semantic information of the Web of data, generally represented with the Resource Description Framework (RDF)[1] data model, is now being considered for real time analysis. This is the case in the Waves project[2], where we provide real-time analytics of RDF data streams for an international company leading innovation technologies for smart water network management. In particular, we are analyzing data captured from potable water networks in major cities in the world, e.g., studying pressure, flow, turbidity, pH, chlore and other chemical measures, in almost real-time. Some of the key goals of this project are to identify malfunctions in these water networks, *e.g.*, water leaks by analyzing flow and pressure measures, to explain their origins leveraging knowledge base enrichment and to predict potential issues within the pipeline system. With more relevant and faster agent interventions on the network, such research and development can have a substantial impact at both the environmental (to limit potable water loss) and economic (to reduce financial costs) levels. In fact, one must bear in mind that worldwide water leaks peaked to 32 billion m^3/year within last years,

---

[1] http://www.w3.org/TR/rdf-mt/.
[2] http://www.waves-rsp.org/.

© Springer International Publishing AG 2017
E. Blomqvist et al. (Eds.): ESWC 2017, Part II, LNCS 10250, pp. 79–93, 2017.
DOI: 10.1007/978-3-319-58451-5_6

90% of them being invisible due to the underground nature of the network, which makes it a burning issue for the 21st Century.

Detecting water leakage could be performed using quantitative data without exploiting the possibilities of semantic web technologies. However, since we aim to explain discovered leaks, taking advantage of RDF technologies (*e.g.*, RDFS, OWL and SPARQL) and functionalities (*e.g.*, data and knowledge integration, reasoning) becomes a necessity. Such scenarios imply the association of expressive schemata, denoted as ontologies, and explicit measured data. Therefore, an intelligent knowledge management system should enable to infer valuable information that can help in providing sound and complete answers to a continuous query processing component or to help in the design of efficient data analytics.

The integration of a reasoning component in Event Stream Processing (henceforth ESP) is a complex task due to the general cost, in terms of computing resources and time, of inferring data using expressive ontologies. In order to address these requirements, we have designed a prototype system based on the following contributions: (i) we present a generic distributed streaming architecture that addresses materialization and query reformulation reasoning services (Sect. 2), (ii) we propose an encoding approach that minimizes system latency and supports inferences (Sects. 4 and 5), (iii) we highlight the efficiency of our compressing approach with results of an experimentation (Sect. 6).

2   Architecture

In Fig. 1, we present an overview of our architecture. Due to the usage of the Apache Kafka [14] and Apache Storm [12] components, we have designed a system capable of ensuring scalability, fault-tolerance, high throughput and low latency properties.

One characteristic of our project is its capacity to handle both static and dynamic data and knowledge. By static, we mean data and knowledge that are rarely updated while the dynamic aspect relates to the notion of streams arriving at a fast pace, potentially thousands of them per second.

The static aspect of our system consists in encoding a set of ontologies and knowledge bases that are specific to the application domain. In the case of the Waves project, the ontologies are addressing the following topics: sensors, *e.g.*, SSN[3] (Semantic Sensor Network), hydrology, *e.g.*, CUAHSI[4] and modeling physical quantities, units of measure, and their dimensions, *e.g.*, QUDT[5]. The system also integrates additional knowledge bases to represent water network geographical aspects. This is supported by the Geonames[6] and DBpedia[7] ontologies. These knowledge bases are stored in our external knowledge base component which is currently handled by the Virtuoso RDF store[8].

---

[3] https://www.w3.org/2005/Incubator/ssn/ssnx/ssn.

[4] http://his.cuahsi.org/ontologyfiles.html.

[5] http://linkedmodel.org/catalog/qudt/1.1/.

[6] http://www.geonames.org/.

[7] http://wiki.dbpedia.org/.

[8] http://virtuoso.openlinksw.com/.

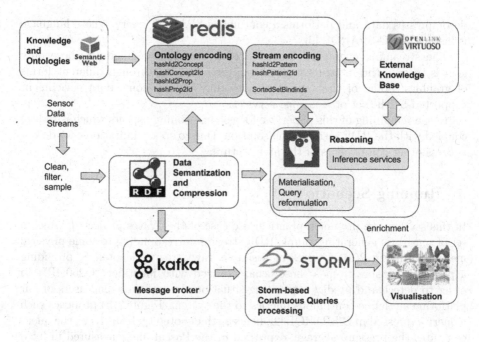

Fig. 1. Architecture overview

The remaining of the architecture is concerned with dynamic, event-based data which are handled by distributed components: Kafka as a distributed, partitioned, replicated commit log service, Storm as the distributed stream processing engine and Redis[9] as a key-value memory store.

A typical scenario in our system is as follows. First, measures are captured from a given sensor network. These streams are cleaned, filtered and possibly sampled before being serialized in a compact RDF format. These data are persisted on-demand to a Redis key-value store and sent to the Apache Kafka message broker. The Kafka component is becoming a standard in streaming processing engines and can be connected to most open source streaming engines (*e.g.*, Storm). Data are fetched from Kafka by a set of distributed nodes which implement the so-called Storm topology, *i.e.*, a network of so-called spouts and bolts. A spout is the source of data streams and can read data from an external framework like Kafka. A bolt is a processing logic unit that performs any kind of processing such as filtering, aggregating, joining, interacting with data stores. Each spout or bolt executes as many tasks across a Storm cluster, and each task corresponds to one thread of execution. Topologies execute across one or more worker processes. Each worker process is a physical Java Virtual Machine (JVM) and executes a subset of all the tasks for the topology. In a Waves topology, each spout subscribes to one stream represented by a Kafka topic, and each bolt

---

[9] http://redis.io/.

decompresses data and performs a continuous SPARQL query, whose language is inspired by C-SPARQL [2].

The streaming engine is connected to a visualization module whose only goal is to ease the interpretation of analyzed streams through different forms of graphics. Some of these visualizations may require some data enrichment supported by the set of reasoning services.

In the remaining of this paper, we focus on reasoning aspects which is tightly coupled with the RDF serialization solution. Due to space limitations, we do not address other components of the architecture.

3   Running Scenario

In this section, we present a practical use case of the Waves project in which a set of sensors is generating simple RDF streams corresponding to some physical measures. In Fig. 2(a), we present a simple, raw stream, denoted S, providing a pressure measure from a sensor characterized with identifier "Q250HP". In order to detect and predict interesting situations in real-time, end-users of our platform can define continuous queries to the system. Figure 2(b) proposes such a query expressed in C-SPARQL [2], henceforth denoted Q. Intuitively, the query computes the pressure average, expressed in the Pascal unit, measured in fixed windows lasting 5 min and sliding every 2 min. Moreover, these averages are only computed for sensors situated in a certain location (a bounding box is specified from ranges of latitude and longitude values) and for a certain sensor type (namely Sensor2). Clearly, this raw stream S does not satisfy the WHERE clause of the C-SPARQL query Q: neither the type of the sensor, the unit of its measure and location are specified in the raw stream. Hence, the result set of Q over S would be empty. We consider that given the messages sent by real-world sensors, such situations are bound to occur frequently.

In fact, sensor "Q250HP" is providing measures in the Pascal unit, is of type Sensor3 and is situated in the bounding box expressed in Q. But these information are only stored in some external knowledge base.

This knowledge base contains two components. An ontology stating that Sensor3 and Sensor4 are sub classes of Sensor2, expressed in a Description Logic [1] formalism as $Sensor3 \sqsubseteq Sensor2$ and $Sensor4 \sqsubseteq Sensor2$. And a set of facts stating that sensor Q250HP provides pressure values expressed in the Pascal unit and is located at latitude 48.59 and longitude 2.75. Thus the data stream, if properly enriched, can satisfy the continuous query Q.

Instead of performing joins at run-time for each incoming events, we prefer to materialize these events with the information that may satisfy a continuous query. Intuitively, the continuous queries are retrieving events from a given set of Kafka topics. Thus it is possible to define possible materialization when a query is associated to a topic. The problem then amounts to define a compact and efficient serialization for the RDF graphs corresponding to the events.

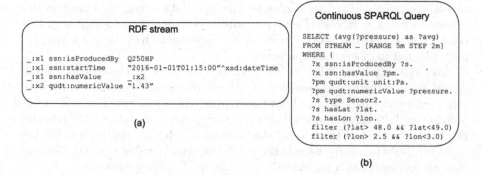

Fig. 2. (a) RDF stream S and (b) continuous query Q

4   Compression Approaches

4.1   Knowledge Base Encoding

With our knowledge base encoding approach, we provide an efficient encoding scheme and data structures to support the reasoning services associated to the terminological and assertional boxes (resp. Tbox and Abox). The input ontology is considered to be the union of all ontologies necessary to operate over one's application domain (*e.g.*, SSN, CUAHSI, QUDT, Geonames and DBpedia). In the current version of our work, we address the ρdf [10] subset of RDFS, meaning that we are only interested in the `rdfs:subClassOf`, `rdfs:subPropertyOf`, `rdfs:domain` and `rdfs:range` constructors. Our Tbox encoding scheme uses our LiteMat system (full details in [5]). Intuitively, it provides a unique, semantic-based identifier to each entry of the Tbox (*i.e.*, class and properties). That is, the identifier of each Tbox element is prefixed by the binary identifier of its super element. This approach enables to represent the class and property hierarchies in a compact way since the identifier of a given class (resp. property) provides all its direct and indirect super classes (resp. properties). Moreover, to capture all inferences related to the class hierarchy, the encoding relies on a classification performed by an OWL reasoner. To support `rdfs:domain` and `rdfs:range` inferences, the property dictionary is extended with the class identifiers that respectively correspond to their domain and range. A final dictionary is generated over instances of the Abox. Once all these dictionaries have been computed and stored in a Redis key/value store, the system can encode the whole Abox, which is then constituted of integer value triples and stored in a Virtuoso instance.

4.2   RDF Distributed Stream Compression

A distributed architecture integrating reasoning services requires low latency to cope with massive real-time streams. However, frequent data transfers between several components (*e.g.*, messaging middleware, data stores, etc.) produce significant network overhead. There are quite various methods to deal with this complex issue and the one we focus on is compression. RDF data are particularly adapted to efficient compression.

Like RDSZ [8] and ERI [6], our approach assumes that events in a given stream share structural similarities, *i.e.*, the RDF graph shapes are similar. We can leverage on this aspect to limit the stream memory footprints. As the compression exploits structural similarities, a new graph (*e.g.*, set of RDF triplets) can be represented on the basis of the previous graph. Our approach breaks up each graph into two parts, namely the graph Pattern and value/variable Bindings that are associated to a graph pattern, hence the PatBin denotation.

In Fig. 3, we present the different steps necessary to generate a pattern signature. This deterministic approach will serve to compare stream graph signature with continuous query signatures in an efficient manner. Considering an arriving graph event corresponding to data stream S of our running example, Fig. 2(a), we use our previously computed property dictionary, Fig. 3(a), to replace property IRIs with integer values. We thus obtain a more compact set of triples (Fig. 3(b)). The compactness of this signature takes benefits from the facts that all events we have encountered correspond to trees. Starting from the root node of our tree, we then sort the graph, in a level-wise manner, according to the property integer values. This order is then used to define a pattern signature (Fig. 3(c)). Intuitively, a signature is composed of property identifiers separated by ':' symbols to delimit properties occurring at the same tree level, '(',')' symbols to describe sub-trees. Note that subjects and objects are not necessary in these signatures since our signature language enables to easily reconstruct the original shape of the tree, *i.e.*, by abstracting subjects and objects with variables.

Correspondences between implicit graph pattern variables and their values (*i.e.*, on triple subjects and objects) are represented by bindings and are sent to Kafka. For a graph N, the bindings are compressed using a differential approach

(a) Property Dictionary extract

```
http://purl.oclc.org/NET/ssnx/ssn#isProducedBy : 144
http://purl.oclc.org/NET/ssnx/ssn#hasValue : 140
http://purl.oclc.org/NET/ssnx/ssn#startTime : 80
http://data.nasa.gov/qudt/owl/qudt#numericValue : 86
```

(b) Intermediary (sorted) pattern

```
_:x1 140 "2016-01-01T01:15:00"^^xsd:dateTime
_:x1 144 Q250HP
_:x1 80  _:x2
_:x2 86  1.43
```

(c) Output pattern signature

```
140:144:80:(86)
```

Fig. 3. Pattern signature process

based on the bindings of the previous graph $N-1$. If the graph N shares some bindings with the $N-1$ graph, then they are replaced by blanks.

Moreover, the mechanism is still not adapted for distributed computing, since encoding the current N graph is based on the bindings of the previously processed $N-1$ graph. However, this implies data exchange between distributed machines if these graphs are processed in different nodes, leading to a network overhead. To solve this, we propose to encode the current N graph based on the bindings of the initially processed graph from which the pattern has been extracted and stored. Hence, we create a context in which we put the pattern, the bindings of the graph from which the pattern has been extracted and the occurring namespaces. All the incoming graphs are encoded based on the context with which they share the same pattern. To guarantee the access to contexts in the distributed infrastructure, we need to store them in a centralized system. Again, Redis has been chosen for storage due to its convenient features (e.g. key-value in-memory store, fast read/write, etc.). Each context created is automatically stored in Redis. The contexts being stored in a centralized system, all the machines have access to compress and decompress operations of RDF graphs. In addition, each machine benefits from its local cache LRU (Least Recently Used) mechanism. That is each machine contains the latest recently used patterns processed by this machine and serves to speed up the contexts read access.

5   Inference Solution

In this section, we present our reasoning approach which is based on a tradeoff between materialization and query reformulation. Although these inference solutions can be used independently, *i.e.*, materialization or query reformulation alone, we highlight that the full potential of the approach is to combine both of them.

5.1   Materialization

The goal of the materialization step is to enrich raw RDF streams in such a way that they can potentially satisfy some given continuous queries. By potentially satisfying a query, we mean that there is a graph homomorphism between a stream and a continuous query graph pattern. It does not necessarily means that a materialized streaming graph pattern actually satisfies the query since some values may not satisfy certain conditions, *e.g.*, filters, of the query. This enrichment is based on retrieving some additional data from external knowledge bases which are stored and possibly encoded in an RDF repository (*e.g.*, the Virtuoso RDF store).

Of course, the task of discovering to what extent a materialization can transform an unsatisfiable raw stream into a potentially satisfiable one, must be performed automatically by the system. That is, the system has to find out a set of sound transformations according to a set of continuous queries and knowledge base axioms. We compute such discoveries using graph matching operations over the graph patterns of RDF streams and continuous queries. This approach is valid since the vocabularies used in these two components correspond to our predefined set of encoded ontologies (Sect. 4.2).

Given the potential high volume of different data stream types, *e.g.*, in our use case, measures such as pressure, flow, chlorine, turbidity, etc., and the number of continuous queries, it is important to propose an efficient discovery approach. Our method considers that streams are submitted to Kafka topics and that these topics are processed to retrieve stream graph patterns. Moreover, the continuous queries (implemented as Storm bolts) are connected to Storm Spouts which are themselves related to Kafka topics. Hence, it is possible to reduce the space search by matching pairs of stream and continuous query graph patterns connected to the same Kafka topics.

Given a Kafka topic T, the graph matching discovery problem amounts to finding if a Stream Graph Pattern SGP is a sub graph of a given continuous query graph pattern CQ, *i.e.*, excluding FILTER, GROUP BY and OPTIONAL clauses and considering group graph patterns related by UNION clauses as individual queries. This search for a sub graph relationship is semantic-aware, meaning that class and property subsumption relationships are taken into account. For instance, with our previously defined ontology, the following situation: $SGP = \_ : x1$ *type Sensor*3, $CQ = \_ : x4$ *type Sensor*2 would correspond to a sub graph relationship due to the *Sensor*3 $\sqsubseteq$ *Sensor*2 axiom. Note that this is not the case for this other example: $SGP = \_ : x1$ *type Sensor*2, $CQ = \_ : x4$ *type Sensor*3.

If SGP is not a sub graph of CQ then we consider that this sort of data streams can not be enriched to satisfy the continuous query. In the case SGP is equal to CQ then no materialization is required since SGP can potentially satisfy CQ out-of-the-box. Finally, if SGP is a sub graph of CQ then the triple-based difference between CQ and SGP is computed to identify the set of triples that are missing in SGP to potentially satisfy CQ. Based on mapping assertions between subject and object identifiers of SGP and CQ, we can instantiate a computed triple set from external knowledge bases. In our running example, this amounts to generating the bold lines of Fig. 4(a). Basically, the unit, location and type of sensor "Q250HP" triples are added to the streams. Note that the sensor type is expressed with the integer value corresponding to its binary encoding: the binary identifiers of *Sensor*2, *Sensor*3 and *Sensor*4 are respectively 101100, 101110 and 101101 which respectively correspond to the 44, 45 and 46 integer values.

The discovery of a graph match is fast due to our compact, deterministic graph signature representation. Nevertheless, it may become a performance bottleneck due to high velocity stream production. To prevent this from happening, the system stores discovered graph pattern correspondences and only searches for new ones when novel stream patterns are recorded in the system and/or when continuous queries are updated or inserted. A discovered graph pattern exactly matches the graph associated to a materialized stream and is expressed as the original graph patterns, *i.e.*, as defined in Sect. 4.2.

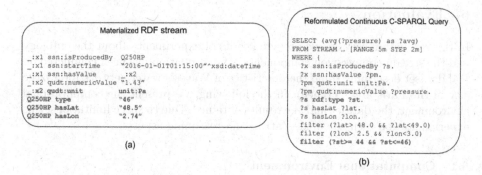

Fig. 4. Materialized RDF stream and reformulated continuous query

5.2   Query Reformulation

The goal of the query reformulation component is to modify the original continuous query such that subsumption relationships are properly addressed. A special attention is given to classes specified in `rdf:type` triples. If any of these classes are at some point a super class in our Tbox then some reformulation is necessary. The system proceeds as follows: in each triple pattern with a `rdf:type` property, replace the class C (object position) with a non previously used variable (denoted V_i). In Fig. 2(b), the `?s type Sensor2` triple is replaced by the triple `?s type ?st` in Fig. 4(b). Then a SPARQL FILTER clause is introduced in the reformulated query on that variable V_i. The goal is to cover all possible sub classes of the original class C. The specification of these classes are performed at the encoding level and hence benefits from the nice properties of our ontology encoding. Due to our encoding approach, we know that sub classes of $Sensor2$ are necessarily included in the "101100" and "101111" identifier range which correspond to respectively to the 44 and 46 values. These lower and upper bound values are easily computed (using two bit shift operations) from the binary version of C's identifier. With this approach, we cover all sub classes of a given class with a single FILTER query line, independently of the length of this class subsumption relationships. The last bold line of Fig. 4(b) represents this filter clause for our running example.

A similar approach is perform for the property hierarchy. It consists of analyzing whether any of the non `rdf:type` properties is at some point a super property. Then the system operates in an identical manner: it replaces the property with a new variable and inserts a FILTER line that restricts the range of accepted property values for that variable.

Note that this approach is particularly efficient when several reformulation (*e.g.*, on classes and properties) are needed in a single query. With our filter approach, a reformulated query grows linearly and not exponentially as is generally the case for standard query reformulation approaches.

6   Evaluation

In this evaluation section, we provide results of experiments about the ontology and the stream compression components. The evaluation has been conducted on real dataset describing some characteristics of Waves's water network for a large city in the Paris area (France). In the following, we present the computational environment, the dataset and the results obtained. Due to space limitations, this experimentation focuses on the PatBin and reasoning aspects.

6.1   Computational Environment

Throughout this experimentation section, we are using two different computational settings. The evaluation concerning the compression have been realized on a laptop with a Windows 8 operating system, equipped with an Intel Core i7 processor (2.90 GHz), 16 GB of RAM, running JDK/JRE 1.8. The ontology encoding evaluation has been performed on a Linux Ubuntu 14.04 distribution with 16 GB of RAM, Intel Core I5 quad-core processor and running a JDK 1.8. We used the HermiT version 1.3.8 as an external reasoner and programmed the encoding solution with Apache Jena 3.0.0. Finally, we are using Apache Spark [15] version 1.5.2 for the encoding of the ABox. The Spark cluster consists of 3 Dell PowerEdge machines equipped with 64 GB of RAM.

6.2   Datasets

For experimentation, we use a real world dataset describing different water measurements captured by sensors. Values of flow, pressure and chlorine are examples of these measurements. These data are provided in CSV format and need to be represented in a semantic model. For this, we are annotating values using three popular ontologies: SSN, CUAHSI-HIS and QUDT. Each sensor observes at least one physical phenomenon or a chemical property, and hence produces timestamped streams containing an observation.

6.3   Results

Knowledge Base Encoding Evaluation. We ran our ontology compression Java program a total of five times and obtained an average of 18.8 s for the merged ontology presented in Table 1. With respect to the low numbers of classes and properties, this duration can be considered rather long. In fact, this can be justified by the rather high expressivity of the resulting ontology which happens to correspond to $\mathcal{SROIQ}(D)$ Description Logics. This expressiveness matches the OWL2 DL ontology language which is known to be the OWL fragment with the highest computational complexity for standard inference services (apart from OWL Full which is undecidable).

Comparatively, The same algorithm is able to encode the DBPedia OWL ontology, which contains over 800 classes and 3000 properties, in less than

Table 1. Compressed ontology in terms of number of classes, object and data type properties

Ontology	#classes	#object pr	#data pr	Duration (sec.)
SSN	117	142	6	–
QUDT	229	69	29	–
CUAHSI extract	103	0	0	–
Merged ontology	449	174	35	18.8
DBPedia	814	3,035	1,310	4.1
Wikidata	213,958	255	98	118

4 s for an expressivity corresponding to $\mathcal{ALCHF}(D)$ DL. The encoding of the Wikidata ontology takes approximatively 2 min. This is mainly due to the large class hierarchy (over 200,000 classes) and not to its expressiveness which corresponds to the $\mathcal{AL}$ DL.

Finally, we provide an evaluation of a data instance encoding which is needed for static knowledge bases. This processing is distributed over our Spark cluster and the measures are presented in Table 2. These measures are about 70% faster than state of the art compression approaches defined over Apache Hadoop [13].

Table 2. Duration and throughput of data instances, triples in $*10^6$, duration in seconds and throughput in triples/sec.

Dataset	#Triples	Duration	Throughput
DBPedia	79.1	282.2	280 943
Wikidata	242.1	1 334.8	181 394

Signature Generation Performance. We used RDSZ results to check the algorithm's compression performance. A specific Java class stores the algorithm statistics in terms of performance and compression rate, thus we made some similar measures for PatBin to ensure a fair comparison. The system time was measured once the input file was parsed as a Java String containing all triples, and a second time right after the compression step. The subtraction of those values gives the compression performance. RDSZ's statistics also provide information about the compression rate, by giving the size of the compressed output (in UTF-8 bytes); therefore we used this method for our algorithm again. Both those measures are presented in Table 3; we performed a series of verification for different input file sizes (using the turtle serialization). We used the basic configuration of RDSZ algorithm, with no specific argument. As we can see, the compression performance is much faster for PatBin; this is mostly due to the fact that we only have to deal with predicates. Indeed, RDSZ must initialize its binding table with both subject and predicates, and verify for each class if it is

Table 3. Signature generation performance for RDSZ and PatBin, time of compression in microseconds, size of he signature in bytes

	RDSZ time	PatBin time	RDSZ size	PatBin size
5 triples	383	1	312	13
10 triples	387	2	370	29
25 triples	394	3	425	89
50 triples	397	6	523	184
100 triples	401	10	750	382

not already present in the table. We have twice less work to do with only the properties. PatBin also has better results in terms of compression rate, which tends to decrease for big input files: this is mostly due to the fact that we used examples files that are represented as big forests, thus having a long signature on several lines.

Graph Matching Performance. The graph matching performance has been performed by checking the equality between a newly compressed string, and an array of stored compressed strings, acting as a cache. We made our evaluation on several sizes of cache, to vary the number of comparison made; we also tested different sizes of files (different numbers of triples) in order to have an output longer or shorter. For each individual evaluation, we took files with the same number of triples, and we also made sure that the input file was not in the cache; this ensured each value in the cache would be verified, and thus the test would not be biased by ending the checking too soon. Both the results in cache and to be checked were (different) compressed results obtained from a C-SPARQL query. The results are displayed in the Fig. 5; since the results have a very high variance, we had to do an average of different results to have valid results. The measures concern only the matching: the signature generation for the file to be matched is not taken in account. Each measure has been identified by a point on its curve, for better visualization. The three measures for PatBin appear mingled with the lower (X) axis, because the computation time is much shorter than RDSZ. In both cases, the matching time increases when we the cache size and/or the triple number. For PatBin, the results are much better: with 25 triples and a cache size of 100 compressed strings, the checking time is only about 19 μs, i.e., two orders of magnitude lower than RDSZ. This proportion cannot be established precisely because of the variance, however the computation times remain much better for PatBin. This is due to the fact that the signature obtained after compression is much more compact that the one of RDSZ, since PatBin does not retain the triples in its signature. We also checked hot and cold performances for cache searching: in both cases, we filled a cache with 1000 random patterns, and checked if a new entry was present in the cache. We verified that the randomness ensures the caches are completely verified in both cases. The cold performances give an execution time of 105 μs for PatBin, and 156 for

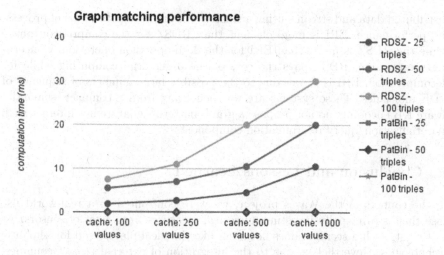

Fig. 5. Materialized RDF stream and reformulated continuous query

RDSZ. For the hot performances, we computed the average of five executions: PatBin is still more efficient, with 95 µs of matching time against 160 for RDSZ.

7 Related Work

We consider two systems that integrate reasoning within a RDF streaming context. IMaRS [3] incrementally maintains a materialization of ontology entailments in a timely manner. The system extends the DRed [9] approach with the use of the window operators and the introduction of an expiration time for each triple. The system does not interact with a query reformulation component, is not distributed and it is recognized that automatically defining efficient expiration time is difficult in a streaming context. Finally, StreamQR [4] proposes a query reformulation solution which is based on the kyrie rewriter. The architecture of the system does not support scalability and interactions with a materialization component have not been considered.

Several systems consider RDF stream compression. The Zstreamy [7] system is presented as a scalable platform for publishing semantic streams on the Web. The compression approach is simply based on a standard Zlib compression. CQELS Cloud [11] addresses the problem of scalable stream processing and proposes a simple dictionary encoding approach reminiscent of RDF stores. RDSZ [8] (RDF Differential Stream compressor based on Zlib) and ERI [6] (Efficient RDF Interchange format) correspond to lossless RDF stream compression approaches. Both take advantage of structural similarities of RDF graph events. ERI proposes a more fine-grained approach to pattern and pattern binding representations. Moreover, ERI does make an extended usage of differential compression as RDSZ does. In general, the compression approaches of the two systems are comparable with RDSZ being slightly more efficient for randomly distributed data and streams using a small set of predicates. In terms of processing performance, ERI is more efficient than RDSZ for the compression phase while the RDSZ is faster than ERI for the decompression operation. Concerning compression, RDSZ pays the cost of the differential computing while for decompression, ERI is slower due to the possibly large numbers of sequence of RDF molecules. These systems are not benefiting from a compact, semantic-aware KB encoding, do not propose a graph pattern signature nor interact with materialization/query reformulation components.

8   Conclusion and Lessons Learned

In the context of the Waves project, we were confronted to a real-world use case that is principally ingesting numerical measures from a set of sensors. At first sight, such a scenario does not seem like the ideal playground for semantic technologies. Nevertheless, due to the integration of external (*e.g.*, Geonames, DBpedia) and domain specific (*e.g.*, SSN, CUAHSI) knowledge bases, as well as RDF related technologies (*e.g.*, SPARQL, RDFS, OWL), we were able to highlight the added value of a semantic approach. The main impact was the ability to explain some network malfunctions via the execution of inference-enabled continuous SPARQL queries. Of course, one of the key learned lesson concerns the impact of reducing latency when reasoning over large event streams. We found out that finding a trade-off between materialization and query reformulation was an important factor in reducing processing latency. But this approach is reaching its full potential with the kind of semantic-aware encoding and compression presented in this work.

As future work, we aim to test Waves's system on diverse IoT contexts and thus emphasize that our approach can be generalized to different use cases. Moreover, we are currently implementing an adaptive query processing engine to guarantee the execution of optimized continuous SPARQL queries. Finally, we will extend LiteMat's inference capabilities with support for RDFS++ (an ontology language supported by the Allegrograph RDF Store), *i.e.*, supporting RDFS as well as `owl:sameAs`, `owl:transitiveProperty` and `owl:inverseOf` ontology constructs.

Acknowledgment. This work has been supported by the Waves project which is partially supported by the French FUI (Fonds Unique Interministériel) call #17.

References

1. Baader, F., Calvanese, D., McGuinness, D.L., Nardi, D., Patel-Schneider, P.F. (eds.): The Description Logic Handbook: Theory, Implementation, and Applications. Cambridge University Press, Cambridge (2003)
2. Barbieri, D.F., Braga, D., Ceri, S., Valle, E.D., Grossniklaus, M.: C-SPARQL: SPARQL for continuous querying. In: Proceedings of the 18th International Conference on World Wide Web, pp. 1061–1062 (2009)

3. Barbieri, D.F., Braga, D., Ceri, S., Valle, E., Grossniklaus, M.: Incremental reasoning on streams and rich background knowledge. In: Aroyo, L., Antoniou, G., Hyvönen, E., Teije, A., Stuckenschmidt, H., Cabral, L., Tudorache, T. (eds.) ESWC 2010. LNCS, vol. 6088, pp. 1–15. Springer, Heidelberg (2010). doi:10.1007/978-3-642-13486-9_1

4. Calbimonte, J.-P., Mora, J., Corcho, O.: Query rewriting in RDF stream processing. In: Sack, H., Blomqvist, E., d'Aquin, M., Ghidini, C., Ponzetto, S.P., Lange, C. (eds.) ESWC 2016. LNCS, vol. 9678, pp. 486–502. Springer, Cham (2016). doi:10.1007/978-3-319-34129-3_30

5. Curé, O., Naacke, H., Randriamalala, T., Amann, B.: LiteMat: a scalable, cost-efficient inference encoding scheme for large RDF graphs. In: 2015 IEEE International Conference on Big Data, Big Data 2015, pp. 1823–1830 (2015)

6. Fernández, J.D., Llaves, A., Corcho, O.: Efficient RDF interchange (ERI) format for RDF data streams. In: Mika, P., et al. (eds.) ISWC 2014. LNCS, vol. 8797, pp. 244–259. Springer, Cham (2014). doi:10.1007/978-3-319-11915-1_16

7. Fisteus, J.A., Garcia, N.F., Fernandez, L.S., Fuentes-Lorenzo, D.: Ztreamy: a middleware for publishing semantic streams on the web. Web Semant. Sci. Serv. Agents World Wide Web **25**, 16–23 (2014)

8. Fernández, N., Arias, J., Sánchez, L., Fuentes-Lorenzo, D., Corcho, Ó.: RDSZ: an approach for lossless RDF stream compression. In: Presutti, V., d'Amato, C., Gandon, F., d'Aquin, M., Staab, S., Tordai, A. (eds.) ESWC 2014. LNCS, vol. 8465, pp. 52–67. Springer, Cham (2014). doi:10.1007/978-3-319-07443-6_5

9. Gupta, A., Mumick, I.S., Subrahmanian, V.S.: Maintaining views incrementally. SIGMOD Rec. **22**(2), 157–166 (1993)

10. Muñoz, S., Pérez, J., Gutierrez, C.: Simple and efficient minimal RDFS. J. Web Sem. **7**(3), 220–234 (2009)

11. Le-Phuoc, D., Nguyen Mau Quoc, H., Le Van, C., Hauswirth, M.: Elastic and scalable processing of linked stream data in the cloud. In: Alani, H., et al. (eds.) ISWC 2013. LNCS, vol. 8218, pp. 280–297. Springer, Heidelberg (2013). doi:10.1007/978-3-642-41335-3_18

12. Toshniwal, A., Taneja, S., Shukla, A., Ramasamy, K., Patel, J.M., Kulkarni, S., Jackson, J., Gade, K., Fu, M., Donham, J., Bhagat, N., Mittal, S., Ryaboy, D.: Storm@twitter. In: Proceedings of the 2014 ACM SIGMOD International Conference on Management of Data, SIGMOD 2014, pp. 147–156 (2014)

13. Urbani, J., Maassen, J., Bal, H.E.: Massive semantic web data compression with mapreduce. In: Proceedings of the 19th ACM International Symposium on High Performance Distributed Computing, HPDC 2010, pp. 795–802 (2010)

14. Wang, G., Koshy, J., Subramanian, S., Paramasivam, K., Zadeh, M., Narkhede, N., Rao, J., Kreps, J., Stein, J.: Building a replicated logging system with apache Kafka. PVLDB **8**(12), 1654–1665 (2015)

15. Zaharia, M., Chowdhury, M., Franklin, M.J., Shenker, S., Stoica, I.: Spark: cluster computing with working sets. In: 2nd USENIX Workshop on Hot Topics in Cloud Computing, HotCloud 2010 (2010)

From Data to City Indicators: A Knowledge Graph for Supporting Automatic Generation of Dashboards

Henrique Santos[1]([⊠]), Victor Dantas[1], Vasco Furtado[1], Paulo Pinheiro[2], and Deborah L. McGuinness[2]

[1] Universidade de Fortaleza, Fortaleza, CE, Brazil
{hos,victordantas2}@edu.unifor.br, vasco@unifor.br
[2] Rensselaer Polytechnic Institute, Troy, NY, USA
pinhep@rpi.edu, dlm@cs.rpi.edu

Abstract. In the context of Smart Cities, indicator definitions have been used to calculate values that enable the comparison among different cities. The calculation of an indicator values has challenges as the calculation may need to combine some aspects of quality while addressing different levels of abstraction. Knowledge graphs (KGs) have been used successfully to support flexible representation, which can support improved understanding and data analysis in similar settings. This paper presents an operational description for a city KG, an indicator ontology that support indicator discovery and data visualization and an application capable of performing metadata analysis to automatically build and display dashboards according to discovered indicators. We describe our implementation in an urban mobility setting.

1 Introduction

While a single agreed upon definition of a smart city may be elusive, many definitions, if not most definitions include some technology and infrastructure that provide a high quality of life for its residents. Determining a desirable quality of life often includes evaluation of city qualities such as: sustainability, safety, inclusiveness, walkability, creativity, and innovation. Cities with high scores on these qualities are often judged as being desirable places to live. Achieving the capability of assessing any of these or other desirable qualities, however, requires two key components: accessing and understanding city's data. Consequently, a city's ability to produce and share relevant data that can be understood and used by a broad range of diverse stakeholders is critical for evaluating and comparing cities and can be viewed as key indicator of a Smart City as well as the ability to derive knowledge from city's data and further use it to power innovation.

Governments are increasingly sharing city data, often with the goal of promoting innovation via societal participation with the use of data. In the context of data sharing, different categories of stakeholders may be identified: designers and software developers may use data to produce public services through the use

© Springer International Publishing AG 2017
E. Blomqvist et al. (Eds.): ESWC 2017, Part II, LNCS 10250, pp. 94–108, 2017.
DOI: 10.1007/978-3-319-58451-5_7

of web and mobile applications; scientists may produce elaborate analysis and studies about the cities; public officers may use the data to improve city administration using data-based decision-making techniques; journalists may use open data to produce more reliable, factually-based and attractive news. The use of knowledge graphs (KGs) as a way of better understanding and analyzing data has proven successful in many cases [2,4,8,14]. They are not simply linked data using an RDF model; they also provide support for knowledge management including explicit provenance encoding capabilities, entity description encodings, potential to connect to and leverage reasoners and so forth.

To obtain measured values for characterizing city's properties, some approaches [5,6,9] have made use of the development and calculation of city indicators. Indicators are metrics that one can use to assess the city level of maturity in a certain field of interest. More than that, well-defined indicators enable the comparison among different cities so one can determine when one city appears to be doing better than another city with respect to certain criteria. However, robust, reusable, and precise calculation plans for indicators have challenges. For example compound indicators require combinations of data that may be unavailable and those data need to be modeled in enough detail so that indicator calculating systems (and humans) can understand enough to know when data is comparable and may be combined. Further enough information about provenance needs to be available so that trust can be ascertained.

This paper tackles the challenge of calculating indicator values from (raw) data, describing work with both city indicators and KGs for city data as a way to automatically build and display dashboards that can be used by a wide range of users in city comparisons. The proposed KG uses OWL ontologies that describe concepts and relations regarding sensing infrastructure, provenance, data acquisition activities, indicators and city entities themselves. Once built, the KG (or a subset of it) can be serialized in the Contextualized CSV format (CCSV [13] - a format that conveys both data and associated metadata) while a reasoner performs inferences to discover indicators inside the indicator ontologies that are suited for the serialized data. Discovered indicators are then serialized themselves in Turtle format. Both serializations are presented to a dashboard generating application, which performs metadata analysis to automatically build and display dashboards according to the discovered indicators. The three main contributions of this paper are (i) the city KG description that enables transparent and explainable indicator values; (ii) the Indicator ontology that can support dashboard visualization; and (iii) a dashboard generating application that works with knowledge from KGs. The rest of this paper is organized as it follows. The next section introduces current approaches to KGs, city indicators, city modeling and data annotation. In Sect. 3, the proposed KG is defined alongside its ontologies, modeling decisions and serialization process. Section 4 describes the dashboard generating application and its metadata analysis that supports building and displaying dashboards in the context of urban mobility, Sect. 5 concludes and discusses future plans.

2   Related Work

Recently, the Knowledge Graph (KG) phrase has been used to define large collections of structured data in a meaningful way. Being more than simply linked data, the semantics encoded in a KG enables tasks that may be challenging in simple linked data RDF models. Metadata faceting, provenance tracking and context-awareness are examples of enhanced features that KGs can support. The term gained popularity with Google KG [14] in a effort to merge Freebase [3] (which also may be considered a KG), Wikipedia and the CIA World Factbook[1] augmented with their search engine's queries and results. Academic KGs are also available including YAGO [2,8] and DBpedia [1].

2.1   City Modeling

The process of modeling a city is complex. The intrinsic complexity of interactions between city entities make it very difficult to map relevant sets of dynamic aspects that are often used to characterize a city. Moreover, these entity interactions, along with the numerous entities and processes, differ from one city to another. Thus, the process of modeling the city is typically use-case centered, where the modeling is performed towards a specified goal. This approach, hence, streamlines the process, identifying which characteristics need to be modeled. The work in [15] proposes a core conceptual model for the Domain Knowledge Model of a Smart City, which originally involves multiple domains and cities. The proposed work aims to support cross-domain and cross-city interoperability by specifying terms from different stakeholders. Ontologies play a big role in enabling cross-city comparison. The Semantic Web has been used in the Open Government Data (OGD) approach to make it possible for cities to share information and knowledge under a common vocabulary. Pushing this further, the GCI (Global City Indicators) Ontology [6] is an effort for the modeling of city entities that covers the concepts used by global indicators using Semantic Web technologies.

2.2   Data Annotation

Data can be encoded in many distinct formats including CSV, XML and NetCDF [12]. In many cases, CSV is a format of choice because of its ease of use by both automated actors and human actors. Human actors often manually enter acquired data in a spreadsheet application (e.g., MS Excel or LibreOffice Calc). Spreadsheets are also capable of exporting content in CSV format. Basically, the CSV format can be seen as a minimalist enabling approach for data interoperability.

Regardless of the format, until our proposed CCSV format [13] we are not aware that any single encoding was able to provide effective mechanisms for annotating data in a way that supports data acquisition as a contextualized data

---

[1] https://www.cia.gov/library/publications/the-world-factbook

point collection. For instance, CSV lacks features for expressing the semantics associated with the data contained in it, so it is challenging to know, in an automated and interoperable way, the meaning of the data enclosed inside a CSV file. For example, it can be difficult to determine if two entries are observationally equivalent (measured under the same conditions, using the same units, in the same area, etc.). Also, different agents may generate data in different formats and standards, making CSV even more difficult to process automatically.

Although there are existing approaches for accessing CSV metadata and also for providing a metadata vocabulary for CSV data, they are typically more concerned with content restrictions, rather than the context in which the CSV data was collected. W3C's recommendations from the CSV on the Web Working Group[2] elaborate on techniques for enabling the access of CSV metadata by describing the content metadata in a separate JSON or RDF/XML file that makes use of RDF vocabulary. To bridge this gap, we proposed the Contextualized CSV (CCSV) [13] as a format that deals with both content and context restrictions of the data points enclosed in it. The CCSV dataset is basically a regular CSV file with a Turtle preamble.

2.3 Indicators

The ISO 37120:2014 [9] is a standard that defines 100 indicators across 17 themes that were evaluated to be a precise way to measure a city's performance of its services and quality of life. The themes span areas including Economy, Education, Health, and Safety. The main goal of this standard is to provide a concise set of well-defined global indicators that any city can use to measure itself. Moreover, cities that adhere to this standard are able to compare themselves, and evaluate how well they are doing in comparison to others. Making use of the ISO standard, the PolisGnosis Project [5] is a final goal of an ongoing effort by the University of Toronto. The project aims the following:

– To provide a description of all the 100 ISO indicators in terms of ontologies for the semantic web;
– To develop an engine capable of performing analysis in order to discover root causes of differences concerning why indicators change over time for a given city and why they are different between different cities.

Until the time of this writing, the PolisGnosis Project has focused largely on the GCI Ontology engineering[3] as a standard to publish the ISO indicator values, while our efforts attempt to also support a broader range of representation challenges including representation and reasoning for data visualization.

3 City Knowledge Graph

City indicators have some requirements that need to be followed when defining them and calculating their values. These requirements ensure that the indicator

---

[2] http://www.w3.org/2013/csvw/wiki/Main_Page
[3] http://ontology.eil.utoronto.ca

is well defined and the calculation process will generate trusted values. We have identified the following requirements:

- Temporal coverage: Indicator values carry more representativeness when calculated taking into account data from a determined time frame. Such an approach enables temporal comparisons such as if a theme of interest had improving or deteriorating performance in one particular year;
- Entities of interest: Indicator definitions relate named entities, thus it is important to provide formal definitions for those entities;
- Provenance: Indicators may refer to a particular set of activities and/or data sources;
- Context: Indicators can also refer to data acquired under certain conditions, making context management also important;
- Location: Indicators values may refer to an specific area within a city or geographic region;
- Visualization: An easy way to visualize the calculated values is desirable.

In order for the KG to fulfill these requirements, we have made use of ontologies that can provide metadata descriptions, domain model and indicators definitions and, where the existing ontologies weren't able to cover, we have created extensions as a new ontology. The following subsections describe our choices.

3.1 Metadata Ontologies

City data production happens in a plethora of different sources and processes. To characterize the diversity and scale of city data produced, we reused ontologies defining the data acquisition concept, which have demonstrated their capability of encoding contextual knowledge for millions of acquired data points that would be otherwise lost during regular data acquisition activities.

VSTO-I [7]. "The Virtual Solar-Terrestrial Ontology - Instrument model" is an ontology[4] that contains concepts that describe entities capable of collecting data (e.g., instruments, detectors and platforms) and activities related to these entities such as a deployment of an instrument on a platform. By making use of this ontology, the KG is able to keep track of all sources of data. The main reused classes are:

- *vstoi* : *Instrument*: A device, mechanism or software that is used to acquire attribute values of entities of interest.
- *vstoi* : *Deployment*: A deployment is an activity of physically installing an Instrument by an agent. More than that, the deployment states that an Instrument is able to start collecting data under certain conditions (calibration, configuration etc.).

---

[4] http://hadatac.org/ont/vstoi#.

HAScO. The Human-Aware Science Ontology[5] is the top metadata ontology in the KG definition. HAScO describes scientific concepts related to data acquisition. With HAScO, it is possible to describe studies, projects and data collection activities like an interview of a subject or an empirical observation. HAScO is the next generation of HASNetO [11] (The Human-Aware Sensor Network Ontology[6]), which is a comprehensive alignment and integration of the VSTO-I sensing infrastructure and the PROV ontology. The KG makes use of the following classes, among others:

- *hasco : Study*: A study is a prov:Activity where steps are performed to prove or disprove an hypothesis.
- *hasco : StudyStep*: A study step is a prov:Activity that composes a study. *hasco : DataAnalysis* and *hasco : DataAcquisition* are examples of it.

HACitO. The Human-Aware City Ontology[7] extends the functionalities of VSTO-I and HAScO to the Smart City context. Figure 1 depicts the main extensions and relationships inside HACitO. HACitO makes it possible for the KG to support data production with full annotation on its origin, when the data is first generated. But, as most of the information systems in a city are legacy systems and cannot be adapted to produce fully annotated data, HACitO describes the manual data annotation data acquisition activity. The goal is to keep track of all the possible metadata involved in that data production process. For that, the ontology defines the class *hacito : ManualDataAnnotation* as a subclass of *hasco : StudyStep*, which is a data acquisition activity by the means of manual data annotation using an annotator software, which in turn is described by *hacito : AnnotatorSoftware*, as a subclass of *vstoi : Instrument*. The annotator is deployed to a legacy data production information system, which is an extension *hacito : InformationSystem* of *vstoi : Platform*.

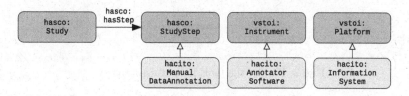

Fig. 1. HACitO ontology

3.2 Indicator and Domain Ontologies

Indicators serve as metrics that provide insight into city performance. They are typically calculations over existing data. The calculated values facilitate quantitative comparisons between different cities, thus enabling city managers to make

---

[5] http://hadatac.org/ont/hasco#.

[6] http://hadatac.org/ont/hasneto#.

[7] http://hadatac.org/ont/hacito#.

decisions informed by current data and also to support data-driven planning. The proposed KG support for indicators is based on the GCI Ontologies [6] and the ISO 37120 Indicator Definitions Ontology[8] for the ISO 37120:2014 indicators. As discussed above, we believe that the GCI Ontology for the ISO 37120 indicators is good for publishing indicator values and comparing cities but is not aimed to support data visualization. To overcome this, the KG is able to hold user-created indicators, To address this, we have developed our QoE Indicators ontology that includes both indicators aimed at representing calculated numerical values but also indicators specifically aimed to support convenient visualization. To address this, we have developed our QoE Indicators ontology[9] that includes both indicators aimed at representing calculated numerical values but also indicators specifically aimed to support convenient visualization, which extends the GCI Ontology.

To make this possible, we have described the QoE indicators using the following data visualization concepts:

– Dimension: An entity value that usually cannot be aggregated, often used for row or columns headings;
– Measure: An entity value that can be used to calculate something, e.g. a sum or medium, often used to support display and plotting.

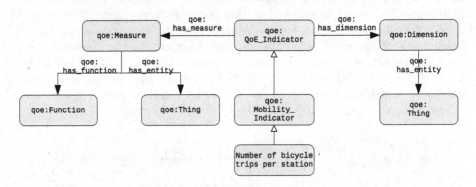

Fig. 2. Part of the QoE indicators ontology

In one example, if one has a bar chart where each bar shows the number of single commuters during a single month of the year, for a total of twelve bars, one for each month, the dimension would be the month and the measure would be sum (or count) of every person who has commuted in a given month. Figure 2 depicts part of the QoE Indicators Ontology. In the middle, the *qoe : QoE_Indicator* is defined by some *qoe : Measure* and some *qoe : Dimension*, each of which has an associated *qoe : Thing*, i.e., the related entity. It is important also to note that the measure has a *qoe : Function*, which states what kind of calculation will be

---

[8] http://ontology.eil.utoronto.ca/ISO37120.owl.
[9] http://hadatac.org/ont/qoe#.

performed over that value. It is possible for an indicator to have more than one dimension and/or measure. For instance, a line chart with two measures would actually display two lines, one for each measure. An interesting case is an indicator with only measures and no dimensions. The resulting data visualization would be just a number.

One of the defined indicators in the QoE Indicators Ontology is 'Number of bicycle trips per station' which defines their associated entities for dimensions and measures using the classes $qoe - m : Bicycle - Share\_Trips$ and $qoe - m : Bicycle - Share\_Station$, respectively. These domain entities are part of the QoE Domain Ontology[10] which is shown in the excerpt on Fig. 3. The QoE Ontologies (Indicators and Domain) are evolving definitions that should be tailored for each city and intended use.

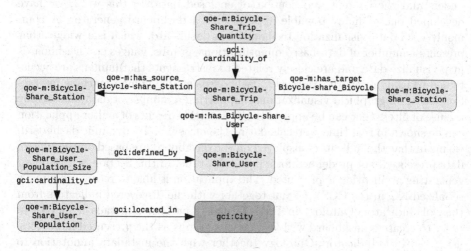

Fig. 3. Part of the QoE domain ontology

3.3   KG Serialization

The KG serialization process is concerned with bringing the data from the KG to a physical file together with all its metadata, making it possible for third-party applications to make use of all the knowledge attached to the data. For that to take place, routines were developed to perform the following:

1. A file is created for every class in the domain ontology with valid instances;
2. For every file, write the instances as CSV registers with each column being a triple in the KG where the instance is the subject;
3. Annotate every register and column using the CCSV format;

---

[10] http://hadatac.org/ont/qoe-m#.

4. Add to the annotations the study, deployment and data acquisitions related to the data in the file;
5. Using the annotated data, discover suitable indicators and export them in Turtle format.

The next section describes how this serialization was performed in the context of urban mobility.

4   Dashboard Application

The serialized KG is a set of files that utilize a common vocabulary, making it possible for third-party applications to interpret the CCSV format and thus understand the content and context of enclosed data. In this work, we have developed one of many possible applications: a dashboard generator. A commonly used data visualization technique is a dashboard, which is a widget that presents a number of data-based quantifications, graphs, gauges etc. Dashboards help visualize data and are closely related to instruments that humans are accustomed to using regularly. Moreover, dashboards enable human-machine interaction based on graphical visualizations, supporting a number of data analyses by the use of filters that can be applied to the data. The results of a filter application can be shown in real-time by recalculating the measures. By dynamic dashboard, we mean that the indicators displayed on the dashboard are based on the type of data presented, not predefined for a particular type. In this Section, a dashboard generating application is presented. The application is able to receive as input a serialized KG in the CCSV format together with the discovered indicators from the QoE Indicators Ontology in Turtle format to perform a metadata analysis in order to create a dashboard with as many as graphs as the presented indicators, using the QoE Indicators Ontology together with the metadata annotation to dynamically configure each visualization.

In this use case, we have worked with data acquired from the bicycle-sharing system in the city of Fortaleza, Brazil. The datasets contained data about the network formed by the usage of the system, where a user is able to grab a bicycle from a station and return it to any other station, including the station where he/she obtained it initially. We obtained two CSV files that described the network:

– Bicycle-share stations: File containing only Bicycle-share stations, each station with an associated id, label and a lat/long.
– Trips performed: File containing all the bicycle-share system journeys. Each journey was presented with an id, an associated user that uses the bicycle, an origin and a destination bicycle-share station.

This data was collected by legacy information systems and most likely manipulated afterwards to clear up unneeded data and for better organization.

4.1   Dataset Characterization and KG Manipulation

In order to load these datasets into the KG, we first had to characterize their metadata in the following aspects:

- Data source: Which ICT system or device generated this data?
- Data acquisition: By which data acquisition activity were they acquired? By an already able to annotate system or manual data annotation performed by an user?
- Study: Are the datasets part of the same study?
- Time frame: When were the datasets generated?

Then, we made use of tools and techniques presented in the work cited in [13], namely the CCSV format and the CCSV-Loader application. Listing 1.1 shows part of the KG[11] after loading the datasets. Due to space restrictions, we present only the metadata related to the trips dataset. The data source is shown in lines 14–22 where both the annotator software and the legacy ICT system are described, while lines 1–5 states that the annotator software was deployed alongside the system at the specified date. Following, lines 6–9 describes the data acquisition activity, referring to the associated deployment. Finally, the dataset is shown in lines 10–13, where PROV-O is used to state from which activity they were generated.

```
1  <deployment-bss>
2     a vstoi:Deployment ;
3     vstoi:hasPlatform <system-bss> ;
4     vstoi:hasInstrument <annotator-01> ;
5     prov:startedAtTime"2016-11-08T14:42:42Z"^^xsd:dateTime .
6  <dataacquisition-trips>
7     a hacito:ManualDataAnnotation ;
8     hasco:hasContext <deployment-bss> ;
9     prov:startedAtTime "2016-11-09T15:13:25Z"^^xsd:dateTime .
10 <dataset-trips>
11    a vstoi:Dataset ;
12    prov:wasGeneratedBy <dataacquisition-trips> ;
13    prov:startedAtTime "2016-11-09T16:07:23Z"^^xsd:dateTime .
14 <annotator-01>
15    a hacito:AnnotatorSoftware ;
16    dc:hasVersion "X.Y"^^xsd:string .
17 <system-bss>
18    a hacito:InformationSystem ;
19    rdfs:label "Bicycle-share information system" ;
20    dc:description "System for managing all the collected data from
                   the bicycle-share system of Fortaleza."@en ;
21    dc:hasVersion "X.Y"^^xsd:string ;
22    dc:subject "bicycle, mobility" .
```

Listing 1.1. Part of the city KG

---

[11] http://hadatac.org/ttl/city_kg-full.ttl

The following step was to serialize the KG. Listing 1.2 shows the CCSV preamble serialization of the $qoe - m : Bicycle - Share\_Trip$ serialization in lines 2–6. The linkage between the trip and associated stations and user are established by the station id and user id, as shown in lines 7–9. Lines 10–13 specifies the id locations for every association. The same was performed for the $qoe - m : Bicycle - Share\_Station$ entity.

```
1 <trips> a vstoi:Dataset; ccsv:hasDataRecord <reg> .
2 <reg>
3     a qoe-m:Bicycle-Share_Trip; dc:identifier <id> .
4     qoe-m:has_Bicycle-Share_User <usr> ;
5     qoe-m:has_source_Bicycle-share_Station <src> ;
6     qoe-m:has_target_Bicycle-share_Station <trg> .
7 <src> a qoe-m:Bicycle-Share_Station; dc:identifier <src_id> .
8 <trg> a qoe-m:Bicycle-Share_Station; dc:identifier <trg_id> .
9 <usr> a qoe-m:Bicycle-Share_User; dc:identifier <usr_id> .
10 <id> ccsv:atColumn 0 .
11 <src_id> ccsv:atColumn 4 .
12 <trg_id> ccsv:atColumn 7 .
13 <usr_id> ccsv:atColumn 1 .
```

Listing 1.2. KG serialization CCSV preamble for $qoe - m : Bicycle - Share\_Trip$

The serialization encompasses a process for indicators discovering. The Listing 1.3 shows a Prolog code we developed to verify if an indicator is suitable for the data, based on its CCSV data annotation. Lines 1–8 shows the transformation of the indicator class into Prolog rules, while lines 10–14 shows the same for the domain classes. Following, lines 16–20 shows the relations in the CCSV data annotation regarding the content of the files. In this case, the CCSV files have data records of trips and stations (line 20). The inference rules are described in lines 22–27. They make use of transitivity to verify if an indicator Y is suitable for a KG X, i.e., if the graph contains the needed data to perform the calculation.

```
1 % indicator ontology
2 indicator(nr_trips_per_station).
3 has_value(nr_trips_per_station,nr_trips).
4 has_value(nr_trips_per_station,nr_station).
5 is_cardinality_of(nr_trips,trip).
6 is_cardinality_of(nr_users,user).
7 is_cardinality_of(nr_stations,station).
8 has_cardinality(X,Y) :- is_cardinality_of(Y,X).
9
10 % domain ontology
11 trip(trip). station(station1). station(station2).
12 has_user(trip,user).
13 has_source_station(trip,station1). has_target_station(trip,station2).
```

```
14 has_station(X,Y) :- has_source_station(X,Y); has_target_station(X,Y).
15
16 % dataset metadata
17 graph(bicicletar).
18 has_dataset(bicicletar,trips). has_dataset(bicicletar,stations).
19 dataset(trips). dataset(stations).
20 has_data_record(trips,trip). has_data_record(stations,station).
21
22 % inference rules
23 refx(X,Y) :- has_station(X,Y).
24 refx(X,Z) :- has_station(X,Y), refx(Y,Z).
25 good_ind(X,Y) :- graph(X), indicator(Y), related(X,Y).
26 related(X,Y) :- has_dataset(X,Z), has_data_record(Z,W), has_value(Y,V)
       , is_cardinality_of(V,U), compatible(W,U).
27 compatible(X,Y) :- refx(X,Y).
```

Listing 1.3. Prolog rules and statements for indicator discovery

Listing 1.4 shows the discovered indicator "Trips by departure station" with its associated dimension and measure entities in Turtle format.

```
1 <indicator01>
2     a qoe:Trips_by_departure_station ;
3     rdfs:label "Trips by departure station"@en ;
4     qoe:dimension <dimension01> ; qoe:measure <measure01> .
5 <dimension01>
6     a qoe:Dimension; qoe:has_entity qoe-m:Bicycle-Share_Station .
7 <measure01>
8     a qoe:Measure; qoe:has_function qoe:Count ;
9     qoe:has_entity qoe-m:Bicycle-Share_Trip .
```

Listing 1.4. Discovered indicators in Turtle

4.2 Dashboard Building

We have developed a dashboard generating application called the Semantic BI (Business Intelligence) Generator, which is able to interact with a number of BI solutions to automatically generate interactive dashboards based on the KG serialization. For this implementation, we focused on Qlik Sense[12] which provides an API for that be used to programmatically create and setup visualizations. First, the user inputs the serialized KG and the indicators files. The tool, then, performs SPARQL queries against the indicators and KG metadata to retrieve: (i) dimension entity id column; (ii) measure entity id column; and (iii) measure calculation function. Figure 4 shows the Semantic BI Generator after the serialized KG files and discovered metrics are loaded. On the left, a preview of the to-be-generated dashboard is presented, while on the right it is possible to

---

[12] http://www.qlik.com/us/products/qlik-sense

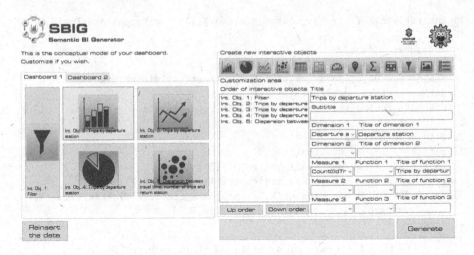

Fig. 4. Semantic BI generator with the discovered indicator

modify or add new visualizations as desired. In this case, the discovered indicator is shown as a bar chart (the one currently selected), while the others have been manually added. Also, note on the right that the columns and function were filled based on the metadata information. After that, the user pushes the generate button and the tool will setup the new dashboard inside the Qlik Sense environment.

Figure 5 shows the generated dashboard. It is possible to see the top left graph showing the dimensions as the bicycle-share stations and the measure counting the number of trips.

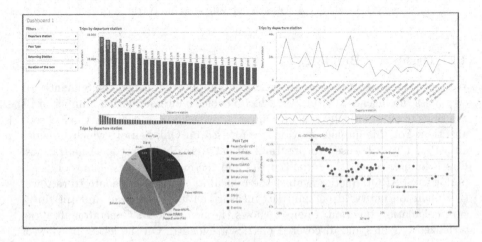

Fig. 5. Generated dashboard

5 Conclusion and Future Work

We have presented an operational description for a city Knowledge Graph that supports automatic generation of dashboards along with an indicator ontology that supports data visualization techniques. To build our KG and to develop our indicator ontology, we have reused many existing ontologies describing identified required metadata. We also proposed an extension to the GCI Ontology focusing on data visualization concepts. A process for KG serialization with indicator discovery was performed as way to foster knowledge interoperability between the KG and third-party applications. The city KG and the QoE Ontology were used in conjunction with the Semantic Business Intelligence Generator, a dashboard generating an application capable of performing CCSV metadata analysis to automatically build rich visualizations. Potentially more importantly, the presented contributions allow users with no previous knowledge about the data (by whom and how it was generated), but who are aware of city entities and processes (that is the case for most field specialists including transportation engineers) to leverage a metadata hierarchy (provided by our ontology choices) to find the right data to be analyzed.

. The research still has room to mature. For instance, we are currently working on an ontology for interactive objects to support the discovery of best suited visualization types based on an indicator definition. Also, we are continuously expanding indicators definitions to support not only data plotting but also calculation procedures (like complex network algorithms) and its associated semantics to an specific KG subset, enabling network data analytics for non-experts. In terms of KG building and metadata management, the Human-Aware Data Acquisition Framework[13] (HADatAc) is being designed and developed as a framework for managing data acquired using a multitude of sources including instruments, sensors, humans, and computer models. Leveraging HAScO and VSTO-I, HADatAc is already being used in support of a number of projects, namely:

- The Jefferson Project [10]: developed in collaboration between IBM, Rensselaer Polytechnic Institute (RPI), and The FUND for Lake George;
- An Urban ecology project led by RPI's Center for Architecture, Science and Ecology supporting large empirical observations and a variety of experiments.
- The Smart City Center at the Universidade of Fortaleza where scientific observations are conducted to understand the use of city resources in support of mass transportation.
- The CHEAR[14] Project where ontologies are being developed to support research on exposure science and child health and also tools and infrastructure for building and maintaining a knowledge graph of related content.

---

[13] https://tw.rpi.edu//web/project/hadatac
[14] https://tw.rpi.edu//web/project/CHEAR

References

1. Auer, S., Bizer, C., Kobilarov, G., Lehmann, J., Cyganiak, R., Ives, Z.: DBpedia: a nucleus for a web of open data. In: Aberer, K., et al. (eds.) ASWC/ISWC - 2007. LNCS, vol. 4825, pp. 722–735. Springer, Heidelberg (2007). doi:10.1007/978-3-540-76298-0_52
2. Biega, J., Kuzey, E., Suchanek, F.M.: Inside YAGO2s: a transparent information extraction architecture. In: Proceedings of the 22nd International Conference on World Wide Web Companion, WWW 2013 Companion, International World Wide Web Conferences Steering Committee, Republic and Canton of Geneva, Switzerland, pp. 325–328 (2013)
3. Bollacker, K., Evans, C., Paritosh, P., Sturge, T., Taylor, J.: Freebase: a collaboratively created graph database for structuring human knowledge. In: Proceedings of the 2008 ACM SIGMOD International Conference on Management of Data, SIGMOD 2008, NY, USA, pp. 1247–1250. ACM, New York (2008)
4. Dong, X., Gabrilovich, E., Heitz, G., Horn, W., Lao, N., Murphy, K., Strohmann, T., Sun, S., Zhang, W.: Knowledge vault: a web-scale approach to probabilistic knowledge fusion. In: Proceedings of the 20th ACM SIGKDD International Conference on Knowledge Discovery and Data Mining, KDD 2014, NY, USA, pp. 601–610. ACM, New York (2014)
5. Fox, M.S.: PolisGnosis Project: Representing and Analysing City Indicators. Working Paper, Enterprise Integration Laboratory, University of Toronto (2015). http://eil.utoronto.ca/wp-content/uploads/smartcities/papers/PolisGnosi.pdf
6. Fox, M.S.: The role of ontologies in publishing and analyzing city indicators. Comput. Environ. Urban Syst. **54**, 266–279 (2015)
7. Fox, P., McGuinness, D.L., Cinquini, L., West, P., Garcia, J., Benedict, J.L., Middleton, D.: Ontology-supported scientific data frameworks: the Virtual Solar-Terrestrial Observatory experience. Comput. Geosci. **35**(4), 724–738 (2009)
8. Hoffart, J., Suchanek, F.M., Berberich, K., Lewis-Kelham, E., de Melo, G., Weikum, G.: YAGO2: exploring and querying world knowledge in time, space, context, and many languages. In: Proceedings of the 20th International Conference Companion on World Wide Web, WWW 2011, NY, USA, pp. 229–232. ACM, New York (2011)
9. ISO: Sustainable development of communities - Indicators for city services and quality of life. ISO 37120:2014, International Organization for Standardization, May 2014. http://www.iso.org/iso/catalogue_detail?csnumber=62436
10. McGuinness, D.L., Pinheiro, P., Santos, H.O., Klawonn, M., Chastain, K.: Semantic support for complex ecosystem research environments. In: AGU Fall Meeting Abstracts 33, December 2015
11. Pinheiro, P., McGuinness, D.L., Santos, H.: Human-aware sensor network ontology: semantic support for empirical data collection. In: Proceedings of the 5th Workshop on Linked Science, Bethlehem, PA, USA (2015)
12. Rew, R., Davis, G.: NetCDF: an interface for scientific data access. IEEE Comput. Graph. Appl. **10**(4), 76–82 (1990)
13. Santos, H., Furtado, V., Pinheiro, P., McGuinness, D.L.: Contextual data collection for smart cities. In: Proceedings of the Sixth Workshop on Semantics for Smarter Cities, Bethlehem, PA, USA (2015)
14. Singhal, A.: Introducing the Knowledge Graph: things, not strings (2012). https://googleblog.blogspot.com/2012/05/introducing-knowledge-graph-things-not.html
15. Zhao, J., Wang, Y.: Toward domain knowledge model for smart city: the core conceptual model. In: 2015 IEEE First International Smart Cities Conference (ISC2), pp. 1–5 (2015)

Ontology-Driven Unified Governance in Software Engineering: The PoolParty Case Study

Monika Solanki[1(✉)], Christian Mader[2], Helmut Nagy[3], Margot Mückstein[3], Mahek Hanfi[3], Robert David[3], and Andreas Koller[3]

[1] Department of Computer Science, University of Oxford, Oxford, UK
monika.solanki@cs.ox.ac.uk
[2] Fraunhofer IAIS, Bonn, Germany
[3] Semantic Web Company, Vienna, Austria

Abstract. Collaborative software engineering environments have transformed the nature of workflows typically undertaken during the design of software artifacts. However, they do not provide the mechanism needed to integrate software requirements and implementation issues for unified governance in the engineering process. In this paper we present an ontology-driven approach that exploits the Design Intent Ontology (DIO) for aligning requirements specification with the issues raised during software development and software maintenance. Our methodology has been applied in an industrial setting for the PoolParty Thesaurus server. We integrate the requirements specified and issues raised by Pool-Party customers and developers, and provide a graph search powered, unified governance dashboard implementation over the annotated and integrated datasets. Our evaluation shows an impressive 50% increase in efficiency when searching over datasets semantically annotated with DIO as compared to searching over Confluence and JIRA.

1 Introduction

In today's dynamic, agile and collaborative environments, software design and development has metamorphosed into a complex social activity, involving teams of software architects, developers, testers and maintainers. In response to this need, several collaborative and social development frameworks for software engineering have been implemented and are in wide use [2].

In collaborative software development, given a set of requirements, typically, several iterations, deliberations and informal discussions are undertaken among the design team members, before a final consensus can be reached, on the features that are to be included in the concrete realisation of the artifact. The requirements and discussions, if at all documented, are recorded as unstructured text, that makes their search and retrieval a cumbersome process. Further, during and after implementation and deployment of the software, several issues may typically arise and get recorded either as part of the maintenance process or in response to new requirements. One critical shortcoming of collaborative

© Springer International Publishing AG 2017
E. Blomqvist et al. (Eds.): ESWC 2017, Part II, LNCS 10250, pp. 109–124, 2017.
DOI: 10.1007/978-3-319-58451-5_8

environments in wide use today such as Atlassian Confluence[1], JIRA[2] and Github[3] is that they provide generic fields such as "issue" and "comment" which encapsulate all discussion types. This makes it extremely difficult to retrieve information relevant to a specific aspect of a requirement. Further, they also do not provide the interfaces needed to associate design requirements with implementation issues in a structured and systematic way. Integrating the requirements with the issues which should be a core functional capability of most collaborative design environments, is not yet well supported.

In this paper we present an ontology-driven approach that captures the knowledge emerging during software design, development, implementation and maintenance and exploits it for *unified governance* of the engineering process. The term "unified governance" was defined in the ALIGNED[4] project and denotes capturing knowledge of the software development lifecycle from various sources and expressing it using a common ontology so that it can be governed, i.e., queried and new knowledge inferred, in a unified way. We exploit DIO (Design Intent Ontology)[5], a content ontology design pattern, that provides a generic mechanism for formally describing the intents or the rationales that emerge or are generated during the processes that underlie modern design decision phases. Our approach conforms to one of the main criteria for systems that record design rationale and decisions, mainly that it should be least disruptive to the designers and the actual design process itself [3,5].

We evaluate our approach within the settings of an industry-driven use case from the Semantic Web Company (SWC henceforth)[6], where customers and developers of the PoolParty Thesaurus Server (PPT henceforth) document their requirements in a tailored version of Confluence and file bug reports or raise design issues using JIRA. By aligning and creating merged repositories of requirements, customer feedback, bug reports, project documentation and mining of web resources we are able to exploit the integrated knowledge to develop a semantic search mechanism for unified governance.

The remainder of this paper is structured as follows: Sect. 2 discusses related work. Section 3 describes our use case scenario. Section 4 presents the requirements that need to be addressed by a unified governance framework. Section 5 describes our knowledge mining and representation approach. Section 6 presents the architecture and implementation. Section 7 outlines our evaluation strategy and finally Sect. 8 presents conclusions.

2   Related Work

Extensive research [4] has shown that capturing design intents is a very difficult problem. Surveys [8,11] on the capture of design rationale and barriers to their

---

[1] https://www.atlassian.com/software/confluence.
[2] https://www.atlassian.com/software/jira.
[3] https://github.com/.
[4] http://aligned-project.eu/.
[5] https://w3id.org/dio.
[6] https://www.semantic-web.at/.

uptake have also been reported. IBIS [9], one of the most widely adopted argumentation models, is a network-oriented model based on capturing the issues, positions and arguments underlying a design deliberation.

Representing design intents or design rationales as ontologies have been explored for various specialised domains such as software engineering [6], ontology engineering (OE) [13], product engineering [14] and aerospace engineering [10]. However there is no generic, domain-independent design intent capture model available as a design pattern that can be specialised for any design rationale capture scenario.

Based on IBIS, one of the first attempts to provide an ontological representation of design rationale was by Medeiros et al. [6]. The authors proposed the Kuaba ontology for the representation of model-based software design. A major limitation of this approach is that the reuse of the design rationale is strongly dependent on the formal model used for artifact design, thereby introducing encoding bias [7].

An argumentation ontology has been proposed as part of the DILIGENT [13] framework for the collaborative building of ontologies. The basic conceptualisation in the ontology has been derived from IBIS and extended to include argumentation-specific entities, particularly those suitable for ontology engineering. The critical limitations of this ontology are that the issues are not interlinked to the requirements.

The fundamental difference between the various semantic and non-semantic models for representing design rationale and DIO, lies in the granularity of the representation. IBIS-based models provide a single conceptualisation for many aspects of a design intent, while DIO provides a finer level of representation, thereby making the specification more expressive, without compromising computability. As an example, typically the terms, "Idea", "Position" and "Solution" have been used to represent the solution to a "Question" or an "Issue", while the term "Argument" has been used for representing the rationales for and against a proposed solution. On the other hand, DIO incorporates the mutually disjoint concepts of "Argument" and "Justification" to represent the distinct rationales as well as the mutually disjoint concepts of "AlternativeSolution" and "MandatedSolution" to represent all solutions and accepted solutions respectively.

Table 1 compares the conceptual abstractions defined in DIO with those in some of the approaches discussed above.

Wikidsmart by zAgile[7] claims to provide "semantic enablement" to Atlassian Confluence by capturing the data in a semantic repository, using "a set of ontologies and metamodels specific to a domain of interest". It is, however, unclear what ontologies are used by the framework's semantic repository to model the information contained in Confluence. Our work follows a similar approach to Wikidsmart in the way that we also extract information from existing Confluence and JIRA installations and express it using semantic technologies. However, we contribute and publish reusable OWL-based ontologies for modeling this

---

[7] http://www.zagile.com.

Table 1. A comparison of the ontological approaches for capturing design intents

	DIO	Kuaba	DILIGENT	RaDEX	ISAA
Available as a design pattern	Yes	No	No	No	No
Generic across domains	Yes	No	No	No	No
Formal representation	Yes	Yes	No	No	No
Ontological serialisation	Yes	No	No	No	No
Explicit provenance metadata	Yes	Limited to agents	Limited to agents	No	No
Traceability wrt. design requirements	Yes	No	No	No	Yes
Traceability wrt. design artifact	Yes	Yes	No	No	Yes
Capturing intent	Yes	No	No	No	Yes
Assumption specification	Yes	No	No	No	
Constraint specification	Yes	No	No	No	No
Heuristics specification	Yes	No	No	No	No
Temporal specification	Yes	Partial	No	No	No

knowledge (DIO and DIOPP), ready to be processed by a wide range of tools of the Linked Data technology stack.

3 Motivating Scenario

The motivation for our work is the current setup at SWC, where Atlassian Confluence is used to support requirements engineering and JIRA is used by team members and SWC customers for issue and change tracking, organising ideas from team members as well as collecting them from customers. Following the agile methodology of software development, the data is recorded in Confluence under headings such as "Requirements", "Goal", "User Story", "Epic" and "Stakeholders". Additional fields such as "Precondition", "Detailed description", "Acceptance criteria & Test scenario" are included to provide further context to the requirements. A single field, "Comment", captures the opinions/discussion carried out by human agents.

SWC collects the requirements for each version of PPT in the designated Confluence space. Requirements are then linked to pages containing epics and user stories. Most of these pages are structured based on standard templates defined by SWC. The outputs from these template-based pages are largely document-centric and require extensive human intervention to synthesise and synchronise them with PoolParty development tasks.

By using ontologies to annotate and provide metadata to the content extracted from Confluence and JIRA, SWC would be able to create merged repositories of requirements, customer feedback, bug reports and project documentation thereby consolidating PoolParty experiences, customer ideas and market needs in order to integrate them into products. This is a key factor for successful development of SWC products and for raising customer satisfaction and enterprise agility. The integrated information would enable the mining of intents that lead to the development in PoolParty. Questions asked by customers will flow faster into the requirements engineering system. The process will help to generate concise reports on distributed business objects and entities relevant for the development processes, and to coordinate the data management and development workflows required to deliver new versions of the evolving PoolParty product.

4 Requirements for Unified Governance

Based on the challenges that need to be addressed to enable the mapping of design intents for unified governance, where intents are derived from requirements in Confluence and issues in JIRA, we identified the following requirements:

- **Enabling Knowledge reuse**: As new agents — humans and software get involved during various phases of design, new processes are introduced. The key motivation behind recording the intents underlying decisions is that the knowledge can be used to inform future agents and processes of the discussions that have happened so far.
- **Shared semantics via annotations**: The framework must facilitate a common understanding and uniform interpretation of the design intents among the participants of the decision making process. The unified representation format could act as a *lingua franca* between various tools and resources that assist in the design and implementation process.
- **Flexibility**: The ontology conceptualisation should be flexible enough to be mapped to current requirement specification environments and primitives. It should facilitate the integration of interfaces for capturing design requirements and the design issues.
- **Traceability**: It must allow forward and backward traceability of final designs as well as initial design requirements from design intents and issues.

5 The Knowledge Representation Framework

A high level representation of the architecture implemented for integrating the requirements and the issues is illustrated in Fig. 1. The core components of the

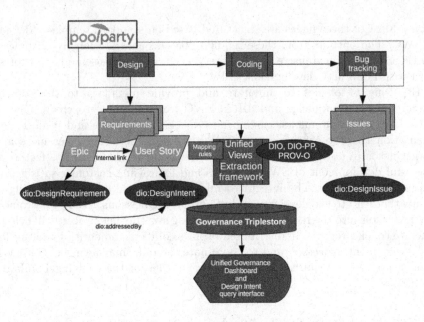

Fig. 1. The knowledge representation and extraction architecture

architecture are the ontologies, the Unified Views extraction pipeline and the unified governance dashboard that provides visual analytics for the results of searches over the integrated datasets.

The DIO content design pattern provides a minimalistic abstraction and defines conceptual, generic entities for the modelling of semantically enriched knowledge required to capture the intents or rationale behind the design of an artifact. DIO is a domain agnostic pattern and most domain specific design rationale models will specialise from it.

Figure 2 illustrates the graphical representation of DIO. It depicts the entities defined for the pattern and their relationships with entities from PROV-O[8].

The axiomatisation of a `DesignIntentArtifact` can be represented in OWL using the Description Logic [1] notation as,

$$\texttt{DesignIntentArtifact} \sqsubseteq ((\exists\texttt{wasAttributedTo.DesignIntent})$$
$$\sqcap (= \texttt{1wasAttributedTo.Agent}) \sqcap (\geq \texttt{1description})$$
$$\sqcap (= \texttt{1version}) \sqcap (= \texttt{1generatedAtTime}))$$

---

[8] http://www.w3.org/ns/prov-o.

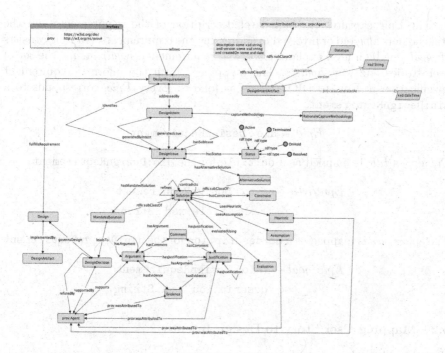

Fig. 2. Graphical representation of DIO

The axiomatisation for a `DesignIssue` captures the intuition that a design issue can have several alternative solution, but just one mandated solution.

$$\text{DesignIssue} \sqsubseteq (\text{DesignIntentArtifact}$$
$$\sqcap (\forall \text{hasAlternativeSolution.AlternativeSolution})$$
$$\sqcap (\forall \text{hasMandatedSolution.MandatedSolution})$$
$$\sqcap (= 1 \text{hasMandatedSolution.MandatedSolution}))$$

For further details, the interested reader is referred to [12].

As DIO is a general purpose design intent ontology, it does not capture attributes specific to PoolParty development. To bridge this gap, we define DIO-PP[9] - an extension to DIO that is specifically aimed at capturing conceptualisations from Confluence and JIRA. Besides DIO-PP, we also define bespoke mappings to DIO entities, for the data recorded during requirements capture.

5.1   Mapping Epic to Design Requirements

The SWC Confluence environment consist of two main parts: The Epic and the User story.

---

[9] https://w3id.org/diopp.

The Epic captures the high level description of the requirements and the stakeholders affected or involved in addressing the requirements. An Epic consists of: *goal* (the high level explanation of the requirement), *requirements*, the set of user stories and *stakeholders*. The mapping[10] from the informal requirement specification in Epic to DIO is done as follows: Every *Epic*, corresponds to a dio:DesignRequirement.

$$Epic \rightarrow \text{dio : DesignRequirement}$$

The epic's *title* is mapped as a dc:title for the dio:DesignRequirement

$$Epic/title \rightarrow (\text{dio : DesignRequirement}$$
$$\text{dc : title xsd : String})$$

The epic's *goal* is mapped as a dc:description for the dio:DesignRequirement

$$Epic/goal \rightarrow (\text{dio : DesignRequirement}$$
$$\text{dc : description xsd : String})$$

5.2 Mapping User Story to Design Intent

Each Epic includes a set of requirements, each of which are represented as a "User story" in Confluence. Each user story captures the design intent. Every *User story*, corresponds to a dio:DesignIntent.

$$User\ story \rightarrow \text{dio : DesignIntent}$$

The user story's *title* is mapped as a dc:title for the dio:DesignIntent

$$User\ story/title \rightarrow (\text{dio : DesignIntent}$$
$$\text{dc : title xsd : String})$$

The user story's *description* is mapped as a dc:description for the dio:DesignIntent

$$User\ story/description \rightarrow (\text{dio : DesignIntent}$$
$$\text{dc : description xsd : String})$$

The use story may include a link to a JIRA issue, if the issue relates to certain aspects of the requirement. This is captured using the predicate dio:generatesIssue

$$User\ story/jira\ issue \rightarrow (\text{dio : DesignIntent}$$
$$\text{dio : generatesIssue dio : DesignIssue})$$

---

[10] indicated by "$\rightarrow$".

The user story includes part of the solution, that addresses the requirement and is the most significant part of the mapping. The user story captures the following elements that contribute to an `AlternativeSolution` and `MandatedSolution` for a `DesignIssue` in DIO: *Preconditions, Detailed Description, Affected Components, Acceptance criteria and test scenarios, Variations, Comments.* The DIO-PP ontology conceptualises these elements.

$$\text{diopp} : \text{Precondition} \sqsubseteq \text{dio} : \text{Assumption}$$
$$\text{diopp} : \text{AffectedComponent} \sqsubseteq \text{dio} : \text{Heuristic}$$
$$\text{diopp} : \text{AcceptanceCriteria} \sqsubseteq \text{dio} : \text{Evaluation}$$
$$\text{dio} : \text{Comment} \sqsubseteq \text{dio} : \text{Argument}$$
$$\sqcup \, \text{dio} : \text{Justification}$$

The user story's *Detailed Description* element is mapped as a `dc:description` for the `dio:AlternativeSolution`

$$User\ story/detailed\ description \rightarrow$$
$$(\text{dio} : \text{AlternativeSolution} \ \text{dc} : \text{description}$$
$$\text{xsd} : \text{String})$$

5.3   Mapping JIRA Design Issues

JIRA is used as the platform by PoolParty customers to highlight design issues related to the software and request new features. Typical properties that need to be ascribed to a design issue documented in JIRA are the issue type, `IssueType`, its status, `Status`, its priority, `PriorityType`, its resolution status, `ResolutionType` and the various agents that are associated with the issue: `Watcher`,`Assignee` and `Reporter`. The DIO-PP ontology defines these entities. The axiomatisation of a `DesignIssue` for PPT can be represented as,

$$\text{DesignIssue} \sqsubseteq ((= 1 \, \text{isOfIssueType.IssueType}) \sqcap$$
$$(= 1 \, \text{hasVersionType.VersionType}) \sqcap$$
$$(= 1 \, \text{hasPriority.PriorityType}) \sqcap$$
$$(= 1 \, \text{hasStatus.IssueStatusType}) \sqcap$$
$$(= 1 \, \text{hasResolution.IssueResolution}) \sqcap$$
$$(= 1 \, \text{hasReporter.Agent}) \sqcap$$
$$(= 1 \, \text{hasAssignee.Agent}) \sqcap$$
$$(\geq 1 \, \text{hasAffectedComponent.Component}) \sqcap$$
$$(\geq 1 \, \text{hasWatcher.Agent}) \sqcap$$
$$(\geq 1 \, \text{hasEnvironment.}\top) \sqcap$$
$$(\leq 1 \, \text{updateDate}))$$

6   Data Extraction Workflow

As highlighted in Sect. 3, SWC collects the requirements and issues for each version of PPT in the designated Confluence space and JIRA respectively. This data needs to be extracted and semantically annotated with DIO and DIO-PP. To perform the data extraction and conversion process, we provide three contributions: (1) an extraction tool which connects to the Confluence and JIRA instances for PPT and converts the contained data into RDF. (2) a Data Processing Unit (DPU) for UnifiedViews[11] which wraps the extraction tool so that it can be included into custom data processing pipelines. (3) a UnifiedViews pipeline, which encompasses data extraction, data annotation using a PPT thesaurus and loading the annotated data into a remote Virtuosos triple store.

6.1   Extraction Tool

The extraction tool connects to the Confluence and JIRA REST APIs and retrieves (meta-)data that should be converted into the RDF output format using DIO-PP. While data from JIRA can be retrieved in structured JSON format, certain information we need to formalize must be retrieved by parsing HTML pages from the Confluence installation. The tool consists of two independent extraction components - one extracts data from Confluence, the other one from JIRA. The Confluence extractor iterates through all pages in a hierarchy below a page named "Requirements PP" in the PPT development space, retrieves the JSON metadata (e.g., history information or creation date) of these pages and parses the HTML content. This data is then used to build an in-memory object model which is serialized as RDF data. Extraction and conversion of JIRA data works in a similar way, except that no HTML parsing needs to be performed because the API provides the information in JSON format.

6.2   Integration into a Processing Pipeline

As the extraction tool produces an RDF document, we can leverage the potential of Linked Data to enrich the extracted data with information coming from other sources. Specifically, we enrich data from Confluence and JIRA with annotation (tagging) information coming from an enterprise thesaurus, allowing to, e.g., improve the search functionality. For the Unified Governance use case we show how to add tagging information by providing a pipeline for the UnifiedViews tool, which is illustrated in Fig. 3. The pipeline consists of seven DPUs: the extraction tool (red box) retrieves data from Confluence and JIRA in RDF format. It is then filtered (blue box below) so that it only contains text literals which serve as an input for annotating. This annotation stage (box below filtering step) connects to the PoolParty Extractor annotation API and returns a RDF graph containing the identified annotations for each resource. Afterwards, the original

---

[11] An Extract-Transform-Load (ETL) tool focused on processing RDF data, http://www.unifiedviews.eu/.

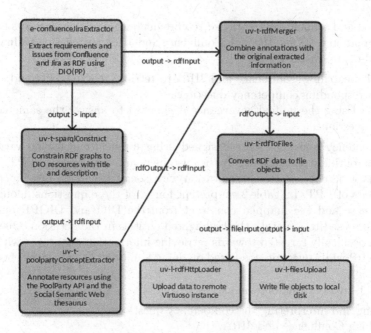

Fig. 3. UnifiedViews extraction pipeline

extracted data is combined with the filtered annotated data and both written to the local file system as well as uploaded to a Virtuoso server.

We ran the extraction process on a subset of the requirements (5394) and issues (184534). This resulted in 215401 triples. The extraction is now run at regular intervals with the resulting triples added to the triple store.

7   Evaluation

In order to assess the benefits of using semantically annotated and integrated requirements engineering datasets for unified governance, we carried out an evaluation that compared the time taken in retrieving requirements and issues from Confluence and JIRA against the time taken when using the graph-search powered unified governance dashboard. Our evaluation involved six participants (members of the consulting team at SWC) with a background in software engineering. All the recruited personnel had reasonable experience in the use of Confluence, JIRA and ontology interpretation.

Confluence and JIRA provide a keyword-driven search facility to retrieve information specific to requirements and issues respectively. The results are returned as natural text. The unified governance dashboard is based on the entities captured in DIO, whose design is governed by competency questions [12], that cover the spectrum of requirements engineering knowledge which should be recorded. In order to measure the accuracy, ease and efficiency of retrieving structured and semantically enriched requirements engineering datasets against the information retrieved through Confluence and JIRA search, our evaluation[12] consisted of two main tasks:

- Task 1: Searching Confluence and JIRA to retrieve requirements and answer the corresponding competency questions.
- Task 2: Using the unified governance dashboard to answer the same competency questions.

Six competency questions were designed using inputs from a wide variety of use cases routinely encountered by the customers and developers of PPT. These questions were chosen after consultation with social scientists, developers and consultants of PPT. In Table 2 we present four[13] of these questions, along with the use case and the mapped construct from the DIO and DIOPP ontology which were used to annotate the highlighted entities in the question. Questions 1–3 are specifically targeted towards retrieving information annotated with DIO and DIOPP while question 4 is based on doing a full text search within both the systems.

Task 1 involved the following steps:

- Reading and interpreting the competency questions.
- Searching Confluence and JIRA.
- Retrieving information leading to answers to the questions.
- Documenting the search strategy used in retrieving the answers and analysing the results.

Task 2 involved the following steps:

- Reading and interpreting the competency questions.
- Using the facet browsing and full text search features of the unified governance dashboard to retrieve potential answers.
- Documenting the search strategy used in retrieving the answers and analysing the results.

The volunteers were divided into two groups, A and B. Group A executed task 1, followed by task 2, while for Group B, it was the other way around. All volunteers were provided with Turtle serialisation of the DIO ontology a week before the evaluation. On the day of the evaluation they were provided with the six competency questions, access to Confluence and JIRA and access to the unified governance dashboard. The screenshots of the dashboard interfaces are presented in Fig. 4. In order to maintain consistency across volunteers, they were advised to answer the questions in the prescribed order and the evaluation was undertaken under supervision. For each of the tasks and for each question, the following observations were recorded: (1) The time taken to answer each question in task 1 and 2. This was the total time taken for searching and analysing the results. (2) The answers for each question in task 1 and 2. (3) The strategy used to answer the question.

---

[12] All evaluation data including participants strategies, timings and the resulting timing analysis graphs has been made available at http://goo.gl/Khlaaf.
[13] Due to space constraint, we report only four here. All six questions can be found in the participant links in the Google doc containing links to the evaluation data.

Table 2. Competency questions used for evaluation

Question	Use case	Mapped DIO/DIOPP concept
List all **stories** that have affected version PoolParty 5.3.1 that were requested by a customer.	It often happens that customers request specific features that should be added or improved in PPT	User story → **dio:DesignIntent**
Which feature caused most **bug reports** in release 5.5.0 (affected Version 5.5.0)?	This goes towards release profiling: finding out which features are weak points of a certain release	Bug reports → **dio:DesignIssue**
What has changed in the last **three years** (2014, 2015, 2016) when it comes to bugs and tasks in the PoolParty support project?	To determine if we are getting better at serving our customers, it is useful to know how the interaction of the customers with our ticketing, system develops over time.	Bug reporting date → **prov:wasGeneratedAt**
Which PoolParty version (starting from PP 2.8 and including bug fix, releases in the respective **minor** or **major** release (e.g. count 5.3.1 for,5.3)) was the one affected with the most bugs that **were classified** as,"blocker", and how would you interpret this?	In order to know if we get better in our development and support, processes, it is useful to know if we are encountering more or less, critical issues than before.	Bug status→ **dio:IssueStatusType**, Bug type → **dio:IssueType**

Figure 5(a – d) illustrates the time taken to answer each of the six questions by four of the participants across the two tasks, while Fig. 5(d) illustrates the average time taken, after discarding the extreme outliers for each participant. Our evaluation shows that the overall time taken to search the integrated datasets annotated with DIO and DIOPP and answer the six competency questions across all participants is 50% less when compared to the time taken to search Confluence and JIRA for the same questions. This is a significant and impressive saving in resources for the SWC, where consultants and developers need to manage a large number of customer requirements and issues on a regular basis. The evaluation also highlighted the weakness of the full text search implementation used in the unified governance dashboard, where for atleast three of the participants, to search answers to questions 5 and 6 took significantly longer as compared to using the full text search provided by Confluence and JIRA. As we had asked the participants to record their search strategies and use of the interface, we were also able to identify pitfalls and bottlenecks in the current design of the unified governance dashboard as an additional evaluation result.

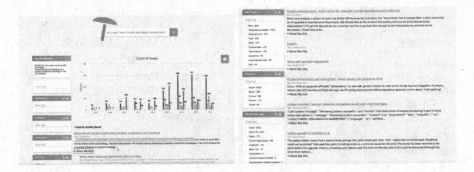

Fig. 4. Unified governance dashboard with full text search and faceted browsing

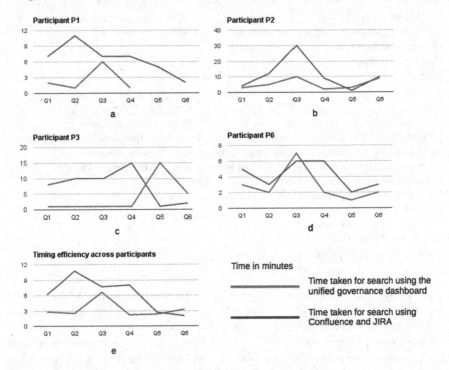

Fig. 5. Timings recorded by four participants and the average time efficiency for all participants and all questions.

8   Conclusions

In this paper, we have proposed a framework that exploits the DIO ontology for integrating the requirements and issues arising during software design, implementation and maintenance. We have evaluated our approach on an industrial software engineering case study for the PoolParty Thesaurus server development and maintenance. We have shown how a minimalistic extension of DIO can be utilised to associate and extract software design rationales with requirements recorded in Confluence and issues submitted in JIRA for PPT. Our evaluation shows an impressive 50% reduction in the time taken to search semantically annotated and integrated datasets when compared to the search provided by Confluence and JIRA. The evaluation also revealed the limitation of the full text search feature currently implemented in the unified governance dashboard. Improving this will be the focus of our immediate future work. We have presented an end-to-end solution that provides significant improvements in productivity, agility and efficiency in software development environments.

In future we aim to build on the results emerging from periodic and detailed empirical evaluation carried out on the semantic search interface for queries with various levels of complexity and implement an ontology-driven, design-intent powered, self learning system for software engineering environments.

Acknowledgments. We are grateful to Prof. Jeremy Gibbons (University of Oxford) for his insightful comments on early drafts of the paper. The research presented in this paper has received funding from the European Union's Horizon 2020 research and innovation programme under grant agreement No. 644055, the ALIGNED project (www.aligned-project.eu).

References

1. Baader, F., Calvanese, D., McGuinness, D.L., Nardi, D., Patel-Schneider, P.F. (eds.): The Description Logic Handbook: Theory, Implementation, and Applications. Cambridge University Press, New York (2003)
2. Bani-Salameh, H., Jeffery, C.: Collaborative and social development environments: a literature review. Int. J. Comput. Appl. Technol. **49**(2), 89–103 (2014)
3. Bracewell, R., Wallace, K., Moss, M., Knott, D.: Capturing design rationale. Comput. Aided Des. **41**(3), 173–186 (2009). Computer Support for Conceptual Design
4. Burge, J.E., Carroll, J.M., McCall, R., Mistrk, I.: Rationale-Based Software Engineering, 1st edn. Springer, Heidelberg (2008)
5. Conklin, E.J., Yakemovic, K.C.B.: A process-oriented approach to design rationale. Hum.-Comput. Interact. **6**(3), 357–391 (1991)
6. Medeiros, A.P., Schwabe, D., Feijó, B.: Kuaba Ontology: Design Rationale Representation and Reuse in Model-Based Designs. In: Delcambre, L., Kop, C., Mayr, H.C., Mylopoulos, J., Pastor, O. (eds.) ER 2005. LNCS, vol. 3716, pp. 241–255. Springer, Heidelberg (2005). doi:10.1007/11568322_16
7. Gruber, T.R.: Toward principles for the design of ontologies used for knowledge sharing. Int. J. Hum. Comput. Stud. **43**(5–6), 907–928 (1995)
8. Horner, J., Atwood, M.E.: Design rationale: the rationale and the barriers. In: Proceedings of the 4th Nordic Conference on Human-Computer Interaction: Changing Roles, NordiCHI 2006, pp. 341–350. ACM (2006)
9. Kunz, W., Rittel, H.W.J., Messrs, W., Dehlinger, H., Mann, T., Protzen, J.J.: Issues as elements of information systems. Technical report (1970)
10. Kuofie, E.J.: Radex: a rationale-based ontology for aerospace design explanation (2010)
11. Lee, J.: Design rationale systems: understanding the issues. IEEE Expert **12**(3), 78–85 (1997)

12. Solanki, M.: DIO: a pattern for capturing the intents underlying designs. In: Proceedings of the 6th Workshop on Ontology and Semantic Web Patterns (WOP 2015), vol. 1461. CEUR-WS.org (2015)

13. Tempich, C., Pinto, H.S., Sure, Y., Staab, S.: An argumentation ontology for DIstributed, loosely-controlled and evolvInG engineering processes of oNTologies (DILIGENT). In: Gómez-Pérez, A., Euzenat, J. (eds.) ESWC 2005. LNCS, vol. 3532, pp. 241–256. Springer, Heidelberg (2005). doi:10.1007/11431053_17

14. Zhang, Y., Luo, X., Li, J., Buis, J.J.: A semantic representation model for design rationale of products. Adv. Eng. Inform. 27(1), 13–26 (2013)

A Hypercat-Enabled Semantic Internet of Things Data Hub

Ilias Tachmazidis[1(✉)], Sotiris Batsakis[1], John Davies[2], Alistair Duke[2], Mauro Vallati[1], Grigoris Antoniou[1], and Sandra Stincic Clarke[2]

[1] University of Huddersfield, Huddersfield, UK
{i.tachmazidis,s.batsakis}@hud.ac.uk
[2] British Telecommunications, Ipswich, UK

Abstract. An increasing amount of information is generated from the rapidly increasing number of sensor networks and smart devices. A wide variety of sources generate and publish information in different formats, thus highlighting interoperability as one of the key prerequisites for the success of Internet of Things (IoT). The *BT Hypercat Data Hub* provides a focal point for the sharing and consumption of available datasets from a wide range of sources. In this work, we propose a semantic enrichment of the *BT Hypercat Data Hub*, using well-accepted Semantic Web standards and tools. We propose an ontology that captures the semantics of the imported data and present the *BT SPARQL Endpoint* by means of a mapping between SPARQL and SQL queries. Furthermore, federated SPARQL queries allow queries over multiple hub-based and external data sources. Finally, we provide two use cases in order to illustrate the advantages afforded by our semantic approach.

1 Introduction

The emerging notion of a smart city is based on the use of technology in order to improve the efficiency, effectiveness and capability of various city services, thus improving the quality of the inhabitants' lives [15]. A fundamental difference between smart cities and similar uses of technology in other areas, such as business, government or education, is the vast variety of the technologies used, the types and volumes of data, and the services and applications targeted [5]. Thus, developing successful smart city solutions requires the collection and maintenance of relevant data in the form of IoT data.

Over the past few years, eight industry-led projects were funded by Innovate UK[1] (the UK's innovation agency) to deliver IoT 'clusters', each centred around a data hub to aggregate and expose data feeds from multiple sensor types. The system that has come to be known as the *BT Hypercat Data Hub* was part of the Internet of Things Ecosystem Demonstrator[2] programme.

Addressing interoperability by focusing on how interoperability could be achieved between data hubs in different domains was a major objective of the

---

[1] https://www.gov.uk/government/organisations/innovate-uk.
[2] https://connect.innovateuk.org/web/internet-of-things-ecosystem-demonstrator/overview.

© Springer International Publishing AG 2017
E. Blomqvist et al. (Eds.): ESWC 2017, Part II, LNCS 10250, pp. 125–137, 2017.
DOI: 10.1007/978-3-319-58451-5_9

programme. Hence, Hypercat [1] was developed, which is a standard for representing and exposing Internet of Things data hub catalogues [6] over web technologies, to improve data discoverability and interoperability. Recent work [13], proposed a semantic enrichment for the core of the Hypercat specification, namely an RDF-based [8] equivalent for a JSON-based catalogue. Other IoT / smart city projects include Barcelona[3], MK:Smart[4] which uses the *BT Hypercat Data Hub* that is Hypercat-enabled but not semantically enriched, and the D-CAT[5] catalogue approach from W3C.

The main objective of this work is to achieve the semantic enrichment [2] of the data in the *BT Hypercat Data Hub* and to provide access to the enriched data through a SPARQL endpoint [11]. Furthermore, adding reasoning capabilities and the ability to combine external data sources using federated queries are important aspects of the implemented system.

The *BT Hypercat Data Hub* provides a focal point for the sharing and consumption of available datasets from a wide range of sources. In order to enable rapid responses, data in the *BT Hypercat Data Hub* is stored in relational databases. In this work, sensor, event, and location databases, i.e., databases containing information about sensor readings, events and locations are used. In order to provide a semantically richer mechanism of accessing the available datasets, the *BT Hypercat Ontology* was developed in order to lift semantically data stored within the relational databases. In addition, data translation through output adapters and SPARQL endpoints was defined. Thus, the semantically enriched data can be queried by accessing the developed *BT SPARQL Endpoint*.

Triplestores contain the information in RDF format combined with a built-in SPARQL endpoint. Thus, triplestores are commonly used for providing SPARQL endpoints. However, as data in the *BT Hypercat Data Hub* is stored in relational databases and this data is frequently updated, a more dynamic solution has been adopted. Thus, instead of copying the existing data into a triplestore, submitted SPARQL queries are dynamically translated into a set of SQL queries on top of the existing relational databases. In this way, a fully functioning SPARQL endpoint is provided, while during query execution, not only the SPARQL query itself is taken into consideration, but also the implicit information that is derived through reasoning over the developed ontology.

This work is organized as follows: Sect. 2 contains background information about the *BT Hypercat Data Hub* prior to its semantic enrichment. Section 3 contains a description of the *BT Hypercat Ontology* which was developed in this work in order to define the semantic representation of existing data. The corresponding mapping of data from a relational database to the semantic representation is described in Sect. 4. The *BT SPARQL Endpoint* is presented in Sect. 5 and the capability to combine information from external data sources by means of federated queries is presented in Sect. 6. Example use cases for the *BT Hypercat Data Hub* are illustrated in Sect. 7, while conclusions and future work are discussed in Sect. 8.

---

[3] http://ibarcelona.bcn.cat/en/smart-cities.

[4] http://www.mksmart.org.

[5] https://www.w3.org/TR/vocab-dcat/.

2   Background

The role of the *BT Hypercat Data Hub* is to enable information from a wide range of sources to be brought onto a common platform and presented to users and developers in a consistent way. Its portal provides a direct interface through which data consumers, such as app developers, can browse a data catalogue and select and subscribe to data feeds that they want to use. In addition, a JSON-based Hypercat [1] machine-readable catalogue, described further below, is also provided (as well as a recently proposed RDF-based Hypercat [13] catalogue). An API enables access to data feeds, secured by API keys, from browsers or within computer programs, while a relational, GIS capable, database enables complex queries that data can be filtered according to a wide range of criteria.

A set of edge adapters enables information coming onto the hub to be converted to a standard format for use inside the platform's core. It also provides a consistent API to end users and developers. The hub provides a consistent approach to integration between data exposed by sensors, systems and individuals via communication networks and the applications that can use derived information to improve decision making, e.g., in control systems. It includes a set of adapters for ingress (input) and egress (output). These are potentially specific to each data source or application feed and may be implemented on a case by case basis. There is therefore a need to translate data between arbitrary external formats and the data formats used internally.

In addition, as mentioned above, a Hypercat catalogue is implemented which is included via the Hypercat API. Hypercat is in essence a standard for representing and exposing Internet of Things data hub catalogues over web technologies, to improve data discoverability and interoperability. The idea is to enable distributed data repositories (data hubs) to be used jointly by applications through making it possible to query their catalogues in a uniform machine readable format. This enables the creation of knowledge graphs of available datasets across multiple hubs that applications can exploit and query to identify and access the data they need, whatever the data hub in which they are held.

From this perspective, Hypercat represents a pragmatic starting point to solving the issues of managing multiple data sources, aggregated into multiple data hubs, through linked data and semantic web approaches. It incorporates a lightweight, JSON-based approach based on a technology stack used by a large population of web developers and as such offers a low barrier to entry. Hypercat allows a server (IoT hub) to provide a set of resources to a client, each with a set of metadata annotations. There are a small set of core mandatory metadata relations which a valid Hypercat catalogue must include; beyond this, implementers are free to use any set of annotations to suit their needs.

3   BT Hypercat Ontology

In our previous work [13], we proposed a semantic enrichment for the core of the Hypercat specification, namely an RDF-based equivalent for a JSON-based catalogue. While Hypercat offers a syntactic first step, providing

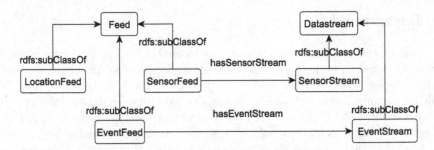

Fig. 1. BT Hypercat Ontology.

semantically enriched data goes further by allowing the unique identification of existing resources, interoperability across various domains and further enrichment by combining internally stored data with the Linked Open Data (LOD) cloud[6]. Data enrichment in the *BT Hypercat Data Hub* is achieved by representing data in RDF using concepts and properties defined in an OWL ontology [9]. Figure 1 shows the top level concepts of the *BT Hypercat Ontology*.

Feed is the top level class for any data feed that is asserted in the knowledge base. It contains the semantic properties of feeds. These include the feed id, creator, update date, title, url, status, description, location name, domain and disposition. There are also subclasses of class *Feed*, namely: *SensorFeed*, *EventFeed* and *LocationFeed* representing feeds for sensors, events and locations respectively.

The modelled data has been incorporated in the *BT Hypercat Data Hub* as one of the following feed types: (a) *SensorFeed*, (b) *EventFeed*, and (c) *Location-Feed*. Practically, each data source can advertise available information through the *BT Hypercat Data Hub* by providing a feed. A feed should be understood as a source of sensor readings, events or locations. Within each feed, data is available through datastreams (a class *Datastream* is defined, which has two subclasses namely: *SensorStream* and *EventStream* representing datastreams for sensors and events respectively). Thus, a given feed may provide a range of datastreams that are closely related e.g., for a weather data feed, different datastreams may provide sensor readings for temperature, humidity and visibility. Considering information about locations, a feed (of type *LocationFeed*) provides information directly by returning locations, namely locations are attached to and provided by a given feed.

A Hypercat online catalogue[7,8] contains details of feeds and information sources along with additional metadata such as tags, which allow improved search and discovery. The developed semantic model enables a semantic annotation and linkage of available feeds and datastreams. The *BT Hypercat Ontology* has been developed and made available with the uri (further details of the *BT Hypercat Ontology* can be found in [14]):

http://portal.bt-hypercat.com/ontologies/bt-hypercat

---

[6] http://lod-cloud.net/.

[7] http://portal.bt-hypercat.com/cat.

[8] http://portal.bt-hypercat.com/cat-rdf.

4   Data Translation

In this section we describe how data that is stored in a relational database within the *BT Hypercat Data Hub*, is made available in RDF.

4.1   RDF Adapter

By defining an ontology, semantically enriched data can be provided in RDF format. Note that prior to the semantic enrichment only XML and JSON formats were available. RDF data is represented in N-Triples format since such a format facilitates both storage and processing of data. Thus, each RDF triple is provided within a single line, in the following format: "<subject> <predicate> <object>.", while a collection of RDF triples is stored as a collection of lines. Note that N-Triples format can easily be transformed into other valid RDF formats, such as RDF/XML. In addition, the generated knowledge base can also be loaded in any given triplestore, namely any given RDF knowledge base, in order to facilitate operations such as query answering. Thus, by following W3C standards interoperability is ensured and the utilization of existing tools and applications is enabled.

The *BT Hypercat Data Hub* includes additional adapters for egress (output) in order to provide data in RDF format. In the following an example of how subject, predicate and object are generated for a *SensorFeed* is presented. Initially, the URI of each *SensorFeed* is generated, namely

<http://api.bt-hypercat.com/sensors/feeds/feedID>

Note that "http://api.bt-hypercat.com/" is the prefix URI for any data provided by the *BT Hypercat Data Hub*. In addition, "/sensors" provides information about the type of the feed (here *SensorFeed*), followed by "/feeds", which indicates that this URI belongs to a resource describing a feed, and finally "/feedID" is an id that uniquely identifies the given feed. For each *SensorFeed*, the *BT Hypercat Data Hub* provides its type, namely:

Subject	<http://api.bt-hypercat.com/sensors/feeds/feedID>
Predicate	<http://www.w3.org/1999/02/22-rdf-syntax-ns#type>
Object	<http://portal.bt-hypercat.com/ontologies/bt-hypercat#SensorFeed>

Note that the generation of other triples follows a similar rationale. However, a detailed description of triple generation for each given concept and property is omitted due to space limitations.

4.2 SPARQL to SQL

In order to develop a *SPARQL to SQL* endpoint, Ontop[9] [3] was used as an external library. Ontop comes with a Protege[10] plug-in that allows the creation of mappings of SPARQL patterns to SQL queries (described below), see Fig. 2. In addition, it provides a reasoner that parses the mappings and the ontology, and handles the translation of SPARQL queries into a set of SQL queries in order to return the corresponding results (for the SPARQL query). A key advantage of using Ontop is that implicit information that is extracted from the ontology through reasoning is taken into consideration. In this way, semantically richer information compared to the knowledge that is stored in the relational database is provided. A description of how mappings can be created is presented below.

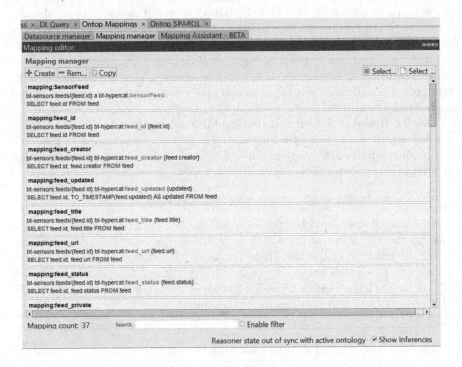

Fig. 2. Protege mapping editor.

In the following, an example of how a SPARQL triple pattern is mapped into a corresponding SQL query is described, and how the retrieved SQL results are used in order to construct RDF triples. *Mapping ID* corresponds to a unique id for a given mapping, *Target* (Triple Template) is the RDF triple pattern to be generated (note that SQL variables are given in braces, such as {feed.id}), and *Source* (SQL Query) is the SQL query to be submitted to the database.

---

[9] http://ontop.inf.unibz.it/.
[10] http://protege.stanford.edu/.

First, the prefixes that are used are defined in order to shorten URIs, for example:

bt-sensors: http://api.bt-hypercat.com/sensors/
bt-hypercat: http://portal.bt-hypercat.com/ontologies/bt-hypercat

Then mappings are defined. For example, the following mapping maps the class *SensorFeed*. Note that class *SensorFeed* is subclass of *Feed*, and thus is a valid assertion, while providing semantically richer information:

Mapping ID	mapping:SensorFeed
Target (Triple Template)	bt-sensors:feeds/{feed.id} a bt-hypercat:SensorFeed .
Source (SQL Query)	SELECT feed.id FROM feed

Note that Fig. 2 contains additional mappings for the class *SensorFeed*.

The following query can be submitted to a *SPARQL to SQL* endpoint in order to retrieve *Feed*s:

```
PREFIX hypercat: <http://portal.bt-hypercat.com/ontologies/bt-hypercat#>
SELECT DISTINCT ?s
WHERE{ ?s a hypercat:Feed . }
```

Thus, Ontop will match the triple pattern "?s a hypercat:Feed" with the mapping "mapping:SensorFeed" since class *SensorFeed* is subclass of *Feed*. An SQL query (see Source) will be submitted to the relational database, while the retrieved *id*s (*feed.id*) will be used in order to generate RDF triples following the triple template (see Target).

Note that the generation of other triples follows a similar rational, while a more detailed description of triple generation for a given concept or property can be found in [14].

5   BT SPARQL Endpoint

In the following, a description of the high level architecture for the developed *BT SPARQL Endpoint* is presented. As shown in Fig. 3, two levels of abstraction are applied. At the lower level, there is a *SPARQL to SQL* endpoint for each relational database in the system, namely each *SPARQL to SQL* endpoint provides a SPARQL endpoint on top of the given relational database. In this way, the system administrator can add or remove a *SPARQL to SQL* endpoint at any time.

At the moment a *SPARQL to SQL* component is supporting the translation of SPARQL queries to PostgreSQL[11] relational databases that contain information about sensors or events. At the higher level, there is only one *SPARQL to*

---

[11] https://www.postgresql.org/.

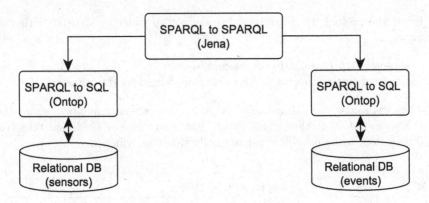

Fig. 3. BT SPARQL endpoint.

SPARQL component (based on the query engine of Apache Jena[12] [4]), which is made available to end users. The underlying functionality indicates that end users submit SPARQL queries to the *SPARQL to SPARQL* endpoint, while the system queries internally all available *SPARQL to SQL* endpoints in order to extract the relevant information from existing relational databases. At any given point, the system administrator can add or remove a *SPARQL to SQL* endpoint depending on the available PostgreSQL databases.

Both *SPARQL to SPARQL* and *SPARQL to SQL* endpoints can be accessed using the *BT SPARQL Query Editor*, which is available for each endpoint. Users can provide the query text, namely the SPARQL query, using a graphic interface. In addition, the *BT SPARQL Query Editor* supports five results formats: HTML, XML, JSON, CSV and TSV.

One of the key advantages of SPARQL queries over SQL queries is that SPARQL queries incorporate semantic reasoning within the returned results. For example, classes *SensorFeed* and *EventFeed* are subclasses of class *Feed*. Thus, the reasoner classifies all objects that belong to either *SensorFeed* and *EventFeed* as *Feed*. The SPARQL query of Sect. 4.2 can also be submitted to a *SPARQL to SPARQL* endpoint in order to retrieve *Feed*s. Note that Ontop supports reasoning over RDFS[13] and OWL 2 QL[14].

6   Federated Querying

As described above, a *Federated SPARQL* endpoint has been added in order to enable federated queries over both the *BT SPARQL Endpoint* and other external SPARQL endpoints that are available through the LOD cloud. Such external

---

[12] https://jena.apache.org/index.html.
[13] http://www.w3.org/TR/rdf-schema/.
[14] https://www.w3.org/TR/owl-profiles/#OWL_2_QL.

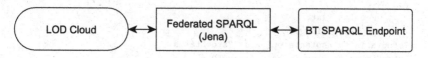

Fig. 4. Federated SPARQL endpoint.

SPARQL endpoints that are part of the LOD cloud are for example: DBPedia[15], FactForge[16], OpenUpLabs[17] and the European Environment Agency[18].

The LOD cloud is expanding and new SPARQL endpoints are added (and removed) allowing for access to new data. Since the *Federated SPARQL* endpoint does not contain any information itself, it serves as a middleware that combines information coming from other SPARQL endpoints, as depicted in Fig. 4.

The *Federated SPARQL* endpoint extends further the functionality of the *BT SPARQL Endpoint* since external SPARQL endpoints can be used in order to retrieve information about events or social and economic information that can be combined with data from the *BT SPARQL Endpoint* for complex data analytics. Examples can be the extraction of data about natural disasters from external datasets combined with related sensor and event data from the *BT SPARQL Endpoint*. Other types of data extracted from external datasets can be, for example, social data related to housing projects and their correlation with sensor and event data from the *BT SPARQL Endpoint*.

Reasoning capabilities and spatiotemporal queries can be combined with external datasets (LOD) in order to retrieve information which is not directly represented in the *BT Hypercat Data Hub*. This can be achieved by means of federated queries spanning over different internal and external SPARQL endpoints.

For example, the following federated query retrieves sensor measurements from the *BT Hypercat Data Hub* related to a specific active bus stop, extracted from an external SPARQL endpoint (OpenUpLabs):

```
PREFIX geo: <http://www.w3.org/2003/01/geo/wgs84_pos#>
PREFIX hypercat: <http://portal.bt-hypercat.com/ontologies/bt-hypercat#>
PREFIX naptan: <{http://transport.data.gov.uk/def/naptan/>
PREFIX skos: <http://www.w3.org/2004/02/skos/core#>

SELECT distinct ?d ?at_time ?western_longitude ?southern_latitude
       ?eastern_longitude ?northern_latitude ?stop ?lat ?long
WHERE {
   SERVICE <http://gov.tso.co.uk/transport/sparql>
   {
       ?stop a naptan:CustomBusStop;
             naptan:naptanCode ?naptanCode;
             naptan:stopValidity ?stopValidity;
             naptan:street"Kingswood Road";
```

[15] http://dbpedia.org/sparql.
[16] http://factforge.net/sparql.
[17] http://gov.tso.co.uk/transport/sparql.
[18] http://semantic.eea.europa.eu/sparql.

```
          geo:lat ?lat;
          geo:long ?long.
    ?stopValidity naptan:stopStatus ?stopStatus.
    ?stopStatus skos:prefLabel "Active"@en.
  }
  SERVICE <http://portal.bt-hypercat.com/BT-SPARQL-Endpoint/sparql>
  {
    ?d a hypercat:Datapoint.
    ?d hypercat:datapoint_at_time ?at_time.
    ?d hypercat:datapoint_western_longitude ?western_longitude.
    ?d hypercat:datapoint_southern_latitude ?southern_latitude.
    ?d hypercat:datapoint_eastern_longitude ?eastern_longitude.
    ?d hypercat:datapoint_northern_latitude ?northern_latitude.
    FILTER (?western_longitude > ?long - 0.1)
    FILTER (?southern_latitude > ?lat - 0.1)
    FILTER (?eastern_longitude < ?long + 0.1)
    FILTER (?northern_latitude < ?lat + 0.1)
  }
  FILTER(BOUND(?d))
}
```

7   Use Cases

This section is devoted to the description of two example use cases of the *BT Hypercat Data Hub*.

7.1   The SimplifAI Project

Urban traffic management and control is a primary concern of any city, and urban traffic transport operators often have at their disposal a disparate variety of real time and historical data, traffic controls (the most common of which are traffic signals) and controlling software. Software systems used for traffic management have a vertical design: they are not integrated at a horizontal level and cannot therefore easily share their data, or exploit data provided from other software/sources.

For achieving a higher level of data integration, and to better capture and exploit real-time and historical urban data sources, the SimplifAI project was carried out by a consortium consisting of the University of Huddersfield, British Telecommunications, Transport for Greater Manchester, and two other SMEs. In particular, the project focussed on exploiting the real-time and historical data sources to pursue better congestion control. As study area, a region of greater Manchester, UK was selected.

The overall concept in the improvement of traffic management was to utilise the semantically enriched data to enable the use of an intelligent function which requires both the integration of traffic data from disparate sources, and the transformation of the data into a predicate logic level, in order to operate. The intelligent function was to create traffic signal strategies in real time to solve challenges caused by exceptional or unexpected conditions.

The initial steps of the SimplifAI project concentrated on the semantic enrichment of traffic data. The raw data was taken from a large number of transport and environment sources and integrated into the *BT Hypercat Data Hub*, using the mapping of Sect. 4. After that, the focus was put on the utilisation of semantic data for generating traffic control strategies.

By enriching semantically the imported data, the unique identification of imported data is enabled. This is orthogonal to the problem solved by planning, as planning can also deal with ad hoc data. However, once the study area expands, using semantically enriched data will allow a systematic way of identifying resources that are mentioned in the generated plans. In addition, federated queries allow the developed system to extract data from the LOD cloud and combine it with data stored in the *BT Hypercat Data Hub* (e.g., the federated query of Sect. 6 combines bus stop information from an external source with internally stored data).

The intelligent function was based on an Automated Planning [7] approach [16], that is able to generate traffic control strategies (actions which change signals at a specified time) to alleviate traffic congestion caused by exceptional circumstances. The initial state of the modelled urban area, and information about available traffic lights and the structure of the network, were provided to the planning approach by the *BT Hypercat Data Hub*. The planner was then executed in order to generate control strategies for a number of test scenarios.

The quality of the strategies output from the planner was evaluated firstly by hand, inspecting the strategies to check that they were sensible, and by simulating their execution using traffic simulation software. Experts verified that strategies are sensible, and follow what would be expected when using "common sense". Simulations confirmed that generated strategies can effectively deal with unexpected conditions better than standard urban traffic control approaches: on average, the area is de-congested 20% faster, and tail-pipe emissions are reduced by 2.5%.

7.2   City Concierge

CityVerve is a Manchester, UK based IoT Demonstrator project, established in July 2016 with a two-year focus on demonstrating the capability of IoT applications for smart cities. One of the use cases of the CityVerve project, City Concierge, is aiming to increase uptake of walking and cycling as a preferred travel mode in Greater Manchester. Currently, Greater Manchester lacks integrated, consistent wayfinding services that can be accessed through a variety of media, including digital and print.

The City Concierge aims to develop a city user interface for the city region, integrating transportation and visitor services, allowing users to make informed choices regarding the way they travel. The scope of the use case includes improvements in the way people navigate around the city with a digital solution in conjunction with physical wayfinding assets, see Figs. 5 and 6.

Currently, it has been established that the *BT Hypercat Data Hub* provides the required infrastructure and functionality in order to enable the City

Fig. 5. Interaction between end users and city wayfinding assets.

Fig. 6. Locations of wayfinding infrastructure.

Concierge. Translating data into RDF enables additional query capabilities such as SPARQL queries on top of the developed system and its combination with the LOD cloud through federated queries. Such queries are vital in order to achieve project's objectives, which include the deployment of IoT and digital software solutions that seek to address current challenges, while having the flexibility for future solutions to be developed on the network deployed as part of the CityVerve project.

8   Conclusion

In this work, the semantic enrichment of the *BT Hypercat Data Hub* has been presented. More specifically, the *BT Hypercat Ontology* has been introduced, which is the basis for the translation of existing data into an RDF representation. In addition, the *BT SPARQL Endpoint* has been implemented as a set of SPARQL endpoints and an additional endpoint, called *Federated SPARQL* endpoint, has been provided in order to allow the execution of federated queries. Moreover, an example federated query illustrates how the *BT Hypercat Data Hub* can be connected to the LOD cloud. Finally, two use cases are illustrating the extended functionality of the system, thus highlighting the benefits of the semantic enrichment.

Future work includes further semantic enrichment of the implemented system. Specifically, current support for SPARQL queries can be extended in order to enable GeoSPARQL queries [10] so as to provide direct access to spatial information that is currently available in the *BT Hypercat Data Hub*. In addition, spatiotemporal reasoning [12] is a prominent direction that could provide richer knowledge by reasoning over data that is coming from both the *BT Hypercat Data Hub* and the LOD cloud.

References

1. Beart, P.: Hypercat 3.00 Specification (2016)
2. Berners-Lee, T., Hendler, J., Lassila, O.: The semantic web. Sci. Am. **284**, 29–37 (2001)
3. Calvanese, D., Cogrel, B., Komla-Ebri, S., Kontchakov, R., Lanti, D., Rezk, M., Rodriguez-Muro, M., Xiao, G.: Ontop: answering SPARQL queries over relational databases. Semant. Web **8**(3), 471–487 (2017)
4. Carroll, J.J., Dickinson, I., Dollin, C., Reynolds, D., Seaborne, A., Wilkinson, K.: Jena: implementing the semantic web recommendations. In: Proceedings of the 13th International World Wide Web Conference on Alternate Track Papers & Posters, pp. 74–83. ACM (2004)
5. d'Aquin, M., Davies, J., Motta, E.: Smart cities' data: challenges and opportunities for semantic technologies. IEEE Internet Comput. **19**(6), 66–70 (2015)
6. Davies, J., Fisher, M.: Internet of things - why now? Jnl Inst. Telecommun. Prof. **7**(3), 36–42 (2015)
7. Ghallab, M., Nau, D., Traverso, P.: Automated Planning: Theory & Practice. Elsevier, Amsterdam (2004)
8. Hayes, P.: RDF Semantics. In: W3C Recommendation (2004)
9. Hitzler, P., Krötzsch, M., Parsia, B., Patel-Schneider, P.F., Rudolph, S. (eds.) OWL 2 web ontology language: primer. W3C Recommendation, vol. 27, October 2009. http://www.w3.org/TR/owl2-primer/
10. Perry, M., Herring, J.: OGC GeoSPARQL-a geographic query language for RDF data. OGC Implementation Standard, September 2012
11. PrudHommeaux, E., Seaborne, A., et al.: SPARQL query language for RDF. W3C Recommendation, vol. 15 (2008)
12. Stravoskoufos, K., Petrakis, E.G.M., Mainas, N., Batsakis, S., Samoladas, V.: SOWL QL: querying spatio-temporal ontologies in OWL. J. Data Semant. **5**, 249–269 (2016)
13. Tachmazidis, I., Davies, J., Batsakis, S., Antoniou, G., Duke, A., Clarke, S.S., Hypercat, R.D.F.: Semantic enrichment for IoT. In: Semantic Technology - 6th Joint International Conference, JIST 2016, Singapore, Singapore, November 2–4, 2016, Revised Selected Papers, pp. 273–286 (2016)
14. Tachmazidis, I., Batsakis, S., Davies, J., Duke, A., Vallati, M., Antoniou, G., Clarke, S.S.: A hypercat-enabled semantic internet of things data hub, Technical report, March 2017. https://arxiv.org/abs/1703.00391
15. Townsend, A.M., Cities, S.: Big Data, Civic Hackers, and the Quest for a New Utopia. WW Norton & Company, Cambridge (2013)
16. Vallati, M., Magazzeni, D., De Schutter, B., Chrpa, L., McCluskey, T.L.: Efficient macroscopic urban traffic models for reducing congestion: a PDDL+ planning approach. In: The Thirtieth AAAI Conference on Artificial Intelligence (AAAI) (2016)

ArmaTweet: Detecting Events
by Semantic Tweet Analysis

Alberto Tonon[1], Philippe Cudré-Mauroux[1], Albert Blarer[2], Vincent Lenders[2],
and Boris Motik[3]([⊠])

[1] eXascale Infolab, University of Fribourg, Fribourg, Switzerland
[2] Science & Technology, C4I, armasuisse, Thun, Switzerland
[3] University of Oxford, Oxford, UK
boris.motik@cs.ox.ac.uk

Abstract. Armasuisse Science and Technology, the R&D agency for the
Swiss Armed Forces, is developing a Social Media Analysis (SMA) system to help detect events such as natural disasters and terrorist activity
by analysing Twitter posts. The system currently supports only keyword
search, which cannot identify complex events such as 'politician dying' or
'militia terror act' since the keywords that correctly identify such events
are typically unknown. In this paper we present ArmaTweet, an extension
of SMA developed in a collaboration between armasuisse and the Universities of Fribourg and Oxford that supports *semantic event detection*. Our
system extracts a structured representation from the tweets' text using
NLP technology, which it then integrates with DBpedia and WordNet in
an RDF knowledge graph. Security analysts can thus describe the events
of interest precisely and declaratively using SPARQL queries over the
graph. Our experiments show that ArmaTweet can detect many complex
events that cannot be detected by keywords alone.

1 Introduction

Twitter[1] is a popular microblogging service. As of late 2016, an estimated 317
million users produce around 500 million messages (or *tweets*) per day that are
broadcast to the users' *followers*. Tweets contain up to 140 characters and cover
almost any topic, including personal messages and opinions, celebrity gossip,
entertainment, news, and more. Current events are widely discussed on Twitter;
for example, around 1.7 M tweets were sent on 7/1/2015 in response to the Charlie Hebdo attacks in Paris. Twitter users often provide live updates in critical
situations; for example, users tweeted 'In Brussels Airport. Been evacuated afer
[sic] suspected bomb.' and 'Stampede now. Everyone running' during the attack
at the Brussels airport on 22/3/2016. Most tweets can be read by unregistered
users, so Twitter can potentially provide a real-time source of information for
detecting newsworthy events before the conventional broadcast media channels.
Thus, the development of techniques for tweet analysis and event detection has

---

[1] http://twitter.com/

© Springer International Publishing AG 2017
E. Blomqvist et al. (Eds.): ESWC 2017, Part II, LNCS 10250, pp. 138–153, 2017.
DOI: 10.1007/978-3-319-58451-5_10

attracted considerable attention. The Natural Language Processing (NLP) community adapted their techniques to tweets [9,17,23], which are short and often use a colloquial style with nonstandard acronyms, slang, and typos. These tools were used to develop numerous approaches to event detection on Twitter, and we survey the NLP tools and the event detection approaches in Sect. 2.

Based on these results, armasuisse Science and Technology—the R&D agency of the Swiss Armed Forces—is developing a Social Media Analysis (SMA) system, which aims to help security analysts detect security-related events. Similarly to previous work [2,15], analysts currently describe the relevant events using keywords, which are evaluated over tweets using standard Information Retrieval (IR) techniques. This approach, however, cannot detect events with complex descriptions. For example, to detect deaths of politicians, an analyst might query the SME system using keywords 'politician die', but this results in both low precision and low recall. For example, the system misses the death of Edward Brooke (the first African American US senator) since, instead of the word 'politician', most tweets about this event contain phrases such as 'the senator' or 'elected to the US Senate'; similarly, the word 'die' is very frequent on Twitter and so the query retrieves mostly irrelevant tweets. To reliably detect such events, one must understand the intended meaning of the query, know which people are politicians, and identify tweets that mention such a person as a subject of a verb 'to die'. Similarly, to match a query for 'militia terror act' to an attack of Boko Haram on a village in Nigeria, one must know that Boko Haram is a militia group and that terror acts include kidnappings and bombings.

In this paper we present ArmaTweet—an extension of SMA to *semantic event detection* developed in a collaboration between armasuisse and the Universities of Fribourg and Oxford. Our system uses NLP technique to extract a structured representation from tweets and integrate it with DBpedia and WordNet in an RDF knowledge graph. Users can thus describe relevant event categories by using *semantic queries* over the knowledge graph. The system evaluates these queries using semantic technologies to retrieve the relevant tweets and passes them to an anomaly detection algorithm to determine whether and how they correspond to actual events. We evaluated our system on the 1%-sample of tweets collected by the Twitter's streaming API during the first six months of 2015. The system detected a total of 941 events across seven different event categories. We evaluated our results using three different definitions of which tweets should be considered relevant to the query. Depending on the selected relevance metric, our system achieved precision between 46% and 67% across all categories. Most of these events could not be detected by the previous version of the system, showing clearly how our approach complements standard keyword search.

2   Related Work

Although analysing tweets is very challenging, initiatives such as the Named Entity rEcognition and Linking (NEEL) Challenge have spurred on the NLP community to develop a comprehensive set of tools including Part-of-Speech

(POS) taggers [17], Named Entity Recognisers [5,22], and dependency parsers [9]. To understand the syntactic structure of tweets, our system must identify dependencies between terms (e.g., identify the subject of a given verb, determine grammatical cases, and so on). We do not know of a Twitter-specific system that provides such functionality, so decided to use the Stanford CoreNLP library [14] that was originally designed to analyse cleaner text.

A recent survey of the methods for event detection on Twitter [7] classifies existing approaches into three groups. The first one contains approaches for detecting *unspecified events*—that is, events of general interest with no advance description. These approaches typically detect trends in features extracted from tweets and/or cluster tweets based on their topic [3,13,29]. Several systems detect breaking news [18,19,26], and one additionally classifies events into predefined types such as 'Sports', 'Death', or 'Fashion' [23]. Some approaches use probabilistic similarity instead of clustering [31]. Analogously to these approaches, we also identify events by detecting trends, but only after semantic queries have been used to identify the tweets matching the user's interests (see Sect. 3).

The second group contains approaches for detecting *predetermined events*, such as concerts [4], controversial events [20], local festivals [10], earthquakes [25], crime and disaster events [12], and disease progression [27]. Such systems are specifically tailored to an event type, and they usually involve training a classifier on manually annotated tweets to learn the correlation of features that identifies tweets talking about an event. The EMBERS system [21] goes a step further by aggregating many sources of information (Twitter, Web searches, news, blogs, Internet traffic, and so on) to detect and *predict* instances of civil unrest.

The third group contains approaches for detecting *specific events*, which typically use IR methods to match a query (i.e., a Boolean combination of keywords) to a database of tweets. Queries are either provided by the users or are learned from the context [2], and recall can be improved by query expansion [15]. These techniques have been combined with geographical proximity analysis to detect civil unrest [30] and model events in Twitter streams [8]. ArmaTweet also identifies tweets using queries provided by users and thus, broadly speaking, falls into this category; however, instead of keyword queries, it uses semantic queries describing the relationships between entities in tweets. The system thus supports queries for specific events (e.g., 'Obama meets Trump') that can be captured using keywords, as well as more complex queries specifying an event *type* (e.g., 'somebody hacks a company') for which a keyword-based approach is not effective. Our system does not rely on a training phase, but requires users to specify their interests precisely by constructing semantic queries. An approach most similar to ours constructs a knowledge graph of events from news articles [24], and the main difference to our work is that it focuses on longer, cleaner texts.

3   Motivation and Methodology

Motivation. Detecting Twitter events using complex descriptions (e.g., based on entities' classes or their relationships) is still very challenging. Consider, for example, the 'politician dying' description from the introduction, and the death of Edward Brooke on 3/1/2015. The event has been widely discussed on Twitter, and running the keyword query 'edward brooke' for that day in SMA returns 121 tweets. This, however, is just a tiny fraction of all tweets produced on that day, and so this event is unlikely to be detected by the techniques for unspecified events (see Sect. 2); for example, the technique by Ritter et al. [23] detected just five completely unrelated events on that day.[2] Moreover, there are no obvious keyword queries: 'die' returns 5161 mostly irrelevant tweets in SMA, 'politician' returns 46 irrelevant tweets, and 'politician die' and 'senator die' return no results (note that SMA uses just 1% of all tweets). Thus, although 'edward brooke' is an effective query, it is unclear how to construct it from description 'politician dying'. Similarly, it is unclear how to exploit classification-based techniques since common features, such as n-grams or bags of words, are unlikely to reflect the semantic information that Edward Brooke was a politician. Other examples of complex events that we consider in this paper include 'politician visits a country', 'militia terror act', or 'capital punishment by country'.

Approach. Since the objective of armasuisse was to detect events with complex descriptions, we depart from statistical and IR approaches and use semantic search instead. In particular, we use natural language processing to associate each tweet with a set of *quads* of the form (*subject,predicate,object,location*), describing who did what to whom and where; any of these components can be empty, which we denote by ×. We also associate with each tweet a set of *entities* whose role (subject or object) in the tweet could not be determined. Subjects, objects, locations, and entities are matched to DBpedia [11], a knowledge base extracted from Wikipedia, and predicates are matched to verb synsets in WordNet [16], an extensive lexicon. Thus, DBpedia and WordNet provide us with a vocabulary and background knowledge for describing complex events. For example, tweets about the death of Edward Brooke are associated with quads of the form (dbr:Edward_Brooke,wnr:200359085-v,×,×), where wnr:200359085-v identifies the synset 'to die' in WordNet, and DBpedia classifies dbr:Edward_Brooke as an instance of yago:Politician110451263. Our simple quad model cannot represent semantic relationships such as appositions, adverbs, dependent clauses, modalities, or causality. While such relationships would clearly be useful, our evaluation (see Sect. 7) demonstrates that our model is sufficient for detecting many kinds of complex event that cannot be detected using keywords only.

Semantic Event Descriptions. To use ArmaTweet, users must first describe formally the events of interest. To facilitate that, the system provides an intuitive and declarative query interface that allows users to query quads in our knowledge graph while exploiting the background knowledge from DBpedia and WordNet. For example, 'politician dying' events can be precisely described by a

---

[2] http://statuscalendar.com/month/201501/ accessed on 14 December 2016.

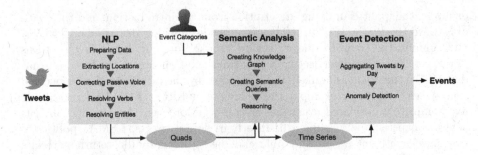

Fig. 1. ArmaTweet architecture

query that identifies all quads in our knowledge graph whose subject is of type yago:Politician110451263, and whose predicate is wnr:200359085-v. As we discuss in Sect. 5, such queries are matched to the knowledge graph in a way that attempts to compensate for the imprecision of natural language analysis. Queries are currently constructed manually, which allows users to precisely describe their information needs. In our future work we shall investigate techniques that can automate, or at least provide some help with, query construction.

System Output. Given a set of tweets and a set of queries describing complex events, ArmaTweet produces a list of events, each consisting of an event date, an event summary, and a set of relevant tweets. The event summary is specific to the event type; for example, for 'politician dying', it identifies the politician in question, and for 'militia terror act', it identifies the militia group and the verb describing the act. Finally, the set of relevant tweets allows the user to validate the system's output, gain additional information, and possibly initiate an appropriate event response. The system currently does not detect long-running events (e.g., political turmoil or health crises)—that is, each event is associated with a single day only. Thus, the same real-world event can be reported as several events having the same summary but occurring on distinct days. Longer-running events are often reported as events with the same summary occurring in close succession, and we shall investigate ways to exploit this in our future work.

System Architecture. Figure 1 shows the architecture of ArmaTweet and its three main components. The *Natural Language Processing* component analyses the tweets' text and extracts the quads and entities, and it is independent of the complex event descriptions. The *Semantic Analysis* component converts the output of the NLP step into RDF, which is then analysed and filtered using the user's event descriptions. The output of this component is a set of *tweet time series*, each consisting of a summary and a set of tweets. Finally, the *Event Detection* component uses an anomaly detection algorithm to extract from each time series zero or more dates that correspond to the actual events. The resulting events and their summaries are finally reported to the user.

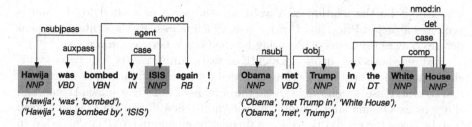

Fig. 2. Data structures produced by OpenIE on two example tweets

To understand the conceptual difference between time series and events, consider the 'militia terror act' event query. Our Semantic Analysis component produces one time series with subject 'Boko Haram' and predicate 'to attack', which contains all tweets talking about attacks by Boko Haram regardless of the time of the tweets. Next, the Event Detection component groups the tweets by time and detects anomalies (e.g., abrupt changes in the number of tweets per day). Since Boko Haram committed several attacks in our test period, our system extracts and reports several events from this particular tweet time series.

Our NLP processing is computationally intensive, but it is massively parallel since each tweet can be processed independently; hence, we parallelised it using Apache Spark. Moreover, we used the state of the art semantic store RDFox[3] to manage and process our knowledge graph. The parts of our system that are independent from the Spark environment (i.e., the core of the NLP component and the queries/rules used for semantic analysis) are available online.[4]

4   Natural Language Processing of Tweets

The NLP component of `ArmaTweet` extracts from tweets in English a set of quads consisting of a subject, predicate, object, and location, and a set of entities that cannot be assigned to a quad. Predicates are matched to verb synsets in WordNet, and the remaining components are matched to DBpedia resources.

Data Preparation. For each tweet, we first prepare certain data structures. Specifically, we first clean the text by removing emoticons and uncommon characters, we substitute # and @ characters with whitespace, and we split Camel-Case words. Next, the OpenIE annotator from the Stanford CoreNLP library [1] transforms the text into *text triples* consisting of a subject, a predicate, and an object; the name 'text triples' emphasises that the components are pieces of text, and not DBpedia or WordNet resources. OpenIE also annotates the mentions of *named entities* (i.e., objects with a proper name) with the entity types (location, organisation, or person); it annotates the text with *part-of-speech (POS)* tags,

---

[3] http://www.cs.ox.ac.uk/isg/tools/RDFox/
[4] http://github.com/eXascaleInfolab/2016-armatweet

which describe the relation of a word with adjacent or related words; and it produces a *(dependency-based) parse tree*, which represents the syntactic dependencies between sentence parts using labelled edges between words.

Figure 2 shows the output of OpenIE on two example tweets. The tweet text is shown in bold. Named entity types are coded using colours: the locations 'Hawija' and 'White House' are shown in green, the organisation 'ISIS' is shown in blue, and the persons 'Obama' and 'Trump' are shown in yellow. The POS tags are shown in italic below the words: 'Hawija' is a singular proper noun (*NNP*), 'was' is a verb in past tense (*VBD*), and 'again' is an adverb (*RB*). Finally, the parse trees are shown as labelled arrows connecting words. The roots of the trees are words without incoming edges—'bombed' and 'met' in this case. Moreover, in the rightmost tree, 'Obama' is the subject of the verb 'met' (denoted by a *nsubj* dependency), while 'Trump' is its direct object (denoted by a *dobj* dependency). Finally, the text triples are shown at the bottom of the figure.

Our NLP component also passes the text to DBpedia Spotlight [6], which identifies entity mentions in the text and associates with each mention an appropriate DBpedia resource. For example, on the example shown in Fig. 2, Spotlight annotates 'Hawija' with `dbr:Hawija`, 'ISIS' with `dbr:ISIS`,[5] 'Obama' with `dbr:Barack_Obama`, 'Trump' with `dbr:Donald_Trump`, and 'White House' with `dbr:White_House`. We chose Spotlight due to its scalability and ease of use. Spotlight is parameterised by a confidence value that regulates the precision of annotation, and a support value used to filter out uncommon entities, and we empirically determined 0.5 and 20, respectively, as values appropriate for our system. Please note that this step is complementary to the named entity recognition of OpenIE: Spotlight provides us with links to DBpedia, whereas OpenIE provides us with high-level entity categories that we use for text analysis.

Location Extraction. We next try to identify the location of the action in text triples by observing that words introducing a *grammatical case* in a sentence that are connected to a location often describe the verb's spatial location. Thus, we first extend each text triple into a *text quad* by specifying the location as unknown. Next, for each text quad where the object is a location (as indicated by entity recognition), we check whether the parse tree contains a grammatical case dependency between a word occurring in the quad's predicate and a word occurring in its object; if so, we move the quad's object to its location. For example, the object of ('Obama', 'met Trump in', 'White House') in Fig. 2 has been classified as a location, and the parse tree contains a grammatical case dependency between the word 'House' occurring in the object and the preposition 'in' occurring in the predicate, and so we treat 'White House' as a location instead of an object. Please note that a location in the subject often does not specify the location of an action; for example, the subject of the sentence 'Oxford is a city' is a location, but 'Oxford' should not be used as a location in a quad since it does not describe the location of an action. We found no clear dependency pattern that could distinguish such cases and reliably extract location from subjects.

---

[5] We abbreviate the actual resource `dbr:Islamic_State_of_Iraq_and_the_Levant`.

Passive Voice Correction. Passive voice can be problematical; for example, in ('Hawija', 'was bombed by', 'ISIS') from Fig. 2, 'Hawija' is the subject and 'ISIS' is the object, which does not correctly reflect the intended meaning of the tweet. To correct such situations, for each text quad, we check whether the predicate contains a word that was classified by the POS tagger as a verb and that has (i) an outgoing *passive auxiliary modifier* dependency (to any other word), (ii) a *passive subject* dependency to a word occurring in the subject, and (iii) an *agent* dependency to a word occurring in the object; if so, we swap the subject and the object. In our example, 'was' is an auxiliary modifier, 'ISIS' is an agent, and 'Hawija' is a passive subject', so we apply the correction.

Entity Resolution. We next match the subject, object, and location of each text quad to the annotations of Spotlight. In case of an exact match we replace the component with the DBpedia resource, and otherwise we replace it with ×.

Verb Resolution. Since Spotlight does not handle verbs, we developed our own approach to verb resolution. First, we identify all verb occurrences in a tweet using POS tags. Next, we lemmatise each verb occurrence—that is, we substitute it with the verb's infinitive form (e.g., 'met' becomes 'to meet', 'bombed' becomes 'to bomb', and so on)—and then we search the tweet's parse tree for any phrasal verb particles connected to the verb's occurrence. Such a dependency indicates that the verb and the particle form an idiomatic phrase (e.g., 'take off' or 'sort out') and should be analysed together, so, whenever we find one, we concatenate the verb with the phrasal verb particle. We finally match the (possibly extended) verb occurrence to a WordNet synset; if several candidate synsets exist, we select the one that is most frequent according to the WordNet's statistics. The output of this part of our system is thus similar to that of Spotlight.

Finally, we resolve the predicates in the quads to the matched verbs. Unlike entities, which we resolved using exact matches, we substitute the predicate of a quad with a matched verb if the former *contains* the latter. This allows us to match 'was bombed by' in Fig. 2 to the synset for 'to bomb'. Again, we replace predicates that could not be resolved with ×.

Quad Output. For each tweet, we return all quads except those containing only × markers. In addition, for each verb that was resolved to the tweet's text but could not be associated with a quad, we also return a fresh quad where the subject, object, and location are empty. Finally, we return the set of all entities that were detected by Spotlight but could not be matched to a quad.

5   Semantic Analysis

The Semantic Analysis component of ArmaTweet integrates DBpedia, WordNet, and the quads in a knowledge graph, and it evaluates complex event descriptions provided by the users. We next discuss the structure of the knowledge graph and the event descriptions, and we describe how we identify the tweet time series.

5.1   The RDF Knowledge Graph for Event Detection

We use RDF as the data model for the knowledge graph. Thus, the RDF versions of DBpedia and WordNet can be imported directly, and we encode tweet information using a simple schema. Each tweet is identified by a URI obtained from the tweet's ID; it is an instance of the `aso:Tweet` class; and data properties `aso:createdAt` and `aso:tweetText` specify the time of the tweet's creation and its text, respectively. A tweet can be associated with zero or more quads, each with at most one `aso:quadSubject`, `aso:quadPredicate`, `aso:quadObject`, and `aso:quadLocation` property value. Finally, a tweet can be associated with zero or more entities whose role in a sentence could not be determined (see Sect. 4).

Figure 3 shows the tweet `ast:551507074258325504` with two quads: one connects `dbr:Edward_Brooke` from DBpedia with the WordNet synset 'to die', and another connects `dbr:Edward_Brooke` with `dbr:Reconstruction_Era` (due to the imprecision of NLP analysis). Finally, the tweet is also directly associated with `dbr:Birmingham`, whose role in the sentence could not be determined.

The time series detected by Semantic Analysis component, each consisting of a summary and a set of tweets, are also stored in the knowledge graph. For example, `_:ts-sp_4344996_1855965` in Fig. 3 is a tweet time series containing all tweets about Edward Brooke dying, and the Event Detection component (cf. Sect. 6) will extract from it zero or more events. Our system currently does not take into account that a person can die only once, and so it can potentially report multiple 'Edward Brooke dies' events. Each time series is classified according to the type of the summary information; currently, this includes subject–predicate (SP), predicate–object (PO), subject–country (SC), predicate–country (PC), and subject–predicate–country (SPC) time series. For example, the time series in Fig. 3 is determined by a subject and a verb, and so it belongs to

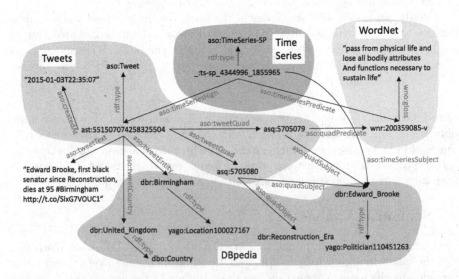

Fig. 3. A fragment of the RDF knowledge graph

the `aso:TimeSeries-SP` class and the values of `aso:timeSeriesSubject` and `aso:timeSeriesPredicate` determine the time series summary.

5.2   Resolving Location in the Knowledge Graph

We observed that the granularity of the event location often varies between tweets; for example, tweets about the Charlie Hebdo attacks refer both to France and Paris. To simplify event detection, we decided to aggregate event information at the country level. Thus, we extend the knowledge graph by resolving references to locations mentioned in tweets to the corresponding countries. For example, the tweet shown in Fig. 3 refers to `dbr:Birmingham` so, since DBpedia contains the information that Birmingham is a city in the UK, we associate the tweet with `dbr:United_Kingdom` using the `aso:tweetCountry` property. Entities in tweet quads are resolved to countries in a similar vein.

5.3   Describing Complex Events and Extracting Time Series

Events of interest are described using conjunctive SPARQL queries that select the relevant quads. For example, queries (1) and (2) describe the 'politician dying' and the 'unrest in a country' events, respectively, where `aso:UnrestVerb` contains all verbs from WordNet that we identified as indicating unrest. The answer variables of each query determine the time series summary.

$$
\text{SELECT ?S wnr:200359085-v WHERE \{ ?Q aso:quadPredicate wnr:200359085-v .} \atop \text{?Q aso:quadSubject ?S . ?S rdf:type yago:Politician110451263 \}}} \quad (1)
$$

$$
\text{SELECT ?P ?C \{ ?Q aso:quadCountry ?C .} \atop \text{?Q aso:quadPredicate ?P . ?P rdf:type aso:UnrestVerb \}}} \quad (2)
$$

We next explain why querying quads is important. In particular, tweets often mention a politician and the verb 'to die', but not in a desired semantic relationship. For example, tweet `ast:551766588421312512` (not shown in Fig. 3) says '@BarackObama @pmharper I'm just trying to get some realization, is school supposed to cause you so much stress&anxiety that you want to die?' and it is annotated with `dbr:Barack_Obama` and `wnr:200359085-v`, but, as one might expect from the text, there is no quad connecting the two resources. The lack of a semantic relationship, however, does not always indicate that a tweet is irrelevant to the event query. For example, tweet `ast:555598764589977600` (not shown in Fig. 3) says 'Edward Brooke, first black US senator elected by popular vote, dies - Reuters', and it is annotated with `dbr:Edward_Brooke` and `wnr:200359085-v`, but, due to the complex sentence structure, the NLP component could not identify the semantic relationship correctly. In fact, our knowledge graph contains 44 tweets with quads matching 'Edward Brooke dies', as well as 111 additional tweets without the semantic relationship.

To exploit the knowledge graph as much as possible but without losing precision, our system proceeds as follows. It creates a tweet time series for each distinct result of a quad query (or, equivalently, for each distinct time series summary), to which it adds as 'high confidence' members all tweets containing a matching quad. Next, for each time series created in this way, the system adds to the time series as 'low confidence' members all tweets mentioning the relevant entities/predicates without the semantic relationship. For example, our system creates a time series for each distinct value of ?S produced by query (1), and this includes _:ts-sp_4344996_1855965 from Fig. 3 that contains tweets ast:551507074258325504 and ast:555598764589977600 as 'high' and 'low confidence' members, respectively. In contrast, no time series is created for dbr:Barack_Obama since our knowledge graph does not contain a quad matching query (1) where ?S is dbr:Barack_Obama. Intuitively, the presence of 'high confidence' tweets raises the importance of the 'low confidence' tweets, which helps compensate for the imprecision of the NLP analysis.

The Semantic Analysis component was realised using the RDFox system, which supports reasoning over RDF datasets using datalog rules. For each time series query, the user must provide the time series name and classify the query according to the summary type, and then the query is converted into a datalog rule that creates the tweet time series and identifies the 'high confidence' tweets. For example, query (1) is named aso:PoliticianDying and classified as a subject–predicate query, and so it is converted into the following datalog rule:

```
[?TS, rdf:type, aso:PoliticianDying], [?TS, aso:timeSeriesSubject, ?S],
[?TS, aso:timeSeriesVerb, wnr:200359085-v], [?TS, aso:timeSeriesHigh, ?TW] :-
    [?TW, aso:tweetQuad, ?Q], [?Q, aso:quadSubject, ?S],
    [?S, rdf:type, yago:Politician110451263], [?Q, aso:quadPredicate, wnr:200359085-v],
    BIND(SKOLEM("ts-sp", ?S, wnr:200359085-v) AS ?TS) .
```
$$(3)$$

This rule uses the datalog syntax of RDFox, which supports calling SPARQL builtin functions in its body. The SKOLEM function is an RDFox-specific extension that creates a blank node uniquely determined by the function's parameters, thus simulating function symbols from logic programming. Thus, for each value of ?S, rule (3) assigns to ?TS a unique blank node that identifies the time series, and its head atoms then attach to ?TS the relevant information and the 'high confidence' tweets. A fixed (i.e., independent from the queries) set of rules then identifies the 'low confidence' members of each time series by selecting tweets that mention all entities/predicates from the time series summary, but without the semantic relationship. For example, for subject–predicate time series, these rules select all tweets that mention the subject and the predicate outside a quad.

6   Event Detection

The Event Detection component accepts as input the tweet time series produced by the Semantic Analysis component, and it identifies zero or more associated events. This is done using the Seasonal Hybrid ESD (S-H-ESD) test [28] developed specifically for detecting anomalies in Twitter data. The algorithm is given a real number p between 0 and 1, a set of time points T, and a real-valued function $x : T \to \mathbb{R}$ that can be seen as a sequence of observations of some value on T where $x(t)$ is the value observed at time $t \in T$. The algorithm identifies a subset

Table 1. Evaluation results by event category

Event category	Type	Total Events	Positive instances by relevance					
			R3		R3+R2		R3–R1	
Aviation accident	SP	84	44	(52%)	51	(61%)	64	(76%)
Cyber attack on a company	PO	129	20	(16%)	42	(33%)	57	(44%)
Capital punishment in a country	PC	153	47	(31%)	67	(44%)	92	(60%)
Militia terror act	SP	220	92	(42%)	125	(57%)	141	(64%)
Politician dying	SP	111	76	(68%)	80	(72%)	85	(77%)
Politician visits a country	SPC	44	29	(66%)	36	(82%)	44	(100%)
Unrest in a country	PC	200	125	(63%)	133	(67%)	148	(74%)
Total:		941	433	(46%)	534	(57%)	631	(67%)

T_a of T of time points at which the value of x is considered to be anomalous, while ensuring that $|T_a| \leq p \cdot |T|$ holds; thus, p is the maximal proportion of the time points that can be deemed anomalous. Roughly speaking, the S-H-ESD test first determines the periodicity/seasonality of the input data; next, it splits the data into disjoint windows each containing at least two weeks of data; finally, for each window, it subtracts from x the seasonal and the median component and applies to the result the Extreme Student Derivative (ESD) test—a well-known anomaly detection technique. Twitter is currently using this technique on a daily basis to analyse their server load. ArmaTweet uses the open-source implementation of this test from the R statistical platform.[6]

To apply the S-H-ESD test, each tweet time series is converted into a sequence of temporal observations by aggregating the tweets by day—that is, the set T corresponds to the set of all days with at least one tweet, and the value of $x(t)$ is the number of (both 'high' and 'low confidence') tweets occurring on the day $t \in T$. We then run the S-H-ESD test with $p = 0.05$—that is, at most 5% of the time points can be deemed anomalous. Moreover, we configured the algorithm to detect only positive anomalies (i.e., cases where the number of tweets is above the expected value), which is natural for event detection.

7   Evaluation

Evaluating ArmaTweet was difficult since there is no ground truth: a list of all relevant events does not exist, so we could not determine the recall of our technique. Thus, we focused on determining the precision and the benefits of semantic event detection. We next present our experimental setup and discuss our findings.

We processed 195.7 M tweets in English collected in the first half of 2015 using Twitter's streaming API (which returns about 1% of all tweets). The NLP component extracted 14.5 M quads from 12.8 M tweets (i.e., 6% of the input). Most quads have two components: 6.2 M quads contain a predicate and an object, and 5 M quads contain a subject and a predicate; the remaining 0.7 M quads have

---

[6] http://github.com/twitter/AnomalyDetection.

three components, and no quads have four components. About 0.5 M quads contain location information. Integrating the quad information with DBpedia and WordNet produced a knowledge graph containing a total of 725.8 M triples, which increased to 800 M triples after applying the semantic analysis rules.

Determining Complex Events. We consulted the Wikipedia page for 2015[7] to identify interesting concrete events, which provided us with a starting point for a series of workshops in which we identified events and event types of interest to armasuisse customers. We eventually settled on the seven complex event categories shown in Table 1. We made sure that our categories cover many different types of event summary (i.e., subject–verb, verb–object, etc.).

Creating Category Queries. For each event category, we constructed a semantic query as follows. We first identified the entities from our example events on Wikipedia (e.g., `dbr:Edward_Brooke`), which we then looked up in DBpedia to identify their types (e.g., `yago:Politician110451263`). Next, we queried our knowledge graph for the verbs occurring together with the sample entities in the tweets. We ranked these verbs by the frequency of their occurrence, and then selected those best matching the event category. Finally, we formulated the category query and tested it on example events. Most queries capture the meaning of the categories directly, apart from the 'Aviation incident' query where, to select useful data, we ask for a subject of type 'airline' and a verb indicating a crash. Creating the queries took about four person-days of an expert in semantic technologies, and optimising this process is the main topic for our future work.

Event Validation. By evaluating the event categories over the knowledge graph and detective events as discussed Sects. 5 and 6, we identified a total of 941 events (see Table 1), which we validated manually—that is, we determined whether the reported event is a positive instance. This, however, turned out to be surprisingly challenging. First, we could often not verify whether the event really happened, so we decided to just evaluate whether the retrieved tweets correctly talk about the event; we justify this choice by noting that detecting 'invented' events could also be very important to security analysts. Second, some events happened in the past (e.g., the anniversary of Robert Kennedy's assassination was widely discussed on Twitter), but we decided to count these as positive instances as well since they are also likely to be of interest. Third, in some cases the retrieved events were only partially relevant to the query, and so we assigned each event one of the following three relevance scores:

- R3 are clear positive instances of the category in question;
- R2 are positive instances where the entity resolution (e.g., `dbr:British_Raj` vs. `dbr:India`) or the subject–object relationships (e.g., 'ISIS attacked X' vs. 'X attacked ISIS') in the event summary are incorrect;

- R1 are events with a 'fuzzy' relationship to the category (e.g., 'ISIS kills X' or 'policeman killed' for the 'Unrest in a country' category); and
- R0 are events with no relevance to the event category.

Results. Table 1 shows the total number of detected events per category and the numbers of positive instances for different relevance scores. As one can see, precision varies considerably across categories. Visits and deaths of politicians could be reliably detected: our NLP component seems very effective on the relevant tweets, and type filtering seems very effective at identifying the appropriate entities. In contrast, detecting cyber attacks is difficult: our query searches for 'company hacked', but the verb 'to hack' often means 'to cut' or 'to manage' so the query retrieved many irrelevant tweets (e.g., about a blogger being stabbed).

A particular problem for ArmaTweet was to correctly differentiate the subject from the object of an action: the approach to passive voice detection we described in Sect. 4 was effective, but should be further improved. Moreover, precision often suffered due acronyms; for example, 'APIs' (i.e., 'Application Programming Interfaces') was resolved to 'Associated Press'. Finally, popular entities posed a particular problem. For example, ISIS appears in a great number of tweets, which increases the likelihood of incorrect event recognition; in contrast, Boko Haram is not that well known and thus seems to be mainly mentioned in tweets reporting terrorist activity. We plan to further investigate ways to 'normalise' the tweet time series based on the 'popularity' of the entities involved.

8 Conclusion

We have presented ArmaTweet—a system developed by armasuisse and the Universities of Fribourg and Oxford for semantic event detection on Twitter. The system represents the tweets' contents in an RDF knowledge graph, thus allowing users to precisely describe the events of interest. The results of our evaluation show that ArmaTweet can detect events such as 'politician dying' and 'militia terror act', which cannot be detected by conventional keyword-based methods. We see two main challenges for future work. First, to help users describe complex events, we will develop adequate user interfaces, as well as investigate ways to extract semantic queries from example tweets. Second, we plan to improve the precision of the NLP component, particularly focusing on the correction of passive voice and the quality of entity resolution in the presence of acronyms.

References

1. Angeli, G., Premkumar, M., Manning, C.: Leveraging linguistic structure for open domain information extraction. In: ACL, pp. 344–354 (2015)
2. Becker, H., Chen, F., Iter, D., Naaman, M., Gravano, L.: Automatic identification and presentation of twitter content for planned events. In: ICWSM, pp. 655–656 (2011)

3. Becker, H., Naaman, M., Gravano, L.: Beyond trending topics: real-world event identification on Twitter. In: ICWSM, pp. 438–441 (2011)
4. Benson, E., Haghighi, A., Barzilay, R.: Event discovery in social media feeds. In: HLTCon 2011, pp. 389–398 (2011)
5. Bontcheva, K., Derczynski, L., Funk, A., Greenwood, M.A., Maynard, D., Aswani, N.: TwitIE: an open-source information extraction pipeline for microblog text. In: RANLP, pp. 83–90 (2013)
6. Daiber, J., Jakob, M., Hokamp, C., Mendes, P.N.: Improving efficiency and accuracy in multilingual entity extraction. In: I-SEMANTICS, pp. 121–124 (2013)
7. Farzindar, A., Khreich, W.: A survey of techniques for event detection in Twitter. Comput. Intell. **31**(1), 132–164 (2015)
8. Gu, H., Xie, X., Lv, Q., Ruan, Y., Shang, L.: ETree: effective and efficient event modeling for real-time online social media networks. In: WI, pp. 300–307 (2011)
9. Kong, L., Schneider, N., Swayamdipta, S., Bhatia, A., Dyer, C., Smith, N.: A dependency parser for tweets. In: EMNLP, pp. 1001–1012 (2014)
10. Lee, R., Sumiya, K.: Measuring geographical regularities of crowd behaviors for Twitter-based geo-social event detection. In: LBSN, pp. 1–10 (2010)
11. Lehmann, J., Isele, R., Jakob, M., Jentzsch, A., Kontokostas, D., Mendes, P., Hellmann, S., Morsey, M., Van Kleef, P., Auer, S., Bizer, C.: DBpedia–a large-scale, multilingual knowledge base extracted from Wikipedia. Semant. Web **6**(2), 167–195 (2015)
12. Li, R., Lei, K.H., Khadiwala, R., Chang, K.C.C.: TEDAS: a Twitter-based event detection and analysis system. In: ICDE, pp. 1273–1276 (2012)
13. Long, R., Wang, H., Chen, Y., Jin, O., Yu, Y.: Towards effective event detection, tracking and summarization on microblog data. In: Wang, H., Li, S., Oyama, S., Hu, X., Qian, T. (eds.) WAIM 2011. LNCS, vol. 6897, pp. 652–663. Springer, Heidelberg (2011). doi:10.1007/978-3-642-23535-1_55
14. Manning, C., Surdeanu, M., Bauer, J., Finkel, J., Bethard, S., McClosky, D.: The stanford CoreNLP natural language processing toolkit. In: ACL, pp. 55–60 (2014)
15. Massoudi, K., Tsagkias, M., Rijke, M., Weerkamp, W.: Incorporating query expansion and quality indicators in searching microblog posts. In: Clough, P., Foley, C., Gurrin, C., Jones, G.J.F., Kraaij, W., Lee, H., Mudoch, V. (eds.) ECIR 2011. LNCS, vol. 6611, pp. 362–367. Springer, Heidelberg (2011). doi:10.1007/978-3-642-20161-5_36
16. Miller, G.A.: WordNet: a lexical database for english. Commun. ACM **38**(11), 39–41 (1995)
17. Owoputi, O., O'Connor, B., Dyer, C., Gimpel, K., Schneider, N., Smith, N.: Improved part-of-speech tagging for online conversational text with word clusters. In: HLTCon, pp. 380–390 (2013)
18. Petrović, S., Osborne, M., Lavrenko, V.: Streaming first story detection with application to Twitter. In: HLTCon, pp. 181–189 (2010)
19. Phuvipadawat, S., Murata, T.: Breaking news detection and tracking in Twitter. In: WI-IAT, pp. 120–123 (2010)
20. Popescu, A.M., Pennacchiotti, M.: Detecting controversial events from Twitter. In: CIKM, pp. 1873–1876 (2010)
21. Ramakrishnan, N., Butler, P., Muthiah, S., Self, N., Khandpur, R., Saraf, P., Mares, D., et al.: 'Beating the news' with EMBERS: forecasting civil unrest using open source indicators. In: KDD, pp. 1799–1808 (2014)
22. Ritter, A., Clark, S., Mausam, Etzioni, O.: Named entity recognition in tweets: an experimental study. In: EMNLP, pp. 1524–1534 (2011)

23. Ritter, A., Mausam, Etzioni, O., Clark, S.: Open domain event extraction from Twitter. In: KDD, pp. 1104–1112 (2012)
24. Rospocher, M., van Erp, M., Vossen, P., Fokkens, A., Aldabe, I., Rigau, G., Soroa, A., Ploeger, T., Bogaard, T.: Building event-centric knowledge graphs from news. J. Web Semant. **37**, 132–151 (2016)
25. Sakaki, T., Okazaki, M., Matsuo, Y.: Earthquake shakes twitter users: real-time event detection by social sensors. In: WWW, pp. 851–860 (2010)
26. Sankaranarayanan, J., Samet, H., Teitler, B.E., Lieberman, M., Sperling, J.: TwitterStand: news in Tweets. In: ACM-GIS, pp. 42–51 (2009)
27. Signorini, A., Segre, A., Polgreen, P.: The use of Twitter to track levels of disease activity and public concern in the US during the influenza A H1N1 pandemic. PLoS ONE **6**(5), 1–10 (2011)
28. Vallis, O., Hochenbaum, J., Kejariwal, A.: A novel technique for long-term anomaly detection in the cloud. In: HotCloud, pp. 1–6 (2014)
29. Weng, J., Lee, B.S.: Event detection in Twitter. In: ICWSM, pp. 401–408 (2011)
30. Zhao, L., Chen, F., Dai, J., Hua, T., Lu, C.T., Ramakrishnan, N.: Unsupervised spatial event detection in targeted domains with applications to civil unrest modeling. PLoS ONE **9**(10), 1–12 (2014)
31. Zhou, X., Chen, L.: Event detection over twitter social media streams. VLDB J. **23**(3), 381–400 (2014)

smartAPI: Towards a More Intelligent Network of Web APIs

Amrapali Zaveri[1,2]([✉]), Shima Dastgheib[1], Chunlei Wu[3], Trish Whetzel[4], Ruben Verborgh[5], Paul Avillach[6], Gabor Korodi[6], Raymond Terryn[7], Kathleen Jagodnik[8,9,10], Pedro Assis[11], and Michel Dumontier[1,2]

[1] Stanford Center for Biomedical Informatics Research,
Stanford University, Stanford, USA
[2] Institute of Data Science, Maastricht University, Maastricht, The Netherlands
amrapali.zaveri@maastrichtuniversity.nl
[3] The Scripps Research Institute, San Diego, CA, USA
[4] T2 Labs, Sunnyvale, CA, USA
[5] Imec – IDLab, Ghent University, Gent, Belgium
[6] Harvard Medical School, Boston, MA, USA
[7] University of Miami, Miller School of Medicine, Miami, FL, USA
[8] Icahn School of Medicine at Mount Sinai, New York, NY, USA
[9] NASA Glenn Research Center, Cleveland, OH, USA
[10] Baylor College of Medicine, Houston, TX, USA
[11] Department of Genetics, Stanford University, Stanford, USA

Abstract. Data science increasingly employs cloud-based Web application programming interfaces (APIs). However, automatically discovering and connecting suitable APIs for a given application is difficult due to the lack of explicit knowledge about the structure and datatypes of Web API inputs and outputs. To address this challenge, we conducted a survey to identify the metadata elements that are crucial to the description of Web APIs and subsequently developed the smartAPI metadata specification and associated tools to capture their domain-related and structural characteristics using the FAIR (Findable, Accessible, Interoperable, Reusable) principles. This paper presents the results of the survey, provides an overview of the smartAPI specification and a reference implementation, and discusses use cases of smartAPI. We show that annotating APIs with smartAPI metadata is straightforward through an extension of the existing Swagger editor. By facilitating the creation of such metadata, we increase the automated interoperability of Web APIs. This work is done as part of the NIH Commons Big Data to Knowledge (BD2K) API Interoperability Working Group.

Keywords: Web API · Web API description · Web services · Linked data · FAIR principles

1 API Interoperability

Workflows for data analysis are increasingly employing cloud-based, Web-friendly application programming interfaces (APIs). Thousands of Web tools

© Springer International Publishing AG 2017
E. Blomqvist et al. (Eds.): ESWC 2017, Part II, LNCS 10250, pp. 154–169, 2017.
DOI: 10.1007/978-3-319-58451-5_11

and APIs are available through Web API registries such as ProgrammableWeb[1], BioCatalogue[2] (specifically for life science data), and cloud platforms such as Galaxy[3]. However, sifting through these and other API repositories to define a linkable toolset pertinent to the workflow is challenging. Discovering relevant APIs often requires a precise combination of matching keywords; and once discovered, the API outputs must be examined to determine whether or not they can be connected together. This task is made more difficult since, in general, there is a lack of rich metadata that precisely describe the APIs, their services, and the data on which they operate. Improvements to this task have been achieved slowly since authoring rich metadata is seen as tedious and unrewarding. Providing easy methods for Web API annotation that integrate with shared terminologies could ease this perception and foster a more discoverable environment for API repositories. In turn, users could more precisely find linkable services that meet their functional requirements as the number of APIs grows.

The problem of authoring coherent, comprehensive, and structured API metadata is gaining attention as a pressing matter due to the aforementioned demand and a lack of work in this area that has fully addressed the issues described. The API metadata problem requires an end-to-end solution – from the specification of metadata elements, to developing metadata templates, to filling out such templates using ontology-based terms, to offering developer-friendly solutions to augment API results. All of this needs to occur in a manner that facilitates discovery, exploration, and reuse. While this problem is admittedly large and complex, our objective here is to carve out specific elements and pilot a lightweight software system for the annotation, discovery, and reuse of what we call smart Web APIs. Our approach is innovative because we address first the problem of API metadata authoring, a task generally disliked in the field, by making it easier to generate useful metadata, and by demonstrating concrete benefits of semantic metadata to developers and users alike.

The overall aim of this project is to undertake a pilot effort that investigates the use of semantic technologies such as ontologies and Linked Data for the annotation, discovery, and reuse of APIs. Linked Data[4] involves the creation of typed links between data from different sources on the Web. It has the properties of being machine-readable, having a meaning that is explicitly defined, being linked to other datasets external to itself, and being able to be linked to from external datasets [4]. The Linked Data principles define the use of Web technologies to establish data-level links among diverse data sources. Linked Data is very useful in cases where exchanges of heterogeneous data are required between distributed systems [2]. Web services can similarly benefit from these principles to facilitate integration and composition [20]. The smartAPI specification employs Linked Data with the aim to connect diverse data sources in pursuit of improved API discovery, interoperability, and reuse.

Our main objective is to develop and evaluate a lightweight software system for the discovery and reuse of smart Web APIs. Smart Web APIs have the advantages that they (i) are easier to discover due to rich semantic annotations, (ii) can be readily connected together without additional data wrangling, and (iii) eliminate data silos by providing Linked Data. Our proposed system will consist of two key components: (a) a coordinated facility for the intelligent annotation of smart Web APIs; and (b) an application to discover smart APIs and how they connect to each other. Essentially, smartAPI helps make APIs FAIR [26]: Findable with the API metadata and the registry; Accessible with the detailed API operations metadata; Interoperable with the responseDataType metadata (profiler); and Reusable with the access to existing APIs stored in an open repository. This work is done as part of the NIH Commons Big Data to Knowledge (BD2K) API Interoperability Working Group (WG)[5] and is available at http://smart-api.info/.

This project has four main contributions:

- Development of the smartAPI metadata specification, based on the results of survey of API metadata guidelines and metadata-in-use in API repositories (Sect. 4).
- Development of an intelligent tool that supports the composition and validation of API metadata that conforms to the smartAPI specification (Sect. 5).
- Development of a profiler that automatically annotates the API response data with semantic identifiers (Sect. 5).
- Development of a repository and smartAPI-conformant API to submit, search, and browse API descriptions (Sect. 5) and obtain field-specific metadata suggestions.

We list the different use cases and projects, specifically in the biomedical domain, that are actively participating in the API Interoperability WG and in the process of annotating (or plan to annotate) their APIs using the smartAPI specification (Sect. 6). We then conclude with a discussion on the future direction of this work in Sect. 7.

2   Related Work

Currently, there exist several challenges in finding relevant APIs as well as reusing those APIs. We discuss both of these challenges in this section. Also, when discussing the annotation and description of Web APIs, we need to distinguish two main groups that interact with these APIs [24]. First, there are annotations targeted at *developers*, with the main aim of facilitating development. Second, there are efforts to describe Web APIs in such a way that *automated clients* can access and compose them. In this section, we will provide a brief overview of both kinds of annotations.

---

[5] https://bd2kccc.org/index.php/working-groups/?v=commons&h=front.

2.1   Challenges of Finding APIs

Finding relevant APIs is a challenging task for developers for diverse reasons. Extensive collections of useful and representative code and data are still lacking [10] despite the quick proliferation of APIs that makes the discovery of resources relevant to individual developers and users difficult [22]. The most visible and accessible APIs are often those that are currently most used, relegating newer and potentially more useful, but less popular, APIs to obscurity [22]. Application frameworks and software libraries often lack proper documentation [9,21], and more sophisticated algorithms need to be developed to facilitate the identification of useful resources [10]. The discovery of relevant APIs can be facilitated by enhancing rich metadata that describe APIs and the services and data associated with them. The smartAPI initiative contributes toward improved discoverability by providing methods that permit simple and intuitive annotation of Web APIs and that are integrated with standard ontologies.

2.2   Challenges of Reusing APIs

Reuse in the context of Web APIs can mean multiple things [24]. First, an API itself is a means to enable reuse of the functionality offered by a certain server. Second, the client-side code for interacting with an API can be (partially) reused across applications. Third, the interface of an API – independent of its implementation – can be (partially) reused by other servers, as is the case with standardized APIs. This third form of reuse is unfortunately rare, since many Web APIs are designed from scratch. The resulting heterogeneity leads to a steep learning curve for the integration of existing Web APIs in applications [10,24], which is the fourth and most common meaning of "Web API reuse". The smartAPI initiative aims to tackle this challenge by developing a profiler that features automatic annotation of API response data. This profiler is integrated with the smartAPI editor to facilitate the semantic annotation of APIs. These features enhance reusability as well as interoperability.

2.3   Annotations for Developers

The XML-based Web Service Description Language (WSDL) provided one of the first models to describe Web services [5,6]. However, WSDL only provides the mechanisms to characterize the technical implementation of Web services; it does not provide the means to capture the functionality of a service. Furthermore, the module source code is generated automatically using a WSDL description, which is then compiled into a larger program. Then, if the description changes, the program no longer works, even if such a change leaves the functionality intact. This prevents WSDL from being used for automatic service discovery at runtime. Furthermore, WSDL is limited by proprietary vendor-specific implementations, being bound to a specific programming language. Swagger[6], on the other hand,

---

[6] http://swagger.io/.

provides an editor for authoring HTTP API documents, and is widely used by API developers[7]. Swagger uses the OpenAPI specification[8], which defines a standard, language-agnostic interface to HTTP APIs. However, each API developer annotates his API in isolation, which results in less interoperable and reusable APIs. The current Web API landscape is hindered by the problem of scalability as every API requires its own hardcoded clients, which only benefits the developers. In particular on the current Web, there is a one-to-many relationship between Web APIs and clients: a single API often has clients for one or more programming languages, but none of these clients work with other APIs. As such, individual clients do not scale with the number of APIs. This makes each API unusually short-lived with a tightly coupled relationship of highly subjective quality. This directly leads to increase in development costs and prevents the design of a more intelligent generation of clients that provide cross-API compatibility [24]. Annotating APIs is an important step in making them accessible for more generic clients.

2.4   Descriptions for Automated Clients

Many approaches for service description exist with different underlying service models. OWL-S [18] and WSMO [16] are the most well-known Semantic Web Service description paradigms. They both allow the description of high-level semantics of services whose message format is WSDL [7]. Though extension to other message formats is possible, this is rarely seen in practice. Semantic Annotations for WSDL (SAWSDL [14]) aim to provide a more lightweight approach for bringing semantics to WSDL services. Composition of Semantic Web Services has been well documented, but all approaches focus on Remote Procedure Call (RPC) interactions and require specific software [19].

In recent years, several description formats for the more lightweight Web APIs have emerged [25]. Several methods aim to enhance existing technologies to deliver annotations of Web APIs. HTML for RESTful Services (hRESTS, [12]) is a microformats extension to annotate HTML descriptions of Web APIs in a machine-processable way. SA-REST [8] provides an extension of hRESTS that describes other facets such as data formats and programming language bindings. MicroWSMO [13,17], an extension to SAWSDL that enables the annotation of RESTful services, supports the discovery, composition, and invocation of Web APIs, but requires additional software.

The description of *hypermedia APIs* is a relatively new field. Hydra [15] is a vocabulary to support API descriptions, but does not directly support automated composition. RESTdesc [23] is a description format for hypermedia APIs that describes them in terms of resources and links. The Resource Linking Language (ReLL, [1]) features media types, resource types, and link types as priorities for description.

With our smartAPI specification, we build upon the already existing widely used OpenAPI specification to provide richer metadata that precisely describes

---

[7] 10M+ downloads according to http://swagger.io/, last accessed Dec 14, 2016.
[8] https://github.com/OAI/OpenAPI-Specification.

the APIs, their services, the data on which they operate, and the data they return. Our smartAPI editor, which is also an extension of the popular Swagger editor, makes it easier to generate useful metadata and indicates which terms are most widely used to annotate Web APIs. The editor also supports suggestion of metadata elements and values along with their usage frequency to the next API provider while she is annotating her API. Furthermore, the *smartAPI profiler* (c.f. Sect. 5), integrated within the editor, provides automatic annotation of the API response data. Finally, the smartAPI registry serves as a repository to save, search, and browse the created API descriptions. Consequently, the smartAPI framework helps to make APIs FÁIR.

3   Survey of API Metadata in the Wild

We conducted a survey of existing metadata repositories and specifications that describe APIs. In particular, the following eight resources were surveyed:

- Repositories:
  - Biocatalogue [3][9], a registry of biological Web APIs with 1,184 entries.
  - Programmable Web[10], a directory of internet-based APIs with over 15,000 API descriptions.
  - Tools & Data Services Registry [11][11], a registry with information about analytical tools and data APIs for bioinformatics with 2, 331 entries.
- Specifications:
  - OpenAPI Initiative[12], created by a consortium of forward-looking industry experts who recognize the immense value of standardizing how HTTP APIs are described.
  - Minimal Information About a Software (MIAS)[13], a key set of minimal fields can that provide maximum value when describing a software.
  - Prototype smartAPI Specification[14], a specification describing semantically annotated Web APIs that facilitates discovery and reuse of Web-based APIs.
  - Semantic Automated Discovery and Integration (SADI)[27][15], a set of design patterns defining the behavior of data retrieval and/or analysis resources that must interoperate on the Semantic Web.
  - schema.org API Reference[16], reference documentations for APIs as described by schema.org.

---

[9] https://www.biocatalogue.org, Accessed April 9, 2016.
[10] http://www.programmableweb.com/apis/directory, Accessed April 10, 2016.
[11] https://bio.tools/, Accessed April 9, 2016.
[12] https://www.openapis.org/, Accessed April 11, 2016.
[13] http://www.softwarediscoveryindex.org/, Accessed April 11, 2016.
[14] http://smart-api.info/website/docs/specification/, Accessed April 11, 2016.
[15] https://rawgit.com/wilkinsonlab/SADI-Specification/master/SADI-W3C-Member-Submission.html#service-metadata, Accessed April 12, 2016.
[16] https://schema.org/APIReference, Accessed April 12, 2016.

We retrieved and listed the metadata elements from each of the resources and also analyzed the degree to which each field was actually employed in practice by its frequency of usage. For instance, in the case of Programmable Web, which contains over 15,000 API descriptions[17], all of the entries use the Title and Description fields. However, only 90% of them supply details about the API provider and the primary category to which the API belongs. Results of the survey are available at https://goo.gl/F4OLnW. Thereafter, we aggregated all the metadata elements from the full set of eight resources to produce a common list of 54 API metadata elements (as discussed in Sect. 4).

4   SmartAPI Metadata Specification

This standard is the result of a survey conducted by the NIH Commons Big Data to Knowledge (BD2K) API Interoperability Working Group of existing metadata repositories and specifications that describe APIs. The smartAPI specification implements the FAIR principles: Findable, Accessible, Interoperable, and Reusable. In particular, we aggregated all the metadata elements from the eight surveyed resources to produce a common list of 54 API metadata elements. We subsequently divided these elements into five categories:

- API Metadata (Table 1[18]): 20 elements
- Service Provider Metadata (Table 2): 6 elements
- API Operation Metadata (Table 3): 12 elements
- Operation Parameter Metadata (Table 4): 10 elements
- Operation Response Metadata (Table 5): 6 elements

The smartAPI Specification includes 21 metadata elements beyond those included in the OpenAPI specification. Examples of the 21 elements are the category to which the API belongs; metadata format and access mode at the API metadata level; the parameter type and parameter value type at the operation parameter level; and the conformance to a specified response profile at the operation response level. The metadata elements marked with a * in the tables are those specific to the smartAPI specification.

Next, we re-evaluated each of the metadata fields according to its applicability and relevance, and further determined whether each MUST, SHOULD, or MAY be included in the API description. The cardinality and datatype of metadata fields were further specified along with a description and example [19]. The smartAPI Specification along with cardinality, datatype, and an example of each metadata element is available at https://websmartapi.github.io/smartapi_specification/.

---

[17] last accessed April 2016.

[18] Note: The tables only contain the elements which are MUST and SHOULD. All other elements can be found on the website https://websmartapi.github.io/smartapi_specification/.

[19] The keywords "MUST", "SHOULD", and "MAY" in this document are to be interpreted as described in RFC 2119 http://www.ietf.org/rfc/rfc2119.txt.

Table 1. smartAPI specification metadata elements: API metadata.

Element	Description	Level	#	Type	Example
Name	A human-readable label for the API	MUST	1..1	URI	MyGene.info API
Access point	The base URI for interacting with the API	MUST	1..1	URI	http://mygene.info/
Description	A human-readable description of the API functionality	SHOULD	0..1	string	MyGene.info Gene Query Web APIs. Learn more at http://mygene.info/.
Response MIME-type	A list of media types the APIs can produce. Can be overridden on specific API calls	SHOULD	0..n	string	application/json
Documentation	Documentation page URL for the API	SHOULD	0..n	URI	http://docs.mygene.info/en/
Version	The version of the API	SHOULD	1..1	string	3.0.0
Terms of service	A document that describes the terms of use for the API	SHOULD	0..1	string	http://mygene.info/terms/
Support*	Indication of whether SSL Support is present or absent	SHOULD	0..1	bool	yes
Authentication mode	Lists the required security schemes to execute this operation	SHOULD	0..1	URI	none

Table 2. smartAPI specification metadata elements: Service Provider Metadata.

Element	Description	Level	#	Type	Example
Responsible organization	The identifying name of the contact person-/organization	MUST	1..1	string	The Scripps Research Institute
Responsible developer	Name of the developer (User ID/Name)	MUST	1..1	URI	http://orcid.org/0000-0002-2629-6124
Contact email	An e-mail address where the provider of the service may be contacted	MUST	1..1	email	cwu@scripps.edu

Table 3. smartAPI specification metadata elements: API Operation Metadata.

Element	Description	Level	#	Type	Example
Operation title*	Title of the operation. A unique identifier of the operation	MUST	1..1	string	q
Operation description	Description of the operation	MUST	1..1	string	Query string. Examples: "CDK2", "NM_052827", "204639_at".
Consumes	A list of MIME types the operation can consume	SHOULD	0..n	string	application/json
HTTP method	The base path on which the API is served, which is relative to the host	SHOULD	0...1	string	GET
Authentication mode	Lists the required security schemes to execute this operation	SHOULD	0..n	string	none
Transfer protocol	The transfer protocol of the API	SHOULD	0..1	string	http

Table 4. smartAPI specification metadata elements: API Parameter Metadata.

Element	Description	Level	#	Type	Example
Operation parameter name	Name of the operation parameter	MUST	1..1	string	q
Operation parameter description	Description of operation parameter	MUST	1..1	string	multiple query terms separated by comma (also supports "+" or white space), but no wildcards, e.g., "q=1017,1018" or "q=CDK2+BTK"
Location	Determines the location of the parameter	MUST	1..1	string	query
Parameter type*	Type of parameter	SHOULD	0..n	string	inputParameter
Parameter value type*	Type of the value of the particular parameter	SHOULD	0..n	string	enterzgene

Table 5. smartAPI specification metadata elements: Operation Response Metadata.

Element	Description	Level	#	Type	Example
Response format	A list of MIME types the APIs can consume	MUST	1..n	string	`application/json`
Example response value	An example of the response value	SHOULD	0..n	integer	200

5 SmartAPI Implementation

smartAPI serves both the API providers and API users. The framework consists of three modules: the editor that facilitates the API metadata authoring for API providers; the searchable API registry where the created API documents are stored and indexed; and the profiler that annotates the API response data.

The smartAPI editor is an extension of the Swagger editor[20], which is widely used by API providers. The Swagger editor uses the OpenAPI specification and provides a framework for creating interactive HTTP API documentation. First, we extended the OpenAPI specification JSON file to incorporate the newly added smartAPI metadata. We extended the auto-completion functionality of the Swagger editor, by suggesting not only the list of predefined metadata and values, but also the values retrieved from the indexed API documents previously created and saved in the registry[21], along with the frequency of their usage. Every new API document added to the registry is indexed using Elasticsearch query engine[22], and their metadata elements and values along with their usage frequency are suggested to the next API provider (Fig. 1b). The conformance level (Required, Recommended, or Optional) of the suggested metadata element is also provided (Fig. 1a).

The smartAPI profiler, shown in Fig. 2a, provides automatic annotation of the API response data, i.e. responseDataType (Fig. 2b). To do this, the API response data (e.g. http://mygene.info/v3/gene/1017) is recursively traversed to provide a keypath/value pair where the keypath consists of one or more labels concatenated together and the value is either a single value or list of strings. The resource annotation is provided by comparing the keypath labels to resource names and synonyms from Identifiers.org[23]. In cases where a match is not found, an example value for the keypath is then compared against resource identifier patterns from Identifiers.org and resulting matches are displayed as suggested annotations. The user may also add his own resource annotation if one does not exist. The annotated API response data is stored in the responseDataType element (Fig. 2b).

---

[20] http://swagger.io/swagger-editor/.
[21] http://smart-api.info/registry/.
[22] https://www.elastic.co/products/elasticsearch.
[23] http://identifiers.org/.

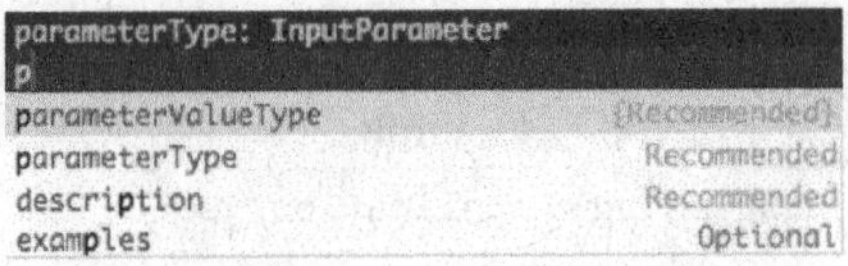

(a) Metadata elements are suggested from the specification, based on the cursor position, along with their conformance level (Required, Recommended, Optional).

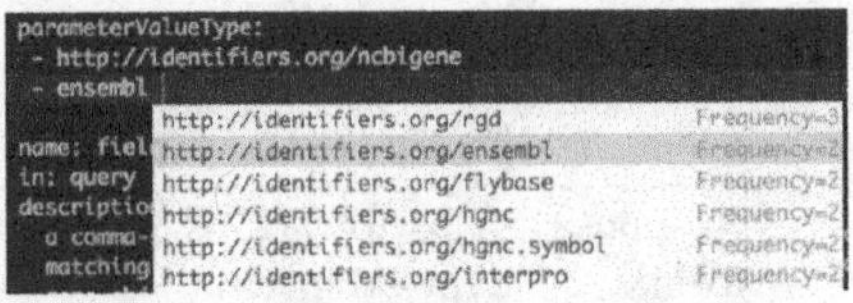

(b) Values for "parameterValueType" are suggested from both identifiers.org and previous APIs saved and indexed in the registry, and sorted by their usage frequency.

Fig. 1. Auto-suggestion functionality for API metadata elements and values.

(a) The API response is annotated using the profiler

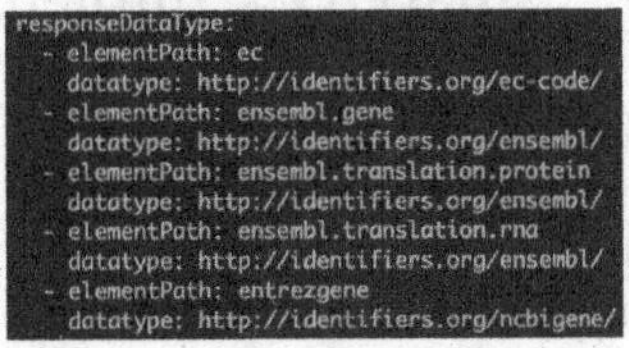

(b) The annotated response data returned by the profiler is inserted to the editor.

Fig. 2. Semantic annotation of the API response (e.g. http://mygene.info/v3/gene/1017) using the smartAPI profiler.

The "parameterValueType" and "responseDataType" elements are added to the specification to semantically annotate the input (parameter) and the output (response) of the API respectively. As shown in Fig. 1b and Fig. 2b, the values of these metadata elements are semantic identifiers from identifiers.org, prefixcommons[24], and other relevant ontologies.

The code, full documentation, and tutorial are available at https://github.com/WebsmartAPI/swagger-editor. A live demo is also available [25].

6 SmartAPI Use Cases

One of the main use cases in which we will examine the usefulness and usability of the smartAPI system is to find and explore connections pertaining to cardiovascular pharmacogenomics. Our use case begins with a set of genes that are differentially expressed in hypertrophic cardiomyopathy (HCM), a leading cause of death among young athletes. HCM arises from genetic defects in close to 20 different genes, although the most common forms of HCM result from mutations in genes encoding proteins of the cardiac sarcomeric apparatus. One concern is that young athletes may be increasing their risk of HCM through

---

[24] http://prefixcommons.org/.
[25] https://www.youtube.com/watch?v=EQpUEiOu1ng&t=3s.

pharmacogenomic interactions. Our objective is to use the smartAPI platform to i) discover which, if any, differentially expressed genes in HCM are targeted by FDA-approved drugs, and ii) identify which HCM genes are also differentially expressed in other published cardiovascular studies. Information about drug targets and pharmacogenomics is already available as Linked Data, through the open source Bio2RDF project[26]. Bio2RDF provides nearly 11 billion Linked Data points from 35 life science databases including DrugBank[27] (a source of drug targets) and PharmGKB[28] (a source of pharmacogenomic interactions). Users of the smartAPI system can gain access to Bio2RDF data by following the Linked Data generated by the MyGene.info and MyVariant.info smartAPIs to Identifiers.org, which in turn will provide links to these Bio2RDF data.

The API Interoperability group is a Working Group[29] in the NIH Commons Framework project. The *NIH Commons* is defined as "an initiative which is essentially a shared virtual space where scientists can work with the digital objects of biomedical research, i.e., it is a system that will allow investigators to find, manage, share, use and reuse data, software, metadata and workflows."[30]. A series of Commons *pilots* has been initiated to develop and test these components in order to understand and evaluate how well they will contribute to an ecosystem that will effectively support and facilitate sharing and reuse of digital objects.

Below, we list the projects that are actively participating in the WG and are in the process of annotating (or plan to annotate) their APIs, specifically in the biomedical domain, using the smartAPI specification:

- **MyGene.info** [28][31] provides Web APIs for both gene queries and gene annotation retrieval. MyGene.info services are being used in Web applications that require querying genes, e.g. BioGPS[32], as well as in an analysis pipeline to retrieve regularly updated gene annotations. MyGene.info has a Swagger-based API document that was loaded into the smartAPI Swagger editor for being validated against the smartAPI specification and saved into the smartAPI registry[33]. The validation process provided a list of missing required, recommended, and optional metadata elements. As a result, "contact" info was added as a required element and the "parameterType", "parameterValueType", and "responseDataType" were recommended. These additions semantically enrich the API document and increase its interoperability with other relevant APIs.
- **MyVariant.info** [28][34] provides simple-to-use Web APIs to query/retrieve variant annotation data, aggregated from many popular data resources.

---

[26] bio2rdf.org.

[27] http://download.bio2rdf.org/release/3/drugbank/drugbank.html.

[28] http://download.bio2rdf.org/release/3/pharmgkb/pharmgkb.html.

[29] https://bd2kccc.org/index.php/working-groups/?v=commons&h=front.

[30] https://datascience.nih.gov/commons.

[31] http://mygene.info/.

[32] http://biogps.org.

[33] http://smart-api.info/registry/.

[34] http://myvariant.info/.

MyVariant.info was modified and saved into the smartAPI registry through the same process as MyGene.info.

- The **National Institutes of Health Library of Integrated Network Cellular Signatures (NIH LINCS) Data Portal**[35] provides access to a diverse array of novel bioassay data that has been curated and packaged with rich metadata for the assay entities. These metadata conform to the NIH LINCS metadata standards[36] that enable integration and interpretation of LINCS data. The LINCS Data Portal API[37] provides programmatic access to all datasets, dataset entities, and metadata within the LINCS Data Portal.

- The **BD2K PIC-SURE HTTP API** facilitates platform-agnostic programmatic access to disparate patient-level heterogeneous datasets to authenticated users. The API provides a selection of methods to access, query, and interrogate data in diverse formats[38]. To test the PIC-SURE API, a demo with the National Health and Nutritional Examination Survey (NHANES) is available online[39]. NHANES is a publicly available epidemiological survey conducted by the US CDC, recording over 1, 100 variables from more than 41, 000 respondents across the US; it is essentially a snapshot of patients' exposomes and phenomes. The exposome is composed of collections of environmental, behavioral, and dietary factors that are associated with health and disease, and phenomes include clinical and physiological phenotypes that are predictive of health.

- The **Alliance of Genome Resources (AGR)**[40] is an initiative formed in 2016 that has the goals of providing better support for the biological sciences via an integration of shared data; standardization of data models and interfaces; and unified outreach to researchers, educators, and the public. The initial members of AGR are the Gene Ontology Consortium[41] and six model organism databases: Saccharomyces Genome Database[42], WormBase[43], FlyBase[44], Zebrafish Model Organism Database[45], Mouse Genome Database[46] and Rat Genome Database[47]. This integration will provide the best visualizations and tools currently in use and allow efficient development of new tools in a collaborative manner. As the project moves toward deeper integration of content and software, we will provide easy-to-use cross-organism queries of the extensive data available in the component resources. The data

---

[35] http://lincsportal.ccs.miami.edu/dcic-portal/.
[36] http://www.lincsproject.org/data/data-standards/.
[37] http://lincsportal.ccs.miami.edu/apis/.
[38] http://bd2k-picsure.hms.harvard.edu.
[39] http://bd2k-picsure.hms.harvard.edu/example-01.html.
[40] http://www.alliancegenome.org.
[41] http://www.geneontology.org/.
[42] http://www.yeastgenome.org/.
[43] http://www.wormbase.org/.
[44] http://www.flybase.org/.
[45] http://www.zfin.org/.
[46] http://www.informatics.jax.org/.
[47] http://rgd.mcw.edu/.

access will be available via an API that will be conformant to the smartAPI specification.

The API Interoperability project is still an ongoing project, and there are a number of BD2K centers that have been actively participating in the WG meetings and have expressed interest in adopting and implementing the smartAPI specification and editor to annotate their APIs. Once we have annotated the APIs using the smartAPI editor, we will store them in the smartAPI registry[48], which will not only provide all of the smartAPI-conformant APIs in one location but will also be integrated into the editor. With this integration, the data and values will be used to suggest related fields and values for new similar APIs during the annotation process (refer to Sect. 5).

7   Conclusions and Future Work

In this paper, we have defined a smartAPI metadata template that contains 54 API metadata elements used to describe an API. Results are reported for a survey of eight resources that were used to identify these API-associated metadata. We constructed the smartAPI metadata template for the validation of API annotations. Additionally, we built a Web application for the intelligent annotation of smartAPIs. Since authoring metadata can be tedious and overwhelming, we developed a software built upon the already existing Swagger editor that will help users describe their APIs by (i) indicating highly used fields, (ii) suggesting commonly used values, and (iii) enabling the discovery and reuse of terms authored by others. Moreover, we developed a profiler for automatic annotation of API response data and integrated that within our editor to enable semantic annotation of APIs, which increases their reusability and interoperability.

Our proposal to facilitate the authoring of rich API metadata is especially significant because of the increased emphasis on providing cloud-based APIs. If left unmanaged, a majority of the APIs will lack the proper metadata needed to find APIs. As sketched out by the participants of the Software Discovery Index Workshop[49], our work begins to explore their roadmap to address challenges facing specifically the biomedical research community in locating, citing, and reusing biomedical software. We believe that the semantic tools and technologies developed in this project will form an important cornerstone in the overall vision of the Commons. As future work, we will assess the ease and utility of authoring smart API metadata for biomedical APIs as well as APIs in other domains. Although we have developed our own API repository (http://smart-api.info/registry/), we expect to be able to export to other repositories that generally have fewer metadata requirements, e.g. ProgrammableWeb. Additionally, our main aim for future work will be to focus on use cases that illustrate our aim of making the APIs interoperable.

---

[48] http://smart-api.info/registry/.
[49] http://www.softwarediscoveryindex.org/.

Acknowledgments. The smartAPI pilot project was funded as a supplement to CEDAR (U41HG000131521). Mygene.info and MyVariant.info are supported by U01HG008473 (from NHGRI). The LINCS Data Portal is supported by grant U54HL127624 awarded by the National Heart, Lung, and Blood Institute through funds provided by the trans-NIH LINCS Program http://www.lincsproject.org/ and the trans-NIH Big Data to Knowledge (BD2K) initiative http://www.bd2k.nih.gov.

References

1. Alarcón, R., Wilde, E.: RESTler: Crawling RESTful services. In: Proceedings of the 19th International Conference on World Wide Web, pp. 1051–1052. ACM (2010)
2. Auer, S., Heath, T., Bizer, C., Berners-Lee, T.: LDOW 2016: 9th workshop on linked data on the web. In: Proceedings of the 25th International Conference Companion on World Wide Web, WWW 2016 Companion, International World Wide Web Conferences, pp. 1039–1040 (2016)
3. Bhagat, J., Tanoh, F., Nzuobontane, E., Laurent, T., Orlowski, J., Roos, M., Wolstencroft, K., Aleksejevs, S., Stevens, R., Pettifer, S., Lopez, R., Goble, C.A.: BioCatalogue: A universal catalogue of web services for the life sciences. Nucleic Acids Res. 38, W689–W694 (2010)
4. Bizer, C., Heath, T., Berners-Lee, T.: Linked data-the story so far. In: Semantic Services, Interoperability and Web Applications: Emerging Concepts (2009)
5. Chinnici, R., Moreau, J., Ryman, A., Weerawarana, S.: Web Services Description Language (WSDL) Version 2.0. Part 1: Core Language. W3C Recommendation (2007). http://xml.coverpages.org/wsdl20000929.html
6. Christensen, E., Curbera, F., Meredith, G., Weerawarana, S.: Web Services Description Language (WSDL) 1.0. W3C Recommendation (2000). http://xml.coverpages.org/wsdl20000929.html
7. Christensen, E., Curbera, F., Meredith, G., Weerawarana, S.: Web Services Description Language (WSDL). W3C Note, March 2001. http://www.w3.org/TR/wsdl
8. Gomadam, K., Ranabahu, A., Sheth, A.: SA-REST: Semantic Annotation of Web Resources. W3C Member Submission, April 2010. http://www.w3.org/Submission/SA-REST/
9. Hsu, S.K., Lin, S.J.: MACs: Mining API code snippets for code reuse. Expert Syst. Appl. 38(6), 7291–7301 (2011)
10. Ishag, M.I.M., Park, H.W., Li, D., Ryu, K.H.: Highlighting current issues in API usage mining to enhance software reusability. In: Proceedings of the 15th International Conference on Software Engineering, Parallel and Distributed Systems (SEPADS 2016). Recent Advances in Computer Engineering Series (2016)
11. Ison, J., Rapacki, K., Ménager, H., et al.: Tools and data services registry: A community effort to document bioinformatics resources. Nucleic Acids Res. 44(1), D38–D47 (2015)
12. Kopecký, J., Gomadam, K., Vitvar, T.: hRESTS: An HTML microformat for describing RESTful Web services. In: Proceedings of the International Conference on Web Intelligence and Intelligent Agent Technology, pp. 619–625. IEEE Computer Society (2008)
13. Kopecký, J., Vitvar, T.: MicroWSMO. WSMO Working Draft, February 2008. http://www.wsmo.org/TR/d38/v0.1/
14. Kopecký, J., Vitvar, T., Bournez, C., Farrell, J.: Semantic annotations for WSDL and XML schema. IEEE Internet Comput. 11, 60–67 (2007)

15. Lanthaler, M., Gütl, C.: Hydra: A vocabulary for hypermedia-driven Web APIs. In: Proceedings of the 6th Workshop on Linked Data on the Web, May 2013. http://ceur-ws.org/Vol-996/papers/ldow2013-paper-03.pdf
16. Lausen, H., Polleres, A., Roman, D.: Web Service Modeling Ontology (WSMO). W3C Member Submission, June 2005. http://www.w3.org/Submission/WSMO/
17. Maleshkova, M., Kopecký, J., Pedrinaci, C.: Adapting SAWSDL for semantic annotations of RESTful services. In: Meersman, R., Herrero, P., Dillon, T. (eds.) OTM 2009. LNCS, vol. 5872, pp. 917–926. Springer, Heidelberg (2009). doi:10.1007/978-3-642-05290-3_110
18. Martin, D., Burstein, M., Hobbs, J., Lassila, O.: OWL-S: Semantic Markup for Web Services. W3C Member Submission, November 2004. http://www.w3.org/Submission/OWL-S/
19. Milanovic, N., Malek, M.: Current solutions for web service composition. IEEE Internet Comput. 8(6), 51–59 (2004)
20. Pedrinaci, C., Domingue, J.: Toward the next wave of services Linked services for the web of data. J-JUCS 16, 1694–1719 (2010)
21. Scaffidi, C.: Why are APIs difficult to learn and use? Crossroads 12(4), 4–4 (2006)
22. Torres, R., Tapia, B., Astudillo, H.: Improving web API discovery by leveraging social information. In: 2011 IEEE International Conference on Web Services, pp. 744–745 (2011)
23. Verborgh, R., Arndt, D., Van Hoecke, S., De Roo, J., Mels, G., Steiner, T., Gabarró Vallés, J.: The pragmatic proof: Hypermedia API composition and execution. Theory Pract. Logic Program. (2016). http://arxiv.org/pdf/1512.07780v1.pdf
24. Verborgh, R., Dumontier, M.: A web API ecosystem through feature-based reuse (2016). CoRR abs/1609.07108, http://arxiv.org/abs/1609.07108
25. Verborgh, R., Harth, A., Maleshkova, M., Stadtmüller, S., Steiner, T., Taheriyan, M., Van de Walle, R.: Survey of semantic description of REST APIs. In: Pautasso, C., Wilde, E., Alarcón, R. (eds.) REST: Advanced Research Topics and Practical Applications, pp. 69–89. Springer, New York (2014). http://link.springer.com/chapter/10.1007/978-1-4614-9299-3_5
26. Wilkinson, M.D., Dumontier, M., Aalbersberg, I.J., et al.: The FAIR Guiding Principles for scientific data management and stewardship. Sci. Data 2 (2016). http://www.nature.com/articles/sdata201618
27. Wilkinson, M.D., Vandervalk, B., McCarthy, L.: The semantic automated discovery and integration (SADI) web service design-pattern, API and reference implementation. J. Biomed. Semant. 2(1), 5–23 (2011)
28. Xin, J., Mark, A., Afrasiabi, C., Tsueng, G., Juchler, M., Gopal, N., Stupp, G.S., Putman, T.E., Ainscough, B.J., Griffith, O.L., Torkamani, A., Whetzel, P.L., Mungall, C.J., Mooney, S.D., Su, A.I., Wu, C.: High-performance web services for querying gene and variant annotation. Genome Biol. 17(1), 91 (2016)

PhD Symposium

Automating the Dynamic Interactions of Self-governed Components in Distributed Architectures

Sebastian R. Bader[✉]

Institute AIFB, Karlsruhe Institute of Technology (KIT), 76133 Karlsruhe, Germany
sebastian.bader@kit.edu

Abstract. The ongoing digitalization and penetration of the Web into each aspect of software development creates new possibilities and challenges. The flexible reuse of components promises to drastically reduce the implementation and maintenance effort. But growing complexity in terms of variety and dynamic changes bring monolithic approaches to their limits. In this paper, an approach is presented which enables components in distributed systems to observe, judge and independently react to dynamic changes in their neighborhood. Reducing the overall complexity to smaller and easier to manage subproblems leads to more flexible and reliable systems. The target is a delegation of decision making to the single components.

Keywords: Distributed systems · Component coordination · Dynamic web

1 Introduction

Initially, the Web started as a Web of documents. Through constant evolution we now see a more and more automatically processable Web of data and services. Semantic technologies allow specifications of characteristics and descriptions both interpretable for humans and machines. They take part to drive the Web from a static information provider to a decentralized interaction platform where data and functionalities are offered, accessed and consumed. The continuously growing number of publicly available Web APIs[1] emphasizes this development.

In addition, the current Web is already a dynamic environment in itself. Resources existing at one point in time can not be taken for granted, as providers stop hosting services, rebrand and relocate APIs. For example, even on the commercial IBM Bluemix platform, Web APIs can disappear without any transparent reason or machine processable information[2]. Contrary, APIs can still be available even if the provider marked them as inactive (e.g. the former Google Web Search API).

---

[1] http://www.programmableweb.com/api-research.

[2] https://concept-insights-demo.mybluemix.net/ Accessed 05.12.2016.

© Springer International Publishing AG 2017
E. Blomqvist et al. (Eds.): ESWC 2017, Part II, LNCS 10250, pp. 173–183, 2017.
DOI: 10.1007/978-3-319-58451-5_12

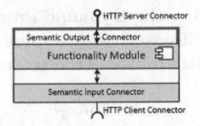

Fig. 1. Self-governed component: Functionality is wrapped by semantic connectors. The input connector independently establishes connections to other components.

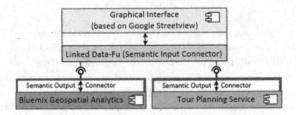

Fig. 2. Prototypical implementation of self-governed components, so far with only one connector (either input or output)

A methodology is outlined to enable distributed Web components to diagnose unforeseen changes and providing adaptive reaction capabilities. Components, as regarded in this work, are software modules which encapsulate a single functionality and are annotated with semantic descriptions. They can be initially designed Linked Data APIs but also occur in the form of translation instances for lifting and lowering the original Web API to Linked Data. The components communicate via RESTful Web APIs and have the ability to select and invoke other Web components autonomously, therefore they are denoted as *self-governed components* in the following.

A self-governed component (Fig. 1) consists of two semantic connectors responsible for its semantic descriptions, the data lifting and lowering, and provision and request to and from, but not limited to, other self-governed components. The offered core functionality is therefore independent of the communication methods and may directly be integrated by code, occur as a remote Web service or a database.

Parts of the concept of self-governed components were implemented for the use case of dispatching field technicians for industrial maintenance. As shown in Fig. 2, a IBM Bluemix analytic service for geospatial data[3] and a commercial tour planning heuristic have been enabled to RESTful communication with RDF through their respective *output connectors*[4]. The *input connector*, realized by a Linked Data-Fu [9] instance, consumes the Web APIs and fills a Google Street View based visualization component.

---

[3] https://console.ng.bluemix.net/catalog/services/geospatial-analytics.
[4] https://github.com/sebbader/BlueWrapper.

For now, the integration instructions are static rules and incrementally executed. In contrast, self-governed components need to be equipped with context-aware reaction capabilities. That includes mechanisms for detecting possible issues, recognizing and evaluating alternative partner components and establishing communication channels.

2   State of the Art

Components are regarded by Morrison [16] as "black boxes" with a strong focus on reusability. He states that in productive systems, the application does not require insights into the functionality of the used components but only on the delivered and consumed information packets. Microservices as described by Thönes [23] further limit the amount of provided functionalities to singular, easy to understand tasks. This drastically reduces the complexity of the necessary descriptions and eases the reuse in unforeseen scenarios.

The Semantic Web Stack [2] constitutes a set of technologies to handle both syntactical and semantical interoperability issues. It thereby defines the architecture of the Semantic Web with central technologies like URIs to identify resources, RDF to encode data, ontologies define meaning and the semantic query language SPARQL.

Semantic descriptions of Web components can be formulated in various languages and ontologies. Currently most important are the Web Service Ontology Language (WSMO) [20], OWL-S [14] and Linked USDL [19]. RESTdesc [26] utilizes a N3 Syntax to specify input and output parameters and how they are connected. Similarly, Dimou et al. [8] combine access information with data mappings in the RDF Mapping language (RML). Verborgh et al. [25] provide a survey on machine-interpretable Web API descriptions.

The types of descriptions can be organized in the categories behavioral, functional and non-functional. The technical details to operate the component are part of the behavioral sections whereas functional statements include basic information on the component's purpose. Non-functional information contain additional details on e.g. prices, provenance or provided Quality of Service.

Also, central registries for Web APIs like RapidAPI[5] or ProgrammableWeb[6] mostly do not provide semantic information, making an automated discovery a hard task. In the approach of Sande et al. [24] for Linked Data sets the data server recognizes other components by dereferencing its existing RDF data or utilizing the Referer Header of incoming HTTP requests and therefore gains knowledge about other data sources.

In order to automatically combine the components, existing approaches [3,21] mostly use centralistic optimization during the design phase. Contrary, Web components are not static elements but can and do change over time. In general, they follow a life cycle as shown by Wittern and Fischer [27]. Mayer et al. [15] extend RESTdesc to cope with a dynamic environment by introducing states.

---

[5] https://www.rapidapi.com/.
[6] https://www.programmableweb.com/category/all/apis.

Similarly, Alaya et al. suggest oneM2M, a IoT approach to gain machine-to-machine (M2M) interoperability with a semantic reasoner [1]. But still, a central organizer with knowledge about the whole network is required.

One way to enable a more flexible way to determine component compositions are policies. The non-functional characteristics of available components are regarded with semantic reasoners [22] in order to match and rank (Palmonari et al. [17]) them against predefined requirements. La Torre et al. [13] propose the dynamic context of the consuming client as a selection criteria for components. Context here is regarded as social media information of e.g. Facebook but also physical data like GPS coordinates. Although only human users have been regarded, it should be possible to transfer the approach to automated components, having context like the location of the hosted server or the company running it.

3   Problem Statement and Contributions

As outlined, self-governed components can perceive spontaneous changes and adjustments. In particular, changes of the produced data model and the technical interaction patterns in addition to modifications of functionality and the component's location are possible without former notice. Together with the rising number of heterogeneous components, the complexity of coordinating the individual elements of Web applications increases further. No central coordinator can guarantee sufficient performance when no specified size limits can be set and potential changes of components can happen at any time.

Although switching from central to distributed architectures can reach supplies the necessary scalability, it does not solve the problem of sudden changes of components. While in the first case a central coordinator was responsible, the adjustments now have to accomplished by the self-governed components themselves. Therefore, this paper focuses to answer the following research question:

Research Question: *How can self-governed components in distributed architectures cope with dynamic and unforeseen changes of other self-governed components which they depend on?*

3.1   Validating and Updating Semantic Descriptions of Web Components

The Web is a dynamic environment where resources and services are not static and may appear, disappear, and change their characteristics spontaneously. If an API or the functionality of an used self-governed component changes, an affected consuming component will only recognize it when its procedures begin to fail.

Even though the provider may announce the modifications upfront, it is usually not done via machine interpretable channels. Self-governed components in critical applications therefore need analytical mechanisms to self-detect such incidents. In the example, the GeospatialAnalytics and the tour planning component provide data in the WGS84 format. An update e.g. could set the tour planning component back to its default settings, where country, address and street name specify a location instead of WGS84 coordinates.

Bhargava and Lingayat [4] try to tackle the topic with local and global monitoring components to discover validations of service level agreements but do not regard any (semantic) descriptions. SHACL [12] on the other hand allows the comparison of incoming Linked Data against expected patterns but can hardly cope with non-functional aspects. Consequently, the following question has to be answered:

RQ 1. *How can the changes in functional, behavioral and non-functional descriptions of self-governed components be validated and, in the case of mismatches, be modified?*

3.2   Spontaneous Connector for Self-governed Components

Although self-governed components share a common stack of technologies like URIs, semantic interfaces and RDF, they still allow nearly endless variations of implementations. A self-governed component which identified a suitable functional input source is not capable to simply consume the API without further specification. Missing data mappings, unknown interface invocation and other behavioral requirements prohibit a plug-and-play like connection. In the described example, the self-governed UI component has to be able to switch from the commercial tour planning to e.g. an instance of the open-source OpenTrip-Planner[7] (see Fig. 3):

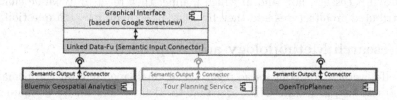

Fig. 3. Exchanging the tour planning component with the similar OpenTripPlanner

For that, Keppmann et al. [11] introduce adjustable "Smart Components" which can change their program code – and thereby also their data sources – during runtime. Nevertheless, these components can not individually customize their connectors to alternative sources. In contrast to that, the system of Bhargava and Lingayat [4] dynamically configures the network but relies on a central coordinator which poses all information about the environment.

This leads to the following research question:

RQ 2. *How can one self-governed component autonomously consume another without formerly specifying connection details and requirements?*

---

[7] http://www.opentripplanner.org/.

3.3   Delegation of Communication Responsibility in Distributed Architectures

In the example, only the self-governed UI component once has to find a replacement for its data provider. No other component is affected as long as the UI component can find a sufficient substitute. This leads to the question how self-governed components know which providers to select in order to gain the required input data.

Cao et al. [6] solve dynamic composition in P2P networks where the participants iteratively create a workflow chain. Although they regard nonfunctional characteristics, the parameters of interest have to be introduced during design time. Additional desired parameter can not be considered. Similarly, Cardellini et al. [7] do not regard the inherent complexity of nonfunctional requirements.

Policies like in [18,22] enable the components to act independently in a surrounding with incomplete information and spontaneously occurring changes. Comprehensible methods have to be developed or adjusted in order to rank available candidates. The consuming self-governed component has to conclude first whether an offering component complies with the defined policies, then select the most appropriate, and establish a connection:

RQ 3. *How can self-governed components in distributed architectures independently derive selection criteria from abstract policies in order to appropriately classify available, alternative self-governed components?*

Summarizing, RQ 1 targets the discovery of evolving problems, RQ 2 examines methods to technically enable reactions and RQ 3 develops approaches to select suitable reactions. In combination, they enable self-governed components to state, if necessary, how and in which manner they react to dynamic changes in distributed architectures and thereby answer the main research question.

4   Research Methodology and Approach

The self-governed components as regarded in this paper rely on the Semantic Web Stack. In particular, URIs are used to identify and locate components, in combination with RDF as the data model, and ontologies to reason about delivered data and semantic descriptions. In addition, this paper only regards self-governed components with RESTful APIs, based on HTTP communication.

The research approach is directly determined by the dependencies between the proposed modules.

The treatment of changing API descriptions (Sect. 4.1) relies on the spontaneous establishment of communication channels (Sect. 4.2) and vice versa. Both parts together will allow to solve the problem of decentralizing coordination responsibility (Sect. 4.3) in order to answer the main research question.

4.1   Validating and Deriving Descriptions of Self-governed Components

The required methods regarding the semantic description of APIs are divided into two tasks. In the first part, component descriptions are regarded as fixed facts. They are compared against the observed data and communication pattern. The aim is to detect inconsistencies caused by e.g. applied upgrades or API changes of the observed self-governed components. On-the-fly reasoning of transferred RDF data allows the recognition of conflicting statements of the received data with their functional and non-functional descriptions. Violations in terms of behavioral aspects usually lead to transaction errors which have to be interpreted separately. Non-functional aspects need a more advanced handling. A mapping of behavior variants to an ontology will be developed to gain inputs for the semantic reasoner.

In addition to methods for validating assumptions, the automated proposal of semantic descriptions is regarded. Independently recognizing component's characteristics is essential to increase the amount of useful and machine interpretable descriptions in the Web as it lowers the effort in both time and necessary skills to deploy a self-governed component. Therefore, benefits for both the consumer (having detailed control mechanisms) and the provider (reducing the manual effort) can be accomplished.

4.2   Spontaneous Connector for Self-governed Components

Consuming components can interact with suppliers on the basis of their semantic descriptions. Data can be easily transmitted in the case of matching demanded and available resources. Nevertheless, in some cases meaningful interactions can still be accomplished although specifications and demands do not fit perfectly. Therefore, the approach will relax the retrieved descriptions of self-governed components. It is to verify whether neglecting parts of stated constraints improves the automated connection of components. Even though the intentional violation will produce mismatches, it has the potential to solve situations where overly restrictive descriptions prevent interactions.

The Linked Data streaming engine Linked Data-Fu as the input connector will serve as the mediator between two components. It is capable to request, process and forward RDF data. Linked Data-Fu will be extended towards an autonomously deciding connector. The challenge is to derive the interaction instructions solely relying on the component's (potentially mismatching) descriptive data and predefined ontological knowledge. For that, Linked Data-Fu provides on the fly semantic reasoning and SPARQL query execution which serve as the foundation for further developments.

4.3   Delegation of Communication Responsibility

The previously developed methods will be combined in a framework for delegating coordination to the component. The framework defines how abstract policies have to be formulated to specify the expected behavior but on the other hand contain enough flexibility to find a matching self-governed component. In general, components are not deployed with exactly the required use case in mind, and therefore have at least slightly divergent descriptions. Consequently, the stated policies need to allow a certain degree of freedom.

The framework also specifies how these user defined policies can be operationalized regarding a faced situation. The derived rules serve to filter and rank the existing alternatives regarding functional and non-functional characteristics and thereby allow decentralized decision making. In order to gain a consistent behavior, the policies together with the derived rules are also transferred to the involved components. Therefore, each self-governed component is capable to configure its neighborhood independently but according to specified manners.

5   Preliminary Results

Currently, the work is in the initial experiment stage. Together with research and industry partners from the STEP project[8], a starting set of self-governed components from various domains is established. This will be the foundation of the a planed Web Component Network (see Sect. 6).

The first conceptual ideas have already been presented at SEMANTiCS 2016[9]. Regarding the domain of industrial maintenance scheduling, the ongoing digitalization increases the requirements for maintenance providers. The heterogeneity of interfaces, changing needs for functionalities and new business models reveal the inadequacy of existing monolithic systems. It was outlined how semantically enriched self-governed components can provide flexible integration into distributed architectures. The next steps are the creation of a testing and development environment. The Web itself is not suitable as conditions can neither be repeated nor sufficiently controlled. On the other hand the Web is the targeted habitat for the investigated self-governed components. Building on the existing Web Service Challenges [5] a sufficiently large set of various self-governed components will be established.

The descriptions of the created components will include incorrect and lacking annotations, syntactic and semantic errors and changes over time. The advantage of a controlled environment is the ability to control the mutations and thereby compare different strategies. The dynamic aspects, as (dis-)appearance and sudden modifications of components, will be implemented by both predefined, repeatable sequences and randomly triggered changes. Similarly to Joshi et al. [10] the system is configurable via seed parameters and creates dynamic but repeatable scenarios.

6   Evaluation Plan

This simulated Web environment will work in the same way and behave following the same dynamic principles as the real Web but at smaller and therefore better treatable scale. The target is a Gold Standard which serves as a testing and evaluating environment for the developed methods but will also be part of the

---

[8] https://www.projekt-step.de/.

[9] http://www.slideshare.net/semanticsconference/sebastian-bader-semantic-technolo gies-for-assisted-decisionmaking-in-industrial-maintenance.

contributions to the research community. The reproducible conditions of the network will allow other researchers to compare their approaches with the results of this research and to further improve the state of the art.

In order to judge the quality for recognition and adjusting of API descriptions of self-governed components (RQ 1) the performance will be measured with this environment.

Combining formerly unconnected Web Components (RQ 2) is a problem also faced in frameworks for service composition. In the described problem the connection effort shall be accomplished independently and self-organized by the component facing a problem. Nevertheless, the performance of tools like Medley [28] can be seen as baseline approaches.

In addition to the proposed evaluation environment, the developed concepts and implementations regarding the delegation of the interaction channels will also be implemented through the industry project STEP[10]. Existing company policies will be provided and automatically translated into technical instructions by the self-governed components. The resulting behavior is then compared to the preferred choices of responsible managers.

7   Conclusion

The variety and amount of available components is a significant advantage of applications running in the Web. A well known set of standard protocols and principles enables the fast reuse of software functionalities. But existing Web based services show that the heterogeneity of their behavior and unforeseeable changes in the implementations require constant manual adjustments. Therefore, a stronger utilization is prohibited.

Self-governed components are one method to enhance the degree of automatization in distributed architectures. The proposed methods target the local decision making to further empower the single components. The ability to independently react on changes minimizes required maintenance of distributed Web applications and decreases the barriers of component reuse. This results in faster deployments, decreasing maintenance efforts and reduced complexity. The additionally provided evaluation environment makes the proposed approaches comparable and opens the paths to continuous improvements.

References

1. Alaya, M.B., Medjiah, S., Monteil, T., Drira, K.: Toward semantic interoperability in oneM2M architecture. IEEE Commun. Mag. **53**(12), 35–41 (2015)
2. Berners-Lee, T., Weitzner, D.J., Hall, W., O'Hara, K., Shadbolt, N., Hendler, J.A.: A framework for web science. Found. Trends Web Sci. **1**(1), 1–130 (2006)
3. Beygelzimer, A., Riabov, A., Sow, D., Turaga, D.S., Udrea, O.: Big data exploration via automated orchestration of analytic workflows. In: ICAC 2013, pp. 153–158 (2013)

---

[10] https://www.projekt-step.de/en/.

4. Bhargava, B., Angin, P., Ranchal, R., Lingayat, S.: A distributed monitoring and reconfiguration approach for adaptive network computing, pp. 31–35. IEEE (2015)
5. Bleul, S., Weise, T., Geihs, K.: The web service challenge-a review on semantic web service composition. Electron. Commun. EASST 17 (2009)
6. Cao, X., Kapahnke, P., Klusch, M.: SPSC: Efficient composition of semantic services in unstructured P2P networks. In: Gandon, F., Sabou, M., Sack, H., d'Amato, C., Cudré-Mauroux, P., Zimmermann, A. (eds.) ESWC 2015. LNCS, vol. 9088, pp. 455–470. Springer, Cham (2015). doi:10.1007/978-3-319-18818-8_28
7. Cardellini, V., D'Angelo, M., Grassi, V., Marzolla, M., Mirandola, R.: A decentralized approach to network-aware service composition. In: Dustdar, S., Leymann, F., Villari, M. (eds.) ESOCC 2015. LNCS, vol. 9306, pp. 34–48. Springer, Cham (2015). doi:10.1007/978-3-319-24072-5_3. http://www.ce.uniroma2.it/publications/esocc2015_xweb.pdf
8. Dimou, A., Verborgh, R., Sande, M.V., Mannens, E., Van de Walle, R.: Machine-interpretable dataset and service descriptions for heterogeneous data access and retrieval, pp. 145–152. ACM Press (2015)
9. Harth, A., Knoblock, C.A., Stadtmller, S., Studer, R., Szekely, P.: On-the-fly integration of static and dynamic linked data. In: COLD 2013, pp. 1–12 (2013)
10. Joshi, A.K., Hitzler, P., Dong, G.: LinkGen: Multipurpose linked data generator. In: Groth, P., Simperl, E., Gray, A., Sabou, M., Krötzsch, M., Lecue, F., Flöck, F., Gil, Y. (eds.) ISWC 2016. LNCS, vol. 9982, pp. 113–121. Springer, Cham (2016). doi:10.1007/978-3-319-46547-0_12
11. Keppmann, F.L., Maleshkova, M., Harth, A.: Semantic technologies for realising decentralised applications for the web of things. In: ICECCS, pp. 71–80 (2016)
12. Knublauch, H., Kontokostas, D.: Shapes Constraint Language (SHACL) (2017). http://www.w3.org/TR/shacl/
13. La Torre, G., Monteleone, S., Cavallo, M., D'Amico, V., Catania, V.: A context-aware solution to improve web service discovery and user-service interaction, pp. 180–187. IEEE (2016)
14. Martin, D., Burstein, M., Hobbs, J., Lassila, O., McDermott, D., McIlraith, S., Narayanan, S., Paolucci, M., Parsia, B., Payne, T.: OWL-S: Semantic markup for web services. W3C Member Submission 22 (2004). 2007–04
15. Mayer, S., Verborgh, R., Kovatsch, M., Mattern, F.: Smart configuration of smart environments. IEEE Trans. Autom. Sci. Eng. **13**(3), 1247–1255 (2016)
16. Morrison, J.P.: Flow-based Programming. In: Proceedings of the 1st International Workshop on Software Engineering for Parallel and Distributed Systems, pp. 25–29 (1994)
17. Palmonari, M., Comerio, M., Paoli, F.: Effective and flexible NFP-based ranking of web services. In: Baresi, L., Chi, C.-H., Suzuki, J. (eds.) ICSOC/ServiceWave -2009. LNCS, vol. 5900, pp. 546–560. Springer, Heidelberg (2009). doi:10.1007/978-3-642-10383-4_40
18. Paoli, F.D., Palmonari, M., Comerio, M., Maurino, A.: A meta-model for non-functional property descriptions of web services, pp. 393–400. IEEE (2008)
19. Pedrinaci, C., Cardoso, J., Leidig, T.: Linked USDL: A vocabulary for web-scale service trading. In: Presutti, V., d'Amato, C., Gandon, F., d'Aquin, M., Staab, S., Tordai, A. (eds.) ESWC 2014. LNCS, vol. 8465, pp. 68–82. Springer, Cham (2014). doi:10.1007/978-3-319-07443-6_6
20. Roman, D., Lausen, H., Keller, U.: Web Service Modeling Ontology (WSMO) (2006). http://www.wsmo.org/TR/d2/v1.3/
21. Sirin, E., Parsia, B., Wu, D., Hendler, J., Nau, D.: HTN planning for web service composition using SHOP2. Web Semant. **1**(4), 377–396 (2004)

22. Speiser, S.: Semantic annotations for WS-Policy. In: ICWS, pp. 449–456 (2010)
23. Thönes, J.: Microservices. IEEE Softw. **32**(1), 116–116 (2015)
24. Sande, M.V., Verborgh, R., Dimou, A., Colpaert, P., Mannens, E.: Hypermedia-based discovery for source selection using low-cost linked data interfaces. IJSWIS **12**(3), 79–110 (2016)
25. Verborgh, R., Harth, A., Maleshkova, M., Stadtmller, S., Steiner, T., Taheriyan, M., Van de Walle, R.: Survey of semantic description of REST APIs. In: Pautasso, C., Wilde, E., Alarcon, R. (eds.) REST Advanced Research Topics and Practical Applications, pp. 69–89. Springer, New York (2014)
26. Verborgh, R., Steiner, T., Van Deursen, D., De Roo, J., Van de Walle, R., Vallés, J.G.: Description and interaction of restful services for automatic discovery and execution. In: International Workshop on AFMS. FTRA (2011)
27. Wittern, E., Fischer, R.: A life-cycle model for software service engineering. In: Lau, K.-K., Lamersdorf, W., Pimentel, E. (eds.) ESOCC 2013. LNCS, vol. 8135, pp. 164–171. Springer, Heidelberg (2013). doi:10.1007/978-3-642-40651-5_13
28. Yahia, E.B.H., Réveillère, L., Bromberg, Y.-D., Chevalier, R., Cadot, A.: Medley: An event-driven lightweight platform for service composition. In: Bozzon, A., Cudre-Maroux, P., Pautasso, C. (eds.) ICWE 2016. LNCS, vol. 9671, pp. 3–20. Springer, Cham (2016). doi:10.1007/978-3-319-38791-8_1

Building and Processing a Knowledge-Graph for Legal Data

Erwin Filtz[⊠]

Institute for Information Business,
Vienna University of Economics and Business, Vienna, Austria
erwin.filtz@wu.ac.at

Abstract. The increasing size and availability of data opens the door for new application areas. Data which has previously been kept separated can be linked and therefore enhanced with additional data from other sources. The linking of data requires a certain data representation such that it can be used in particular domains. In this paper we describe the problem of data representation and search within data exemplified by the legal domain. We propose an approach to represent the legal data (legal norms and court decisions) of Austria and show how this data can be used to build a legal knowledge graph, usable in various applications for lawyers, attorneys, citizens or journalists.

1 Introduction

In recent years more and more data in various domains has become available publicly and for free. The usefulness of such data varies widely and depends on the structure as well as a maybe existing standardized representation of the data. In the legal domain information is usually provided in legal documents such as laws or court decisions, containing the respective information as well as links to related documents. Laws link to each other (e.g. exceptions defined in another law) and court decisions are linked to laws and previous court decisions which have been taken into account in the particular case. All links are highly eligible for representation as linked data in RDF. Although the information is available, the look-up of legal information may be tricky when being interested in the legal situation of specific circumstances or investigating a particular case as a legal professional. The desired information is spread over various data sources, which follow different access and pricing policies.

Information provided by legal information systems (LIS) is often accessed with simple, keyword-based search interfaces and presented as a simple list of hits based on particular search terms, maybe enhanced with meta information about the document (law, court decision,...) to allow a first evaluation whether a result might be interesting or not. The manual process of information retrieval that is very time consuming and when searching for the wrong or not optimal key words, the results might be overwhelming.

Research performed in the area of legal semantics shows that this is a hot topic and will be approached both from an information systems and legal perspective due to the author's legal background. Previous research is typically

© Springer International Publishing AG 2017
E. Blomqvist et al. (Eds.): ESWC 2017, Part II, LNCS 10250, pp. 184–194, 2017.
DOI: 10.1007/978-3-319-58451-5_13

tailored to a particular subdomain of the law or country [23,33,34]. Specialized laws often also reference other laws and it is therefore appropriate to focus on a jurisdiction, e.g. with a focus on Austria. A legal information system providing related information in a common knowledge graph enhanced with semantics would be beneficial for legal and non-legal professionals for information search and argumentation at court or to understand the evolution of legislation over time.

The remainder of this research proposal is structured as follows: In Sect. 2 we will describe the state of the art and related work. The problem statement and contributions follow in Sect. 3. The methodology and approach to this problem are described in Sect. 4. First results are shown in Sect. 5 and the evaluation plan is outlined in Sect. 6. Section 7 concludes the paper.

2   State of the Art

Applying semantics in the legal domain is not new and was a hot topic around the turn of the millennium until 2008 (e.g. [1,5,7,9,15,18,28,29]). However, with the advances in information technology and the semantic world we think that this is still a very interesting topic and should receive appropriate treatment.

The representation of legal information as natural language text is not optimal, therefore another forms of representations have been proposed, for instance legal ontologies, which can be seen as explicit specifications of conceptualizations [19]. In the area of the creation of legal ontologies Gangemi investigated ontology design patterns that are typical for the legal domains and shows some examples [18]. For exchanging information between legal knowledge systems formats like the legal knowledge interchange format (LKIF) [21] or LegalRuleML [2] have been suggested. Using the Resource Description Framework (RDF) as means to represent legal information has been investigated by Ebenhoch, who describes the challenges of legal resource description points and out that the key approach to enrich legal data is enhancing it with metadata [15]. Saias et al. describe the problem of missing semantics in legal information retrieval systems and propose an ontology to enrich the legal data with semantics based on the Portuguese legal system [28]. Winkels et al. describe the need of semantics in a legal context from the practical point of view of the Dutch Tax and Customs Administration, who have to deal with legal information from various sources and formats and developed a parser to automatically detect the identity of sources and references to other legal documents [32]. RDF is also used to describe a particular subdomain of law by Rodríguez-Doncel et al. to express software licenses [26].

A summary of existing legal ontologies is provided by Breuker et al. [11], listing 23 ontologies and categorizing them by application (general language for expressing legal knowledge, information retrieval,...), type (knowledge representation), roles (understanding a domain, searching,...), character (general vs domain-specific), ontology construction (manual, automatic,...) and language.

Legal data needs not only be stored in an appropriate way, but also being searchable. A very suitable way to represent relations between data is using graphs, which has already been investigated in the semantic world [10]. Furthermore, graphs allow the application of already well-known and researched graph algorithms for search and traversal. Mimouni et al. present a solution to graph query legal documents not only on their intertextual relationships but also taking content descriptors into account [25].

Recommender systems provide the user with related information based on particular metrics, for instance similarity. They can also be used in the legal domain to show which information is related or similar to currently displayed information. Drumond et al. describe the requirements and architectural design for such a system [14]. Zeleznikow et al. apply game theory from the economics domain to the Australian family law in order to provide negotiation support as litigation is usually a zero-sum game [34]. A legal recommender system has also been invented by Winkels et al. for Dutch case law[1] and provides the user with related information for the searched case [33].

The related work clearly shows that there is research done in this field. Because of the fact that legal systems vary from country to country and also the prevalent legal system, there is no common ontology or recommender system available. Research is tailored to the specifics of a particular legal area, system or both. Nonetheless, whereas axiomatization of laws and norms has received considerable attention, case law and a semantic, graph-based representation of court decisions, cases and their links, has not yet been tackled systematically. We believe that a systematic approach to fill this gap could complement the existing aforementioned efforts of enhancing law by semantics and enable new research directions.

3   Problem Statement and Contributions

In Austria, legal information is provided free of charge by the legal information system (RIS)[2], which is operated by the Austrian Federal Chancellery, containing information about legislation, law gazettes and case law limited to decisions by the respective supreme courts (Supreme, Constitutional, Administrative, etc.). In addition to RIS, information about the law, comments on decisions and additional information contained in legal commentaries are provided by some companies by paid subscription. These platforms have in common that they allow searching for keywords, specific laws or court decisions and, depending on the data provider, may also offer some related information. However, the search and assessment process of the result takes a long time and requires a lot of manual browsing and reading until a legal professional is able to come to the final conclusion whether or not this particular search result is of relevance for a specific case.

Although legal professionals usually know the law and the most important decisions, in non-trivial cases the look-up of additional information is essential. The RIS can be accessed by everybody with internet access, it is mainly used by

---

[1] https://www.rechtspraak.nl/.
[2] "Rechtsinformationssystem des Bundes", http://ris.bka.gv.at (14.11.2016).

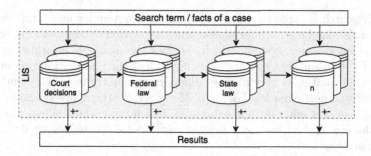

Fig. 1. Interlinked legal documents, laws and decisions, contributing to a legal case.

legal professionals. Hence, time matters as it generates costs for their clients and the information retrieval process should be kept as short as possible without a negative impact on the quality of the outcome. Therefore, a software supported search and assessment system would be beneficial.

Figure 1 shows a generic use case. All information is contained in a LIS and can be queried. The search results are marked as applicable/positive or contradicting/negative for a case. Furthermore, also clustering the results would enhance the explanatory power of the results, shorten the search process and allow focusing on the matters of fact.

To outline the problem with an example: A client of a legal professional had a car accident and is involved in legal proceedings. Searching in the RIS for judgments containing the term "Auto" (car) provides information about several application areas. The results contain court decisions dealing with: (*i*) auto-completion function of search engines, (*ii*) Observation of a suspect with a GPS unit attached to his car, (*iii*) several cases dealing with the assignment of rights and defects liability involving a car, (*iiii*) court judgments having an accident as a fact of the case and many others. This example illustrates the problem on a specific case, of course refining the search terms would be the first approach to get a more fine-grained result. Enhancing the results with semantics and relations to the search terms would be very beneficial for all users of such legal information systems, save a lot of time by classifying the results and facilitating argumentation before a court. This requires contextualised, unambiguous entity recognition and appropriate linkage of similar cases, which is not present in the current LIS.

The legal domain has some special properties, for instance the focus is not just on information retrieval but also on question answering [4], which implies having semantics of the texts available [28]. Dealing with legal documents requires a transformation from natural language text into a more structured format. In the last decade, several efforts were taken to represent legal documents in a more formal way by focusing on XML [7–9,30] and also in RDF [1,15,24]. Moreover, several research projects have been carried out so far, for instance ship certification (CLIME) [31] or tax law related (E-POWER) [6] legal documents.

Our research is targeted towards proposing a system for legal and non-legal professionals searching for information and the relation of the found information in a legal information system, which requires the available information to be represented in a suitable way and allowing to link the information properly.

Legal information is very diverse, differs from country to country and might be specific to a particular legal domain. Moreover, it is typically available in natural language text and not in a structured data format which allows processing out of the box and without any further restrictions. Therefore, we deduct our *main hypothesis*:

Legal information available in systems such as the Austrian legal information system RIS or European legal databases can be structured and enhanced by semantics to support unambiguous and useful interlinking of legal cases in a legal knowledge graph, that helps legal professionals, along with suitable graph traversal and summarization techniques.

Due to the fact that the legal domain is very broad and can be divided into many, very specialized subdomains, we split the hypothesis into two problem areas, which have to be tackled to achieve the ultimate goal of having a legal information system providing legal and non-legal professionals with all required case-related information and turning the information retrieval into a "one stop shop". Therefore, we have discovered two problem areas:

P1 *(Research Problem 1) Representation of legal information.* Legal information is distributed over various sources and typically represented as natural language text. Although all required information about a law or court decision is mentioned in the text, there is no metadata about this specific source of information available. The metadata is expected to be different for different kinds of legal information (for instance laws and court decisions) and will require semantic alignment in case the same metadata fields are used to represent information differently.

In general, a legal system can be classified in several ways e.g. into areas of law (civil, criminal,...), but this classification might not always be satisfying, for instance being to fine- or coarse-grained. Therefore, all related legal information has to be enhanced with additional information. Figure 2 shows how a law is displayed in the RIS, in particular the rights of the legal owner. It contains structured information (bold printed), for instance, "Abkürzung" (abbreviation), the actual law text or "Schlagworte" (keywords). These kind of information can be parsed easily as it is already available in a structured format. Additional information might still be incorporated in the actual law text and not covered by the provided keywords.

Figure 3 shows the search results for the keyword "Auto" (car), which presents all court decisions containing the search term. A legal professional knows the circumstances of the case and for what court ("Gericht") to look and it is also possible to restrict the search to a particular court, but the results still have to be browsed and their usefulness evaluated manually.

Inkrafttretensdatum   **Außerkrafttretensdatum**
01.01.1812

Abkürzung
ABGB

Index
20/01 Allgemeines bürgerliches Gesetzbuch (ABGB)

Text

Rechte des Eigenthümers.

§ 362. Kraft des Rechtes, frey über sein Eigenthum zu verfügen, kann der vollständige Eigen
unbenützt lassen; er kann sie vertilgen, ganz oder zum Theile auf Andere übertragen, oder unbeding!

Schlagworte
Eigentümer, Eigentum, Teil

Fig. 2. Excerpt of the law on rights of legal owners in Austria from RIS

Nr.		Geschäftszahl	Datum	Gericht	Typ	Kurzinformation
1		6Ob26/16s	30.03.2016	OGH	RS	Den Betreiber eine wenn er trotz
2		Bsw35623/05 Bsw649/08 B...	02.09.2010	AUSL EGMR	RS	Im Zusammenhan Vorhersehbarkeit d
3		3Ob1/09g 4Ob195/10w 6Ob...	23.06.2009	OGH	RS	1. Ob eine Interze: ausschließende ec

Fig. 3. Court decisions for search term "Auto" (car) in the Austrian RIS

C1 *Contribution to P1.* The classification of a legal system can be made by
several aspects. Usually, legal systems use different codes of law, which can
serve as a classification basis. However, terms might be used comprehen-
sively in several different codes of law and searching for a keyword contained
in both codes, for instance, Fig. 2 shows an article in the civil code about
"Eigentum" (legal ownership). The same term also appears several times in
the criminal code.

Besides the classification using the respective code a legal norm is contained
in, we can also use the provided keywords by the RIS to assign legal norms
and judicial decisions to appropriate categories. For instance, the example
shown in Fig. 2 can be assigned to the civil code or to the categories mentioned in the keyword section, which then might contain also legal norms
and court decision with the same keyword. Therefore, we will be able to
classify the legal norms and decisions based on the information they provide.

P2 *The search of legal information is difficult.* Graphs are a convenient and
intuitive way to display relations between information. In the legal domain,
the relation and the importance of a relation is case-dependent. The meaning of words and expressions is often ambiguous. Therefore, search engines
for legal information need to be context-aware and consider this fact in the
applied search algorithm. The search for a keyword and all related documents will result in an unmanageable list of results which has to be analyzed
manually. Current LIS and legal databases partly provide the option to sort
the results based on relevance, date or different types of legal documents,

for instance, decisions (with subcategories of different courts), legal norms, commentaries, journals etc. However, the assessment of the result is still up to the user of such an information system.

C2 *Contribution to P2.* The additional benefit of a graph-based semantics-enhanced information system lies in the availability of graph metrics to adjust the search process. Graph metrics describe the structure of it, for instance the degree distribution of a graph for finding the nodes with the highest number of incoming or outgoing links. The betweenness centrality indicates how central a particular node is within the network based on the number of shortest paths going through this node. The closeness centrality relates to the length of the shortest paths and therefore to how close a node is to the other nodes. These and other graph metrics are useful to find important nodes in a graph and to optimize the search algorithm as well as to find the most important related documents as already shown by Hulpus et al. [22]. Based on a proper representation of an information system, this would be beneficial to build summary graphs which enable the user to capture the structure very intuitively and choose which subgraph might seem worth a further inspection.

The number of nodes in a graph is highly varying. Depending on the context, it is possible to classify nodes based on certain properties and merge them in a way such that parent categories can be formed which contain all the information of the merged nodes. A graph size reduction it more comprehensible and in the case of a LIS it does not overwhelm the user with an unmanageable number of results but provides a summary and the user can choose which category seems more promising and investigate it further. As a first starting point, we will use a breadth-first search algorithm to find all related information before building the summary graph. That is done because a BFS algorithm stepwise follows all related information and is expected to provide a more general overview. On the other hand, a depth-first search algorithm can be used when you already know a given source and target node in a graph, which is not the case when building summary graphs. Therefore, a breadth-first search is expected to be more suitable to solve this problem.

4   Research Methodology and Approach

We decided to apply the iterative research methodology Action Research (AR) as described by Checkland and Holwell [12]. This means that we start with a literature review and evaluate the already proposed ontologies in the legal domain. We then will either choose an appropriate one and see which adaptions need to be made to meet our requirements. If no appropriate ontology can be found or the adaption process is not successful we will develop and apply our own ontology. The adapting and/or development process will go hand in hand with continuous checks whether the ontology is still applicable when adding new data sources and continued as long as new and different legal data sources are added which leads to a continued assessment and adaption process. All iterations targeted to a specific task are continued until a satisfying result has been found or we came to the conclusion that this task cannot be solved. The advantages of an iterative research methodology are that the return to a previous step in a sequence of tasks is possible and we are not restricted by previously taken decisions.

In terms of the graph-based search algorithms we will start with a small subset of the legal domain and apply a breadth-first search (BFS) to navigate through the graph. Furthermore, we will calculate several graph metrics like the number of nodes and edges, graph density and centrality scores. The content of nodes with high scores will be compared to the keywords mentioned in the related documents. Subgraphs of DBPedia and a financial transaction network serve as comparison graphs as we already gained knowledge on how a BFS algorithm performs on these graphs.

Furthermore, this thesis is related to two different research projects, which will also serve as a source for ideas of how to approach problems arising as well as for result comparison. The law-related project is called DALICC and deals with the machine-readable representation and comparison of software licenses. The GraphSense project focuses on the processing and analysis of large graphs. Therefore both projects will serve as input combined with the author's mainly legal background.

5   Preliminary Results

We surveyed suitable graph search algorithms such as different graph search algorithms [13,16] as well as summarization methods [22,27], which we already applied in different use cases, for instance finding the k-shortest paths in different datasets [17,20]. We plan to leverage the already gained knowledge in the proposed PhD combined with the author's legal background. Furthermore, our research group has already gained preliminary expertise in advanced search features within the Austrian RIS and their limitations [3] as well as formats and structure of legal data.

6   Evaluation Plan

The results of this research project will be evaluated continuously. The evaluation of the results for the representation of legal data as linked data will be based on existing ontologies and their capability to represent the Austrian legal system. Suitable ontologies will be found by evaluating existing ontologies and their likeliness to applicable to our specific research project. We will start with a small subset of legal documents and try to represent it with the chosen ontology. Throughout the research project we will continuously add different types of legal documents to check whether the new information can still be represented. In terms of the performance and scalability we will have to investigate which sizes of the input and output graphs are feasible for processing in terms of processing time and scalability. Furthermore, the size of a summary graph must be comprehensible such that it can be captured intuitively. We will compare this with the sizes and complexity of legal graphs we can construct from the available case law knowledge extracted from the LIS. The performance of the algorithm will be evaluated by the time it needs to perform the search and memory consumption.

7 Conclusion

Our research addresses the problem of data representation of different domains exemplified by the legal domain. We outlined the problem of a proper representation of domain-specific legal data involving laws, regulations and court decisions, as well as their associated meta-data and graph-based search problems. The motivation, research problems and approach are described in this paper. The future work consists of an analysis of already existing ontologies and evaluation whether they can be taken as a basis for our work.

Acknowledgement. This thesis is supervised by Axel Polleres and funded by the Austrian Research Association (FFG) under the scope of ICT of the Future program (contracts # 849906 and # 855396).

References

1. Amato, F., Mazzeo, A., Penta, A., Picariello, A.: Building RDF ontologies from semi-structured legal documents. In: 2008 International Conference on Complex, Intelligent and Software Intensive Systems (2008)
2. Athan, T., Boley, H., Governatori, G., Palmirani, M., Paschke, A., Wyner, A.: OASIS LegalRuleML. In: Proceedings of the Fourteenth International Conference on Artificial Intelligence and Law, ICAIL 2013, pp. 3–12. ACM, New York (2013). http://doi.acm.org/10.1145/2514601.2514603
3. Bauer, S.: Evaluation of search engines in the context of RIS. Master's thesis, Vienna University of Economics and Business, Vienna, Austria (2016)
4. Benjamins, V.R., Casanovas, P., Breuker, J., Gangemi, A.: Law and the semantic web, an introduction. In: Benjamins, V.R., Casanovas, P., Breuker, J., Gangemi, A. (eds.) Law and the Semantic Web. LNCS (LNAI), vol. 3369, pp. 1–17. Springer, Heidelberg (2005). doi:10.1007/978-3-540-32253-5_1
5. Biagioli, C., Francesconi, E., Passerini, A., Montemagni, S., Soria, C.: Automatic semantics extraction in law documents. In: Proceedings of the 10th International Conference on Artificial Intelligence and Law, ICAIL 2005, pp. 133–140. ACM, New York (2005). http://doi.acm.org/10.1145/1165485.1165506
6. Boer, A., van Engers, T., Winkels, R.: Using ontologies for comparing and harmonizing legislation. In: Proceedings of the 9th International Conference on Artificial Intelligence and Law, pp. 60–69. ACM (2003)
7. Boer, A., Hoekstra, R., Winkels, R., van Engers, T., Willaert, F.: Proposal for a Dutch legal XML standard. In: Traunmüller, R., Lenk, K. (eds.) EGOV 2002. LNCS, vol. 2456, pp. 142–149. Springer, Heidelberg (2002). doi:10.1007/978-3-540-46138-8_22

8. Boer, A., Hoekstra, R., Winkels, R., Van Engers, T., Willaert, F.: Metalex: legislation in XML. In: Legal Knowledge and Information Systems (Jurix 2002), pp. 1–10 (2002)

9. Boer, A., Winkels, R., Vitali, F.: MetaLex XML and the legal knowledge interchange format. In: Casanovas, P., Sartor, G., Casellas, N., Rubino, R. (eds.) Computable Models of the Law. LNCS (LNAI), vol. 4884, pp. 21–41. Springer, Heidelberg (2008). doi:10.1007/978-3-540-85569-9_2

10. Bonstrom, V., Hinze, A., Schweppe, H.: Storing RDF as a graph. In: Proceedings of the IEEE/LEOS 3rd International Conference on Numerical Simulation of Semiconductor Optoelectronic Devices (IEEE Cat. No.03EX726), pp. 27–36, November 2003

11. Breuker, J., Casanovas, P., Klein, M.A.C., Francesconi, E.: The flood, the channels and the dykes: managing legal information in a globalized and digital world. In: Law, Ontologies and the Semantic Web: Channelling the Legal Information Flood, vol. 188, p. 3 (2009)

12. Checkland, P., Holwell, S.: Action research: its nature and validity. Syst. Pract. Action Res. 11(1), 9–21 (1998). http://dx.doi.org/10.1023/A:1022908820784

13. Dijkstra, E.W.: A note on two problems in connexion with graphs. Numer. Math. 1(1), 269–271 (1959). http://dx.doi.org/10.1007/BF01386390

14. Drumond, L., Girardi, R.: A multi-agent legal recommender system. Artif. Intell. Law 16(2), 175–207 (2008). http://dx.doi.org/10.1007/s10506-008-9062-8

15. Ebenhoch, M.P.: Legal knowledge representation using the resource description framework (RDF). In: 12th International Workshop on Database and Expert Systems Applications, pp. 369–373 (2001)

16. Eppstein, D.: Finding the k shortest paths. SIAM J. Comput. 28(2), 652–673 (1999). http://dx.doi.org/10.1137/S0097539795290477

17. Filtz, E., Savenkov, V., Umbrich, J.: On finding the k shortest paths in RDF data. In: Proceedings of the 5th International Workshop on Intelligent Exploration of Semantic Data (IESD 2016) co-located with the 15th International Semantic Web Conference (ISWC 2016), 18 October 2016, Kobe, Japan (2016)

18. Gangemi, A.: Design patterns for legal ontology constructions. In: LOAIT 2007, pp. 65–85 (2007)

19. Gruber, T.R.: A translation approach to portable ontology specifications. Knowl. Acquisition 5(2), 199–220 (1993)

20. Haslhofer, B., Karl, R., Filtz, E.: O bitcoin where art thou? Insight into large-scale transaction graphs. In: Martin, M., Cuquet, M., Folmer, E. (eds.) Joint Proceedings of the Posters and Demos Track of the 12th International Conference on Semantic Systems - SEMANTiCS2016 and the 1st International Workshop on Semantic Change & Evolving Semantics (SuCCESS 2016) co-located with the 12th International Conference on Semantic Systems (SEMANTiCS 2016), 12–15 September 2016, Leipzig, Germany. CEUR Workshop Proceedings, vol. 1695. CEUR-WS.org (2016). http://ceur-ws.org/Vol-1695/paper20.pdf

21. Hoekstra, R., Breuker, J., Di Bello, M., Boer, A., et al.: The LKIF core ontology of basic legal concepts. In: LOAIT, vol. 321, pp. 43–63 (2007)

22. Hulpus, I., Hayes, C., Karnstedt, M., Greene, D.: Unsupervised graph-based topic labelling using DBPedia. In: Leonardi, S., Panconesi, A., Ferragina, P., Gionis, A. (eds.) Sixth ACM International Conference on Web Search and Data Mining, WSDM 2013, 4–8 February 2013, Rome, Italy, pp. 465–474. ACM (2013). http://doi.acm.org/10.1145/2433396.2433454

23. Lenci, A., Montemagni, S., Pirrelli, V., Venturi, G.: Ontology learning from Italian legal texts. In: Proceedings of the 2009 Conference on Law, Ontologies and the Semantic Web: Channelling the Legal Information Flood, pp. 75–94. IOS Press, Amsterdam (2009). http://dl.acm.org/citation.cfm?id=1563987.1563995
24. McClure, J.: The legal-RDF ontology. A generic model for legal documents. In: LOAIT, pp. 25–42 (2007)
25. Mimouni, N., Nazarenko, A., Paul, E., Salotti, S.: Towards graph-based and semantic search in legal information access systems. In: Twenty-Seventh Annual Conference on Legal Knowledge and Information Systems (JURIX 2014), pp. 163–168. Jagiellonian University, Krakow, Poland, December 2014. https://hal.archives-ouvertes.fr/hal-01121968
26. Rodríguez-Doncel, V., Villata, S., Gómez-Pérez, A.: A dataset of RDF licenses. In: Hoekstra, R. (ed.) Legal Knowledge and Information Systems - JURIX 2014: The Twenty-Seventh Annual Conference, Jagiellonian University, Krakow, Poland, 10–12 December 2014. Frontiers in Artificial Intelligence and Applications, vol. 271, pp. 187–188. IOS Press (2014). http://dx.doi.org/10.3233/978-1-61499-468-8-187
27. Ruchansky, N., Bonchi, F., García-Soriano, D., Gullo, F., Kourtellis, N.: The minimum Wiener connector problem. In: Proceedings of the 2015 ACM SIGMOD International Conference on Management of Data, Melbourne, Victoria, Australia, 31 May – 4 June 2015, pp. 1587–1602 (2015). http://doi.acm.org/10.1145/2723372.2749449
28. Saias, J., Quaresma, P.: Semantic enrichment of a web legal information retrieval system. In: JURIX, pp. 11–20 (2002)
29. Valente, A.: Types and roles of legal ontologies. In: Benjamins, V.R., Casanovas, P., Breuker, J., Gangemi, A. (eds.) Law and the Semantic Web. LNCS (LNAI), vol. 3369, pp. 65–76. Springer, Heidelberg (2005). doi:10.1007/978-3-540-32253-5_5
30. Winkels, R., Boer, A., Hoekstra, R., et al.: Metalex: an XML standard for legal documents. In: Proceedings of the XML Europe Conference, London, UK (2003)
31. Winkels, R., Boer, A., Hoekstra, R.: CLIME: lessons learned in legal information serving. In: ECAI 2002: 15th European Conference on Artificial Intelligence, 21–26 July 2002, Lyon France: Including Prestigious Applications of Intelligent Systems (PAIS 2002): Proceedings, vol. 77, p. 230. IOS Press (2002)
32. Winkels, R., Boer, A., de Maat, E., van Engers, T., Breebaart, M., Melger, H.: Constructing a semantic network for legal content. In: Proceedings of the 10th International Conference on Artificial Intelligence and Law, ICAIL 2005, pp. 125–132. ACM, New York (2005). http://doi.acm.org/10.1145/1165485.1165505
33. Winkels, R., Boer, A., Vredebregt, B., van Someren, A.: Towards a legal recommender system. In: JURIX, pp. 169–178 (2014)
34. Zeleznikow, J., Belluci, E.: Family-Winner: integrating game theory and heuristics to provide negotiation support. In: Legal Knowledge and Information Systems: JURIX 2003: The Sixteenth Annual Conference, p. 21. IOS Press (2003)

Ontology Matching Algorithms for Data Model Alignment in Big Data

Ruth Achiaa Frimpong[(✉)]

University of South Australia, Adelaide, SA, Australia
ruth.frimpong@mymail.unisa.edu.au

Abstract. Big Data commonly refers to large data with different formats and sources. The problem of managing heterogeneity among varied information resources is increasing. For instance, how to handle variations in meaning or ambiguity in entity representation still remains a challenge. Ontologies can be used to overcome this heterogeneity. However, information cannot be processed across ontologies unless the correspondences among the elements are known. Ontology matching algorithms (systems) are thus needed to find the correspondences (alignments). Many ontology matching algorithms have been proposed in recent literature, but most of them do not consider data instances. The few that do consider data instances still face the big challenge of ensuring high accuracy when dealing with Big Data. This is because existing ontology matching algorithms only consider the problem of handling voluminous data, but do not incorporate techniques to deal with the problem of managing heterogeneity among varied information (i.e., different data formats and data sources). This research aims to develop robust and comprehensive ontology matching algorithms that can find high-quality correspondences between different ontologies while addressing the variety problem associated with Big Data.

Keywords: Big Data · Ontology matching · Data heterogeneity · Alignment

1 Introduction

Big Data is the "new oil", the substance that is expected to drive the information economy of tomorrow. It is commonly considered to have three main dimensions: Volume, Velocity and Variety. Volume refers to the problem of dealing with large data sets; Velocity refers to the problem of dealing with real-time streaming data where it may not be possible to store all data for later processing and; Variety refers to the need to deal with many different data sources and data formats [9]. Big Data applications and projects are everywhere and companies prepare for the future where they cannot survive without the information gleaned from a variety of data sources. The problem of managing heterogeneity among such varied information resources is increasing. For instance, most database research

© Springer International Publishing AG 2017
E. Blomqvist et al. (Eds.): ESWC 2017, Part II, LNCS 10250, pp. 195–204, 2017.
DOI: 10.1007/978-3-319-58451-5_14

and self-assessment reports recognise that the question of semantic heterogeneity, that is, how to handle variations in meaning or ambiguity in entity interpretation, remains open [1]. Ontologies are often used as a model for knowledge representation and can help in overcoming these heterogeneities [16].

Ontologies play a prominent role for many applications, such as database integration, peer-to-peer systems, e-commerce, semantic web services and social networks [7]. They are a practical means to conceptualise what is expressed in a machine readable format. An ontology usually provides a vocabulary characterising a domain of interest and a specification of the meaning of terms in that vocabulary [5]. In open or evolving systems, such as dynamic Big Data analysis environments, different parties adopt different ontologies that typically need to be merged for analysis to be performed. Information cannot be processed across ontologies if the correspondences or semantic mappings between the elements are unknown. Manually finding such correspondences is time-consuming and prone to error. The success of the Semantic Web and other applications is dependent on the development of algorithms to assist in this process, also called *ontology matching*, which is the main focus of the research.

We seek to develop a robust and comprehensive ontology matching system that can find high-quality correspondences between different ontologies and also resolve the variety problem associated with Big Data.

The remainder of this paper is organised as follows. Section 2 reviews the related literature. The research problems and expected contributions are presented in Sect. 3. Section 4 outlines the research methodology and approach. Sections 5 and 6 discuss the preliminary test and evaluation plan respectively. The paper is finally concluded in Sect. 7.

2   State of the Art

This section discusses the recent developments in ontology matching (OM).

Given two or more ontologies, the OM problem is to find correspondences between them. OM has been extensively studied as an essential technology to achieve interoperability over ontologies [10]. Existing work on OM is mostly focused at the schema level, that is, finding correspondences between two schemas such as database schemas and ontologies. However, in recent literature, the ability to compare different ontologies with the objective of identifying similar instances which refer to the same real-world entity is drawing more research interest. To execute the matching process, OM systems (algorithms or tools) are used. Some of the challenges the OM systems face can be found in [17]. These matching systems are basically developed by selecting a matching strategy and combining two or more of the multitude of matching techniques (see [Chaps. 5–7 of [5]] for details). However, quality mappings cannot be obtained if these techniques are not selected and combined appropriately.

Ontology matching systems such as COMA [3] and Lily [18] find correspondences between ontologies at the schema level and do not consider instance data, hence, they are unable to identify important mappings which may be needed for future analysis. The few systems, such as Falcon-AO [8] (see survey in [14]), that do consider data instances during matching still cannot deal with voluminous and high-variety datasets properly. Existing systems such as GLUE [4] use iterative matching (IM) techniques to speed up the ontology matching process. IM finds the instance correspondences in multiple loops; only a fraction of instances are matched in each iteration, which are then used as sources for matching the remaining instances in the next iterations. Although these techniques are useful when matching large datasets, IM is likely to propagate errors of mismatched instances. To avoid this problem, other systems such as PRIOR+ [12] use the parallel workflow strategy. In this strategy, several matchers are executed independently on the ontologies. The results produced by the individual matchers are combined by some aggregation and extraction methods to obtain the final alignments. Several aggregation and extraction methods have been proposed [5], including: Max/Min method, which returns the maximum or minimum similarity value of individual matchers; Weighted method, which computes a weighted sum of similarity values of individual matchers; and Sigmoid method, which combines the results of the individual matchers using a sigmoid function. The weighting methods require that a threshold be set manually and hence, are unable to adapt to different matching tasks. For example, when the selected matcher changes or their number increases. Employing weighting methods as the only strategy to aggregate individual matchers and extract alignments has been identified in the literature [17] to be ineffective and may result in ambiguous, inconsistent and inaccurate (low-quality) mappings: especially for matching systems that adopt both syntactic and semantic techniques. To solve the above challenges, we propose a novel ontology matching system that uses a dynamic weighted method in addition to user knowledge (reasoning) to find high-quality correspondences in Big Data. The main idea behind the system is to maximize the utilization of available data instances of ontologies.

3   Problem Statement and Contributions

3.1   Problem Statement

The following examples illustrates variety problems associated with Big Data and some other ontology matching problems. The first example shows a problem that occurs when mappings are identified based solely on information in the ontologies, neglecting external background knowledge. The second scenario presents the variety problem associated with Big Data (i.e., different representation of the same attribute). The last example shows the importance of matching at both the schema and instance levels.

The diagrams in Figs. 1 and 2 represent ontologies for two companies A and B. These ontologies consist of concepts, instances, attributes and relations. Green rectangles are concepts, yellow diamonds represent relations between different concepts, attributes are the pale blue hexagons and instances are displayed as pink rounded rectangles.

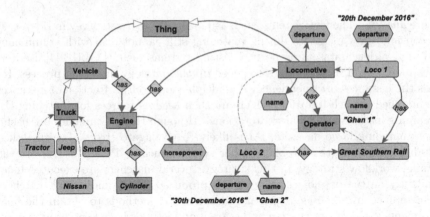

Fig. 1. Ontology A

Assume we would like to find correspondences between Ontology A (Fig. 1) and Ontology B (Fig. 2). Schema level matching would establish a match between Locomotive in ontology A and Train in Ontology B. However, in the real world, a locomotive is an engine of a train hence, matching Locomotive to Train is a false match. In such cases the use of external resources such as WordNet or reference ontologies can help reduce the mismatch.

Secondly, the attribute values associated with the *departure* attributes in ontologies A and B use different representations (i.e., full month name vs. short number form). Using syntactic (e.g. string-based) techniques, the attribute values would not match. However, these values represent dates in the real world hence techniques such as logical and numerical methods need to be combined appropriately with other syntactic techniques to resolve such varieties in data.

Lastly, using semantic and syntactic techniques (such as logical deduction, terminological structure, name-based, external resources), the Truck (Fig. 1) and Lorry (Fig. 2) will result in a match. This is because in the real world, Lorry and Truck are synonyms; Australia and United States of America call a motor vehicle designed to transport cargo as Truck while United Kingdom and Ghana refer to the same entity as Lorry. However, in this case, matching Truck to Lorry would be a mismatch. This is because the instances of Truck are similar to the instances of the Lorry, Autobus and Car hence, Truck would match Lorry, Autobus and Car in this scenario.

Additional problems that current matching systems cannot handle include:

- the inability to incorporate missing type information of instances. This is important since not all ontologies will be at the same level of granularity, which makes it difficult to find appropriate matches.
- the inability to effectively resolve mismatches, inaccuracies and inconsistencies between heterogeneous data.

To solve the identified problems, the following research questions will be addressed:

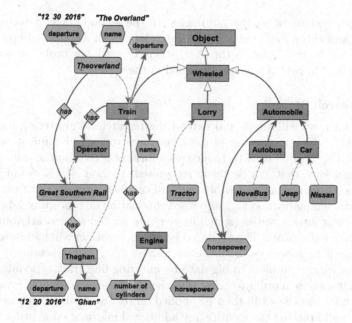

Fig. 2. Ontology B

1. **Big data has varieties and masses of data, how can we use it in ontology matching?**
2. **How can high-quality mappings be extracted without introducing ontological mismatches?**
3. **How can alignment inconsistency, ambiguity and inaccuracy in ontology matching be addressed?**

3.2 Contributions

The expected contributions of the research include:

- a novel instance-based technique that uses mining association rules to compute similarities between instances of different ontologies.
- the development of an effective aggregation and extraction technique which will incorporate user knowledge and a dynamic weighted method to combine the alignments resulting from the individual matchers.
- an effective filtering (alignment improvement) technique to identify the inconsistent and inaccurate mappings when matching Big Data.

4 Research Methodology and Approach

The research is divided into three stages. The first stage will investigate and develop algorithms that will incorporate data instances, identify missing type information, and partition the ontologies. In the second stage, a technique to aggregate and extract alignments will be developed. An alignment improvement technique will be developed in the third stage. Overall, the problems and questions identified in Sect. 3 will be addressed in these stages.

4.1   Research Stage 1

To begin with, we will adopt and extend the hierarchical clustering method of Typifier [11], which makes use of data instances to recover implicit subtypes. Python NLTK[1] will be used to reduce each form of a term in the data to some standardised form that can be easily recognised: Python NLTK would perform activities such as normalisation of date and number formats, term extraction, tokenisation, lemmatisation and stop word elimination on concept or label names to reduce their dissimilarities as well as generate pseudo schema attributes used by the Typifier algorithm. The extended type info. identifier will handle arbitrary formats (such as string vs. actual date format using data instances) to help address the variety problem in big data by inferring fine grain type information from data instances in order to bring the schema to the same level of granularity. The type info. identifier will then be applied on real world data such as OAEI[2], DBpedia[3] and Freebase[4] to identify any additional information in order to avoid class mismatch. Furthermore, we will empirically select an effective clustering algorithm that clusters instances in addition to schemas and adopt it as our ontology partitioner by evaluating their validity and runtime performance. The preliminary results of this evaluation is shown in Sect. 5.

In addition, the parallel workflow strategy will be adopted to design a matcher. The matcher will mainly incorporate existing instance-based, name-based, language-based, graph-based, taxonomy-based and model-based techniques (see [5]). The matcher will consists of the following:

Structural matcher compares the structure of entities in the data
Label matcher compares the similarity between strings (text)
Instance matcher computes the similarity between data instances
Semantic matcher uses external resource and logic to identify schemas and instances that are semantically related

The matcher architecture is illustrated in Fig. 3. This matcher is similar to that of PRIOR+ with the difference being an integration of a semantic and instance-based matcher. The semantic matcher will employ external background knowledge and will be integrated with existing logic-based techniques [19] while the instance-based matcher will employ link-key extraction method. The matcher will then be tested on the partitioned ontologies. The output of this matcher will be a series of alignments (correspondences).

---

[1] https://pypi.python.org/pypi/nltk.
[2] http://oaei.ontologymatching.org OAEI is an annual ontology matching competition that provides authoritative test and evaluations of ontology matching techniques or algorithms.
[3] http://www.dbpedia.org.
[4] http://www.freebase.com/.

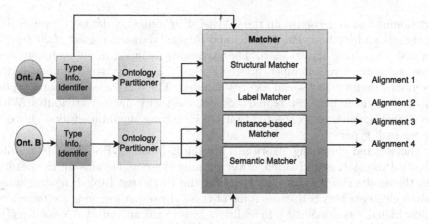

Fig. 3. Proposed architecture for stage 1

4.2   Research Stage 2

The second stage of this research will investigate and develop techniques that will help in producing quality mappings. An aggregation and alignment extraction technique will be developed by integrating user knowledge (similar to ALCOMO[5]) with weighting [5] techniques. This technique will enable us to select the best correspondences produced by the individual matchers. The aggregation and alignment extraction technique will be implemented on the set of alignments produced by the matcher. The results will be evaluated using standard measures such as precision, recall and F-measure by comparison with OAEI gold standard.

4.3   Research Stage 3

This stage will investigate reducing ambiguity to address alignment inconsistency and inaccuracy in ontology matching. An alignment improvement technique will be developed by adopting and extending existing approaches such as [15] using SAT solvers [6] and some reasoning techniques [13]. This will ensure consistency and accuracy in ontology matching by reducing ambiguity of data instances and helping to identify sets of exact and non-exact correspondences. The alignment improvement technique will be evaluated on the result of stage 2.

5   Preliminary Results

Matching and validating large ontologies is very difficult. However, when the ontologies are partitioned, it becomes easier for the single modules to be matched. Relatively, partitioning Big Data makes it easier to be used in an ontology matching process. In order to obtain an effective ontology partitioner, an ontology

---

[5] http://web.informatik.uni-mannheim.de/alcomo/.

partitioning test was carried on the OAEI-IM'10[6] ontology datasets using Scikit clustering[7] and Python. The Person1 and Person2 datasets consist of 2000 entities and 9000 RDF triples each. The attributes included in the ontology are: name, surname, street number, address, suburb, postcode, state, date of birth, age, phone number and social security number. The ontologies contain 4 types, namely Person, Suburb, State and Address, each with distinct attributes. With this simple schema, it is expected that the clustering should identify 4 clusters, one for each type.

The selected clustering algorithms are the K-Means, DBSCAN, Mean shift, Affinity Propagation and Birch. After running these algorithms on the ontologies, the results showed that only K-means and Birch (see Table 1) identified the desired clusters. This is because some of these clustering methods are unable to handle high dimensional data. In addition, Mean Shift and DBSCAN are density-based clustering methods that require a (nearly) continuous density function, hence, cannot yield any useful results in discrete scenarios. Affinity Propagation has high time and memory complexity, hence, is limited to clustering small sized datasets. Although Birch produced the desired clusters, the identification of its threshold parameter is complex and data specific; therefore, will be inappropriate to use for partitioning large datasets. K-Means also require a threshold parameter to be identified. However, the parameter selection for K-Means is relatively simple and it is expected that, the structure of the ontologies and the information from the missing type identifier will enable a reasonable estimate of this parameter. We will suggest K-Means as good for partitioning simple datasets but experiments need to be performed on complex datasets.

Currently, we are performing a similar evaluation on hierarchical clustering methods. The idea is to hierarchically cluster the instances to infer hidden structure in the ontology which can help to further partition a large-grain ontology into finer distinctions that may help find better matches and even guide the future evolution of the ontology.

6   Evaluation Plan

To evaluate our matching algorithms, we will adopt the benchmark tests from OAEI ontology matching campaign 2016. We will begin the evaluation by checking if the datasets have the different type of heterogeneity[8] (e.g., class heterogeneity, attribute-type heterogeneity) and if not, we will introduce specific heterogeneity using the framework in [2]. This framework permits the user to upload the source ontology and input heterogeneities between the source ontology and target ontology based on his/her knowledge and actions. We will then follow the evaluation criteria of OAEI, calculating the precision, recall and f-measure of each test case. The results of our algorithms will be compared with that of other OAEI participants. Finally, our matching system will be submitted

---

[6] http://oaei.ontologymatching.org/2010/im/.

[7] http://scikit-learn.org/stable/modules/clustering.

[8] Heterogeneities such as RDF, EXCEL and DB have been left aside.

Table 1. Ontology clustering results

Clustering algorithm	Validity (Number of clusters)	Runtime performance(s)
K-Means	4	1.17[a]
DBSCAN	0	0.15
Mean Shift	1	0.4
Affinity Propagation	92	2.09
Birch	4	1.33

[a]This is for multiple runs using inertia to select best result (0.117 per run on average)

to the OAEI'18 workshop/conference for its performance to be compared with top-ranked systems that will be participating in OAEI'18 campaign.

7   Conclusion

This research will result in the development of algorithms that will improve upon the existing state of the art in ontology matching by using techniques from different fields such as information retrieval and graph matching. The novelty lies in the individual matchers as well as the aggregation and alignment extraction method. For instance, to deal with class mismatch, a technique to infer fine grain type information from data instances will be incorporated into the matching process to bring the schema to the same level of granularity. A dynamic weighted sum in addition to user knowledge will then be applied to select the final mappings which will be improved using SAT solvers to eliminate ambiguity and inconsistencies.

Acknowledgement. This research is supported by the Data to Decisions Cooperative Research Centre (D2D CRC). The thesis is supervised by Prof. Markus Stumptner and Dr. Wolfgang Mayer.

References

1. Agrawal, R., Ailamaki, A., Bernstein, P.A., Brewer, E.A., Carey, M.J., Chaudhuri, S., Doan, A., Florescu, D., Franklin, M.J., Garcia-Molina, H., Gehrke, J., Gruenwald, L., Haas, L.M., Halevy, A.Y., Hellerstein, J.M., Ioannidis, Y.E., Korth, H.F., Kossmann, D., Madden, S., Magoulas, R., Ooi, B.C., O'Reilly, T., Ramakrishnan, R., Sarawagi, S., Stonebraker, M., Szalay, A.S., Weikum, G.: The Claremont report on database research. ACM Sigmod Rec. **37**(3), 9–19 (2008)
2. Chowdhury, N.A., Dou, D.: Evaluating ontology matchers using arbitrary ontologies and human generated heterogeneities. In: Meersman, R., et al. (eds.) OTM 2012. LNCS, vol. 7566, pp. 664–681. Springer, Heidelberg (2012). doi:10.1007/978-3-642-33615-7_15

3. Do, H.H., Rahm, E.: COMA: a system for flexible combination of schema matching approaches. In: Proceedings of the 28th International Conference on Very Large Data Bases (VLDB 2002), Hong Kong, China, pp. 610–621. VLDB Endowment (2002)
4. Doan, A., Madhavan, J., Domingos, P., Halevy, A.: Learning to map between ontologies on the semantic web. In: Proceedings of the 11th International Conference on World Wide Web (WWW 2002), pp. 662–673. ACM, New York (2002)
5. Euzenat, J., Shvaiko, P., et al.: Ontology Matching, 2nd edn. Springer, Heidelberg (2013)
6. Gomes, C.P., Kautz, H., Sabharwal, A., Selman, B.: Satisfiability solvers. Foundations of Artificial Intelligence, Chap. 2, vol. 3, pp. 89–134. Elsevier (2008)
7. Hu, W., Chen, J., Zhang, H., Qu, Y.: How matchable are four thousand ontologies on the semantic web. In: Antoniou, G., Grobelnik, M., Simperl, E., Parsia, B., Plexousakis, D., Leenheer, P., Pan, J. (eds.) ESWC 2011. LNCS, vol. 6643, pp. 290–304. Springer, Heidelberg (2011). doi:10.1007/978-3-642-21034-1_20
8. Hu, W., Qu, Y.: Falcon-AO: a practical ontology matching system. Web Seman. Sci. Serv. Agents World Wide Web 6(3), 237–239 (2008)
9. Knoblock, C.A., Szekely, P.: Exploiting semantics for Big Data integration. AI Mag. 36(1), 25–38 (2015)
10. Li, J., Wang, Z., Zhang, X., Tang, J.: Large scale instance matching via multiple indexes and candidate selection. Knowl.-Based Syst. 50, 112–120 (2013)
11. Ma, Y., Tran, T., Bicer, V.: Typifier: inferring the type semantics of structured data. In: Proceedings of the 29th International Conference on Data Engineering (ICDE 2013), Brisbane, Australia, pp. 206–217. IEEE (2013)
12. Mao, M., Peng, Y., Spring, M.: An adaptive ontology mapping approach with neural network based constraint satisfaction. Web Seman. Sci. Serv. Agents World Wide Web 8(1), 14–25 (2010)
13. Meilicke, C., Stuckenschmidt, H., Tamilin, A.: Reasoning support for mapping revision. J. Logic Comput. 19(5), 807 (2008)
14. Otero-Cerdeira, L., Rodríguez-Martínez, F.J., Gómez-Rodríguez, A.: Ontology matching: a literature review. Expert Syst. Appl. 42(2), 949–971 (2015)
15. Pührer, J., Heymans, S., Eiter, T.: Dealing with inconsistency when combining ontologies and rules using DL-programs. In: Aroyo, L., Antoniou, G., Hyvönen, E., Teije, A., Stuckenschmidt, H., Cabral, L., Tudorache, T. (eds.) ESWC 2010. LNCS, vol. 6088, pp. 183–197. Springer, Heidelberg (2010). doi:10.1007/978-3-642-13486-9_13
16. Schneider, T., Hashemi, A., Bennett, M., Brady, M., Casanave, C., Graves, H., Gruninger, M., Guarino, N., Levenchuk, A., Lucier, E., Obrst, L., Ray, S., Sriram, R.D., Vizedom, A., West, M., Whetzel, T., Yim, P.: Ontology for big systems: the ontology summit 2012 communiqué. Appl. Ontology 7(3), 357–371 (2012)
17. Shvaiko, P., Euzenat, J.: Ontology matching: state of the art and future challenges. IEEE Trans. Knowl. Data Eng. 25(1), 158–176 (2013)
18. Wang, P., Zhou, Y., Xu, B.: Matching large ontologies based on reduction anchors. In: IJCAI, Barcelona, pp. 2343–2348 (2011)
19. Zhang, W., Zhao, H., Mei, H.: A propositional logic-based method for verification of feature models. In: Davies, J., Schulte, W., Barnett, M. (eds.) ICFEM 2004. LNCS, vol. 3308, pp. 115–130. Springer, Heidelberg (2004). doi:10.1007/978-3-540-30482-1_16

Ontology-Based Data Access Mapping Generation Using Data, Schema, Query, and Mapping Knowledge

Pieter Heyvaert[(⊠)], Anastasia Dimou, Ruben Verborgh, and Erik Mannens

IDLab, Ghent University - imec, Ghent, Belgium
pheyvaer.heyvaert@ugent.be

Abstract. Ontology-Based Data Access systems provide access to non-RDF data using ontologies. These systems require mappings between the non-RDF data and ontologies to facilitate this access. Manually defining such mappings can become a costly process when dealing with large and complex data sources, and/or multiple data sources at the same time. This resulted in different mapping generation tools. While a number of these tools use knowledge from the original data, existing Linked Data, schemas, and/or mappings, they still fall short when dealing with complex challenges and the user effort can be high. In this paper, we propose an approach, together with an evaluation, that discovers and uses extended knowledge from existing (Linked) Data, schemas, query workload, and mappings, and combines it with knowledge provided by the mapping process to generate a new mapping. Our approach aims to improve the mapping quality, while decreasing the task complexity, and subsequently the user effort.

1 Introduction

Nowadays, Linked Data is materialized using RDF, which uses schemas (ontologies and vocabularies) to provide annotations, and is queried using the SPARQL query language [1]. However, Linked Data applications have the need to access data that is available in non-RDF formats [2]. Ontology-Based Data Access (OBDA) systems provides such access where an ontology mediates between the raw data and its consumers [6]. This access requires a mapping between the data schema of the non-RDF data and the ontologies. Subsequently, the aforementioned applications use OBDA systems to access non-RDF data, as if dealing with RDF data. However, manually defining such mappings can become a costly process when dealing with large and complex data sources [3,4], and/or multiple data sources at the same time [5]. This resulted in the development of (semi-)automatic mapping generation tools. Such tools reduce the user effort during the mapping process by reducing the required user interaction. This process

The described research activities were funded by Ghent University, imec, Flanders Innovation & Entrepreneurship (AIO), the Research Foundation – Flanders (FWO), and the European Union.

E. Blomqvist et al. (Eds.): ESWC 2017, Part II, LNCS 10250, pp. 205–215, 2017.
DOI: 10.1007/978-3-319-58451-5_15

takes as minimum input the raw data and outputs a mapping that maps this data to RDF triples. The process' tasks include selecting the appropriate classes, predicates, and datatypes, and matching them with the data.

Existing tools for single scenarios only need limited user interaction, but fail on scenarios involving non-trivial data schemas. Automatic tools are developed for use cases where only a mapping for a single scenario is required and where no mappings for subsequent, similar scenarios need to be created (hereinafter referred to as single-scenario use cases). They have a low task complexity [6], as the required user interaction is limited. Subsequently, the mapping process requires a low user effort. In most cases, these tools only use the original data schema and ignore other knowledge available in existing (Linked) Data [7,8], Schemas (data schemas, ontologies and vocabularies), the Queries that will be executed on the new RDF data (query workload) [2], and Mappings [9] (DSQM). With existing knowledge, we refer to knowledge that is available before the mapping process. Although these tools are able to generate a promising mapping for simple scenarios, they fail on scenarios with more complex data sources involving non-trivial data schemas (hereinafter referred to as complex challenges) [2].

Semi-automatic tools are developed for both single-scenario uses cases and use cases where multiple mappings need to be created in the same domain (hereinafter referred to as multi-scenario use case). They require more user interaction, such as writing SQL queries [10] or validating a suggested mapping [11], which might improve the generated mapping. However, it increases the required user effort. Despite using more existing knowledge compared to automatic tools, such as mappings and Linked Data, they neglect the query workload.

In this work, we present our semi-automatic approach to improve the quality of single-scenario mappings, while decreasing the user effort, compared to the state of the art. Our approach discovers and uses a more extended set of DSQM knowledge compared to existing tools to deal with complex scenarios, and reduces task complexity. Furthermore, the approach is not limited to a specific data format. The remainder of the paper is structured as follows. In Sect. 2, we elaborate on the state of the art. In Sect. 3, we discuss the research questions and the corresponding hypotheses. In Sect. 4, we explain the methodology and approach. In Sect. 5, we give preliminary results, followed by the evaluation plan in Sect. 6. In Sect. 7, we conclude the paper.

2   State of the Art

In this section, we elaborate on the state of the art for OBDA mapping generation and evaluation.

2.1   Mapping Generation Tools

Existing automated tools require no user interaction and use the data, data schema, and/or target ontology, but neglect the query workload and existing

DSQM knowledge. These tools are based on the direct mappings approach[1], in which tables are mapped to classes, data attributes are mapped to datatype properties, and foreign keys to object properties. They generate a new schema, called a bootstrap ontology, based on the database schema. When an existing ontology, called a target ontology, needs to be used, alignment between the target and bootstrap ontology is required afterwards. D2RQ [12] generates additional rules to tackle more complex challenges not considered by direct mappings. The alignment between the bootstrap and target ontology is done with schema matching tools, such as LogMap [13]. MIRROR [14] and Ontop [15] are similar to D2RQ. However, MIRROR extends the mappings by using information in the databases to determine, e.g., subclass-of relationships and $m{:}n$ relationships. Ontop updates the mappings using T-Mappings [16], which use knowledge embedded in the target ontology. Furthermore, D2RQ, MIRROR, and Ontop neglect the actual data, the query workload, and existing DSQM knowledge. AutoMap4OBDA [17] uses both the data and data schema, together with the target ontology. The ontology alignment is done by the tool itself and does not require an external schema matching tool. However, it neglects the query workload and existing DSQM knowledge. While the aforementioned tools work with relational databases (RDBs), Gloze [18] and JTOWL [19] apply a similar approach for XML and JSON files, respectively.

Existing semi-automated tools use the data, data schema, target ontology, existing mappings, existing Linked Data, and user interaction to create and improve the mapping, but not all information in the mappings is used and the query workload is neglected. BootOX [10] is the only semi-automatic tool for RDBs that applies the direct mappings approach and uses a bootstrap ontology, while IncMap and Karma are not. BootOX deals with more complex mappings than the ones tackled by direct mappings, due to the user interaction. IncMap [11] creates an IncGraph, based on the graph used in the Similarity Flooding Algorithm [20], to represent the data schema and the target ontology. The calculation of the weights of the graph is dynamic and allows to incorporate user feedback to improve results. Karma [21] is different because it uses existing DSQM knowledge to suggest mappings to the user. During a multi-scenario use case information in the previous mappings (called the semantic model) is reused, such as the classes, properties, and how these are related to each other [9]. However, they do not utilize the other information available via the mappings to tackle more complex challenges. If the different scenarios are in different domains previous mappings will have a limited usability. They use graph patterns found in existing Linked Data to determine how the classes and properties are related to each other [8]. This allows support for single-scenario use cases, because for these cases the tool is not able to use previous mappings to get that knowledge. However, the quality of mappings generated by using existing mappings is higher, because they provide a more coherent semantic model than the small graph patterns of the Linked Data. Furthermore, none of these tools validate the correct use of properties and classes.

---

[1] https://www.w3.org/TR/rdb-direct-mapping/.

The use of mapping process knowledge in the tools is limited to either the original data, schema and/or ontology, while the query workload is neglected. Furthermore, besides the use of a target ontology for alignment with the bootstrap ontology, only Ontop, AutoMap4OBDA, IncMap, and Karma use knowledge provided by the ontologies to improve the mapping. Karma is also the only tool that uses existing mappings (semantic models and existing data with their corresponding classes and properties) and Linked Data to provide improvements.

Nevertheless, these RDB tools still fall short when tackling more complex challenges, e.g., when subclasses are grouped in a single table and need to be separated, and in real-life scenarios with more complex queries [2,17].

2.2 Mapping Generation Evaluation

Liu and Li [6] propose a task model to evaluate the task complexity. This complexity is the aggregation of any intrinsic task characteristic that influences the task's performance. The task is in our case the mapping process. The model's components that contribute to the complexity are input (e.g., data, procedures, guidance, and random events), goal/output, process (e.g., steps and actions), time, and presentation (e.g., format and task compatibility). Decreasing the task complexity results in decreasing the required user effort, because less is required from the user (e.g., input, actions, and time).

In most cases, when tools are accompanied with an evaluation, they only assess a limited set of the mapping process' aspects: time required to transform the mapping suggestions to the correct ones [11]; W3C Direct Mapping Test Cases[2] [14]; precision, recall, and CPU time when using the semantic model [9] and Linked Data [8]; or number of user actions [21]. However, they only give a limited insight about task complexity and/or mapping quality, and not every tool is compared with the results of other tools. Furthermore, during these evaluations there is a lack of clear test case descriptions and performance indicators for the mapping quality, and they are different for each tool. Therefore, Pinkel et al. [2] developed a mapping generation quality benchmark for Relational-to-Ontology Data Integration scenarios (RODI). Mappings tools are evaluated by assessing the generated mapping's quality, i.e., a comparison between triples generated via the mapping, as a result of given queries, and the expected triples. However, as RODI is focused on automatic tools, it does not provide a formalized way to evaluate the task complexity, which is required when dealing with semi-automatic tools. Furthermore, it works only when the target ontology is used by the tool, while a combination of ontologies might provide better annotations. It is not suited to evaluate tools for other formats, as it only works for RDBs.

3 Problem Statement

In our approach, we aim to improve the mappings of single-scenario use cases, i.e., the precision and recall of the query-answering of the resulting Linked Data

---

[2] https://www.w3.org/2001/sw/rdb2rdf/test-cases/.

improves, by using DSQMs, because the DSQMs might contain knowledge that already tackles these challenges and DSQMs have proven benefits [8,9]. However, we aim to use an extended set of knowledge compared to previous efforts to improve the mappings to tackle these complex challenges. Therefore, we need to discover existing DSQM knowledge, i.e., find the relevant DSQM knowledge that is already available before the mapping process. This leads to the following main research question: *can we improve the (semi-)automatic generation of new single-scenario mappings using existing DSQM knowledge?* To answer this question, we need to answer these subquestions:

- How can we (semi-)automatically discover existing DSQMs that are relevant to the mapping process?
- How can we (semi-)automatically integrate the discovered DSQM knowledge with the DSQ knowledge of the mapping process to generate a new mapping?

  These research questions lead to the following hypotheses:

- Using existing DSQM knowledge improves the quality of a new single-scenario mapping compared to the state of the art.
- Using existing DSQM knowledge decreases the task complexity of the mapping process compared to the state of the art.

4 Research Methodology and Approach

Based on the research questions, we need to tackle two aspects: (semi-)automated discovery of relevant existing DSQM knowledge (Sect. 4.1) and (semi-)automated us of this knowledge to generate mappings (Sect. 4.2). Both can be addressed separately. The first aspect is not tackled by any of the other tools. For the second aspect, we exploit all options where existing tools are limited to only a subset of the possibilities. The knowledge of the new mapping process is combined with relevant existing knowledge (Fig. 1). This results in an initial mapping. Subsequently, user feedback on the mapping, collected via a user interface, is used to improve it. Furthermore, our approach an be used for heterogeneous formats.

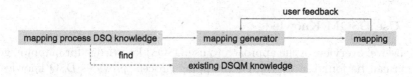

Fig. 1. Overview of the approach

4.1 Discover Existing DSQM Knowledge

A high-level overview of the approach to discover the relevant existing DSQM knowledge can be found in Fig. 2. The mapping process provides DSQ knowledge (bottom elements). Furthermore, we have existing DSQM knowledge (top elements). In the ideal scenario all the DSQM knowledge is at hand. However, this might not always be the case. To address this, we infer knowledge from other knowledge (dashed arrows): based on the original data the data schema can be reconstructed to a certain extent, as done in our previous work [22]; based on the classes and properties used in a mapping [9], Linked Data [8], and/or queries, you can derive the used ontologies. To discover the relevant knowledge we employ algorithms to measure the similarity of other knowledge components (methods a-e in Fig. 3). For example, if two data schemas of data sources about persons are similar then the classes (e.g., foaf:Person) and properties (e.g., foaf:name) of the existing mapping become candidates to be reused for the new mapping (b). Another example is comparing the query workload (e). If the two query workloads contain a query that searches for the graph patterns ?s a foaf:Person. ?s foaf:name 'John Doe'., then both mappings will be similar as both will need to annotate entities with the class foaf:Person and annotate them with their name using foaf:name.

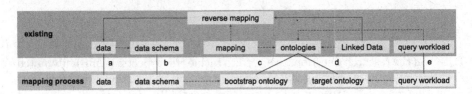

Fig. 2. Discover existing DSQM knowledge

In our approach, we aim to collect as much knowledge as possible to improve the mapping. First, we infer the knowledge that is not at hand. Then, we calculate the similarity measures between the knowledge from the mapping process and the existing knowledge. Finally, based on results of the similarity measures, we select the most relevant knowledge components.

4.2 Use DSQM Knowledge

A high-level overview of the approach to use DSQM knowledge for mapping generation can be found in Fig. 3. The mapping process provides DSQ knowledge (left elements). The mapping consists of the semantic model and 'extra rules' (middle elements). The semantic model contains how the used classes and properties are related to each other, and how classes and properties are mapped to the data fractions. The 'extra rules' represent the mapping rules that are needed to tackle mapping challenges that cannot be solved using the direct mappings approach. Furthermore, we have existing DSQM knowledge (right elements).

A target ontology can be used together with the data (a) or the data schema (b). However, requiring the user to provide a target ontology is not always straightforward and a combination of classes and properties from different schemas might result in a better annotation of the data. Therefore, we can only use the data (c) and/or data schema (d). However, in this case information from the existing DSQM knowledge is required to complete the mapping. This information can come from mappings (α), ontologies (β), Linked Data (γ), and/or query workloads (δ). When using mappings, we need some information to be adjusted to take into account the specifics of the new mapping process. The ontologies contain classes, properties, and how they are related to each other. However, no information about how they are related to data fractions is provided. This is the same for Linked Datasets and queries.

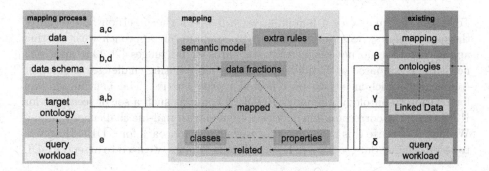

Fig. 3. Use DSQM knowledge

In our approach, we aim to execute the aforementioned methods separately. Subsequently, we merge each result to improve the mapping. During each merge, we generate adequate mapping rules, while assuring correct use of the ontologies.

5   Preliminary Results

In previous work [23], we developed the RMLEditor. It is a graphical user interface (GUI) that enables non-Semantic Web experts to create their own mappings while limiting the need to understand the underlying mapping language, which is RML [24], or the used (Semantic Web) technologies. In our approach, we aim to use it as a starting point to receive user feedback after each mapping generation iteration. Besides the RMLEditor, we also developed the RMLWorkbench [25]. It is a GUI to support data owners to administrate their Linked Data generation and publication workflow. This includes the data in heterogeneous formats and mappings, which are both used by our approach. Therefore, the RMLWorkbench offers a GUI to administrate the different elements of our approach, while hiding the implementation from the user. Furthermore, we have looked into different modeling approaches to generate mappings [26]. They help describing how different elements of the DSQM knowledge can be used for the generation.

We aim to use the approaches separately or combined in our approach. Finally, we developed a tool [22] to effectively perform data analysis on hierarchical data sources to identify RDF terms. This is needed when the schema of the data that needs to be mapped is not available, which might be the case for JSON and XML.

6   Evaluation Plan

Our hypotheses state that the use of DSQM knowledge improves the quality of the mappings, while decreasing the task complexity. Therefore, two aspects need to be evaluated: the quality of the mapping and the task complexity.

6.1   Mapping Quality

To assess the quality of a mapping, we will assess the precision and recall of the query results, because it allows us to compare our approach with existing approaches that do not use certain standards or languages [2]. Our approach needs to be evaluated with different scenarios representing challenges of different complexity, which needs to be reflected in the queries. The benchmark tool for relation-to-ontology mappings RODI [2] contains such a set of scenarios for RDBs with the corresponding queries, designed with real-life challenges in mind. We intend to reuse this benchmark for testing our approach for RDBs. However, to test it against data sources in heterogeneous formats, we need to extend RODI.

6.2   Task Complexity

During our evaluation of the task complexity, we want to apply the model by Liu and Li [6] to our approach, by evaluating each aspect during the mapping process. The input consists mainly of the information and knowledge that needs to be provided by users, e.g., the data, data schema, target ontology, query workload, and the information required during the process. The output is a mapping. The process is defined by the required user actions. The time is the duration to perform these actions. The presentation is defined by the GUI used to complete the actions. To have a mapping process with a low complexity, the input, the actions, and the time to complete these actions needs to be decreased, and the GUI needs to fit the actions. While previous evaluations only analyzed a single component, we want to evaluate all of them to know the complete impact of the mapping process on the task complexity. As RODI is developed for automatic tools, it does not take into account the task complexity. Therefore, we need to extend RODI to also evaluate the different aspects of the task complexity.

7   Conclusion

The main differences of our approach with the state of the art is that we discover relevant DSQM knowledge, and use an extended set of DSQM knowledge, including the found DSQMs, for the generation of OBDA mappings. Challenges for the former include finding the correct similarity metrics and combining these metrics when comparing multiple elements of knowledge. Challenges for the latter include determining how to merge the different knowledge and how to ensure that resulting mapping is valid regarding, e.g., ontology definitions. Furthermore, our approach is not limited to RDBs. It can also be used for JSON and XML data. If we can validate the hypothesis, then users will have a method that requires less user effort to generate higher quality OBDA mappings, and subsequently, they will have access to higher quality OBDA systems. Even more, Linked Data applications will have access to a larger amount of non-RDF datasets, allowing them to utilize RDF-based techniques on these non-RDF datasets.

References

1. Bizer, C., Heath, T., Berners-Lee, T.: Linked data - the story so far. Int. J. Semantic Web Inf. Syst. **5**(3), 1–22 (2009). doi:10.4018/jswis.2009081901
2. Pinkel, C., Binnig, C., Jiménez-Ruiz, E., Kharlamov, E., May, W., Nikolov, A., Skjæveland, M.G., Solimando, A., Taheriyan, M., Heupel, C., et al.: RODI: Benchmarking Relational-to-Ontology Mapping Generation Quality. Semant. Web (2016, pre-print). http://content.iospress.com/articles/semantic-web/sw268
3. Kharlamov, E., Hovland, D., Jiménez-Ruiz, E., Lanti, D., Lie, H., Pinkel, C., Rezk, M., Skjæveland, M.G., Thorstensen, E., Xiao, G., Zheleznyakov, D., Horrocks, I.: Ontology based access to exploration data at Statoil. In: Arenas, M., Corcho, O., Simperl, E., Strohmaier, M., d'Aquin, M., Srinivas, K., Groth, P., Dumontier, M., Heflin, J., Thirunarayan, K., Staab, S. (eds.) ISWC 2015. LNCS, vol. 9367, pp. 93–112. Springer, Cham (2015). doi:10.1007/978-3-319-25010-6_6
4. Kharlamov, E., Solomakhina, N., Özçep, Ö.L., Zheleznyakov, D., Hubauer, T., Lamparter, S., Roshchin, M., Soylu, A., Watson, S.: How semantic technologies can enhance data access at siemens energy. In: Mika, P., Tudorache, T., Bernstein, A., Welty, C., Knoblock, C., Vrandečić, D., Groth, P., Noy, N., Janowicz, K., Goble, C. (eds.) ISWC 2014. LNCS, vol. 8796, pp. 601–619. Springer, Cham (2014). doi:10.1007/978-3-319-11964-9_38
5. He, B., Patel, M., Zhang, Z., Chang, K.C.-C.: Accessing the deep web. Commun. ACM **50**(5), 94–101 (2007)
6. Liu, P., Li, Z.: Task complexity: a review and conceptualization framework. Int. J. Ind. Ergon. **42**(6), 553–568 (2012)
7. Ramnandan, S.K., Mittal, A., Knoblock, C.A., Szekely, P.: Assigning semantic labels to data sources. In: Gandon, F., Sabou, M., Sack, H., d'Amato, C., Cudré-Mauroux, P., Zimmermann, A. (eds.) ESWC 2015. LNCS, vol. 9088, pp. 403–417. Springer, Cham (2015). doi:10.1007/978-3-319-18818-8_25
8. Taheriyan, M., Knoblock, C.A., Szekely, P., Ambite, J.L.: Leveraging linked data to discover semantic relations within data sources. In: Groth, P., Simperl, E., Gray, A., Sabou, M., Krötzsch, M., Lecue, F., Flöck, F., Gil, Y. (eds.) ISWC 2016. LNCS, vol. 9981, pp. 549–565. Springer, Cham (2016). doi:10.1007/978-3-319-46523-4_33
9. Taheriyan, M., Knoblock, C.A., Szekely, P., Ambite, J.L.: Learning the semantics of structured data sources. Web Semant. Sci. Serv. Agents World Wide Web **37**, 152–169 (2016)

10. Jiménez-Ruiz, E., Kharlamov, E., Zheleznyakov, D., Horrocks, I., Pinkel, C., Skjæveland, M.G., Thorstensen, E., Mora, J.: BooTOX: practical mapping of RDBs to OWL 2. In: Arenas, M., Corcho, O., Simperl, E., Strohmaier, M., d'Aquin, M., Srinivas, K., Groth, P., Dumontier, M., Heflin, J., Thirunarayan, K., Staab, S. (eds.) ISWC 2015. LNCS, vol. 9367, pp. 113–132. Springer, Cham (2015). doi:10.1007/978-3-319-25010-6_7

11. Pinkel, C., Binnig, C., Kharlamov, E., Haase, P.: IncMap: pay as you go matching of relational schemata to OWL ontologies. In: Proceedings of the 8th International Conference on Ontology Matching, pp. 37–48. CEUR-WS.org (2013)

12. Bizer, C., Seaborne, A.: D2RQ - treating non-RDF databases as virtual RDF graphs. In: Proceedings of the 3rd International Semantic Web Conference (2004)

13. Jiménez-Ruiz, E., Grau, B.C.: LogMap: logic-based and scalable ontology matching. In: Aroyo, L., Welty, C., Alani, H., Taylor, J., Bernstein, A., Kagal, L., Noy, N., Blomqvist, E. (eds.) ISWC 2011. LNCS, vol. 7031, pp. 273–288. Springer, Heidelberg (2011). doi:10.1007/978-3-642-25073-6_18

14. de Medeiros, L.F., Priyatna, F., Corcho, O.: MIRROR: automatic R2RML mapping generation from relational databases. In: Cimiano, P., Frasincar, F., Houben, G.-J., Schwabe, D. (eds.) ICWE 2015. LNCS, vol. 9114, pp. 326–343. Springer, Cham (2015). doi:10.1007/978-3-319-19890-3_21

15. Calvanese, D., Cogrel, B., Komla-Ebri, S., Kontchakov, R., Lanti, D., Rezk, M., Rodriguez-Muro, M., Xiao, G.: Ontop: answering SPARQL queries over relational databases. Semant. Web 8(3), 471–487 (2017)

16. Rodrıguez-Muro, M., Calvanese, D.: Dependencies: making ontology based data access work in practice. In: Proceedings of the 5th Alberto Mendelzon International Workshop on Foundations of Data Management (2011)

17. Sicilia, Á., Nemirovski, G.: AutoMap4OBDA: automated generation of R2RML mappings for OBDA. In: Blomqvist, E., Ciancarini, P., Poggi, F., Vitali, F. (eds.) EKAW 2016. LNCS (LNAI), vol. 10024, pp. 577–592. Springer, Cham (2016). doi:10.1007/978-3-319-49004-5_37

18. Battle, S.: Gloze: XML to RDF and back again. In: Jena User Conference (2006)

19. Yao, Y., Wu, R., Liu, H.: JTOWL: a JSON to OWL convertor. In: Proceedings of the 5th International Workshop on Web-scale Knowledge Representation Retrieval & Reasoning, pp. 13–14. ACM (2014)

20. Melnik, S., Garcia-Molina, H., Rahm, E.: Similarity flooding: a versatile graph matching algorithm and its application to schema matching. In: Proceedings of the 18th International Conference on Data Engineering, pp. 117–128. IEEE (2002)

21. Knoblock, C.A., Szekely, P., Ambite, J.L., Goel, A., Gupta, S., Lerman, K., Muslea, M., Taheriyan, M., Mallick, P.: Semi-automatically mapping structured sources into the semantic web. In: Simperl, E., Cimiano, P., Polleres, A., Corcho, O., Presutti, V. (eds.) ESWC 2012. LNCS, vol. 7295, pp. 375–390. Springer, Heidelberg (2012). doi:10.1007/978-3-642-30284-8_32

22. Heyvaert, P., Dimou, A., Verborgh, R., Mannens, E.: Data analysis of hierarchical data for RDF term identification. In: Li, Y.-F., Hu, W., Dong, J.S., Antoniou, G., Wang, Z., Sun, J., Liu, Y. (eds.) JIST 2016. LNCS, vol. 10055, pp. 204–212. Springer, Cham (2016). doi:10.1007/978-3-319-50112-3_15

23. Heyvaert, P., Dimou, A., Herregodts, A.-L., Verborgh, R., Schuurman, D., Mannens, E., Van de Walle, R.: RMLEditor: a graph-based mapping editor for linked data mappings. In: Sack, H., Blomqvist, E., d'Aquin, M., Ghidini, C., Ponzetto, S.P., Lange, C. (eds.) ESWC 2016. LNCS, vol. 9678, pp. 709–723. Springer, Cham (2016). doi:10.1007/978-3-319-34129-3_43

24. Dimou, A., Sande, M.V., Colpaert, P., Verborgh, R., Mannens, E., Van de Walle, R.: RML: a generic language for integrated RDF mappings of heterogeneous data. In: Proceedings of the 7th Workshop on Linked Data on the Web (2014)
25. Dimou, A., Heyvaert, P., Maroy, W., De Graeve, L., Verborgh, R., Mannens, E.: Towards an interface for user-friendly linked data generation administration. In: Proceedings of the 15th International Semantic Web Conference: Posters and Demos (2016)
26. Heyvaert, P., Dimou, A., Verborgh, R., Mannens, E., Van de Walle, R.: Towards approaches for generating RDF mapping definitions. In: Proceedings of the 14th International Semantic Web Conference: Posters and Demos (2015)

Engaging Librarians in the Process of Interlinking RDF Resources

Lucy McKenna[✉]

ADAPT Centre, Trinity College Dublin, Dublin, Ireland
lucy.mckenna@adaptcentre.ie

Abstract. By publishing metadata as RDF and interlinking these resources with other RDF datasets on the Semantic Web, libraries have the potential to expose their collections to a larger audience, increase the use of their materials, and allow for more efficient user searches. Despite these benefits, there are many barriers to libraries fully participating in the Semantic Web. Increasing numbers of libraries are devoting valuable time and resources to publishing RDF datasets, yet little meaningful use is being made of them due to lack of interlinking. The goal of this research is to explore the barriers faced by librarians in participating in the Semantic Web with a particular focus on the process of interlinking. We will also explore how interlinking could be made more engaging for this domain.

Keywords: Engagement · Interlinking · Library · Linked Data · Semantic Web

1 Introduction

The Semantic Web (SW) is a Web of Data where the relationships between data are defined in a common machine readable format [1,2]. These relationships are known as Linked Data (LD), which describes a set of principles for publishing and interlinking data on the web [3].

From the perspective of a library, participating in the SW could greatly influence metadata quality and information discovery. By freeing metadata from library databases and sharing it on the SW, libraries could make their resources more visible, leading to an increase in the use of library data and, as such, an increase in the number of library patrons [4]. Publishing to the SW would also allow libraries to share their metadata with greater ease, thus enhancing metadata accessibility and quality. This could lead to a reduction in the amount of time spent creating metadata, reducing library costs [5]. In addition, the process of interlinking RDF resources with those emerging from other cultural heritage institutions and beyond could allow researchers to be directed to a web of related data based on a single information search [3].

Despite these benefits, relatively few libraries are fully participating in the SW as a result of the many barriers faced by librarians in engaging with SW and

© Springer International Publishing AG 2017
E. Blomqvist et al. (Eds.): ESWC 2017, Part II, LNCS 10250, pp. 216–225, 2017.
DOI: 10.1007/978-3-319-58451-5_16

LD technologies [6,7]. Although increasing numbers of libraries are publishing metadata in RDF [3,8], few have successfully interlinked their data with other RDF resources - a central aspect of the SW.

2 Motivation

The goal of this research is to explore the barriers faced by librarians in participating in the SW, with a particular focus on the process of interlinking. Increasing this group's engagement in the process of interlinking will also be explored. It was decided to focus on this area due to the values provided by increased library participation on the SW. These values are twofold; firstly, as described above, use of LD has the potential to open up library collections and increase metadata quality. Secondly, librarians and library metadata has much to offer the SW in terms of data quality and credibility.

2.1 The Potential Role of Librarians in SW Development

It could be argued that LD generation could be conducted by technical experts or via crowd-sourcing, rather than by librarians. However, librarians have been successfully working in areas of information access and knowledge discovery for centuries and, thus are already ideally placed to play a leading role in this domain.

Librarians are experts at using controlled authorities and vocabularies when creating bibliographic metadata. This allows for consistent identification and linking of similar concepts and entities across records, resulting in more efficient catalogue searches. Libraries have developed many reliable authorities for controlling forms of names, titles and subjects, and countless controlled vocabularies for describing subjects, genres, languages, and locations. A number of these resources are already available as LD, thus, rather than duplicating what has already been created, these resources could be used to consistently identify concepts and entities across the SW [9]. Being familiar with the use of these authorities and vocabularies, librarians already have the expertise to establish this.

Since anyone can publish RDF metadata and interlink datasets on the SW, as the Web of Data grows, there will be an increased need to identify who completed these tasks in order to establish the degree of metadata credibility. As authoritative sources of information, it is believed that LD generated by librarians will be treated with increased credibility over that generated by non-authoritative sources [9–11]. Therefore, it is likely that LD generated by librarians will be used with increased frequency [10,11].

From the above it can bee seen that librarian's have the expertise to evolve the SW into a rich and trustworthy information network [12,13].

2.2 Challenges Faced by Libraries in Participating in the SW

As mentioned, RDF datasets are being published by a growing number of libraries [3,8], yet few are integrating these datasets with those emerging from other organisations [9], possibly because this is one of the most challenging areas of LD implementation [14]. This is also likely due to the fact the tools required to complete such data integration are limited [14] and that little usability testing of these tools has been completed with users, or potential users, of the SW who do not have a technical background [11]. As one of the fundamental prerequisites of the SW is the existence of large amounts of meaningfully interlinked resources [15], it is key that institutions not only publish RDF datasets but also interlink their data with others.

In 2015 the Online Computer Library Center (OCLC) conducted a world-wide survey investigating the use of LD in libraries. Of the 79 responses received, 112 LD projects were reported, the benefits of which included exposing data to a larger audience, enhancing the library's metadata, improving search accuracy, and combining LD datasets [6]. Barriers to using LD included; difficulty establishing links, lack of authority control, difficulty learning how to implement LD, lack of information outlining useful applications of LD in libraries, and difficulties incorporating LD generation into existing workflows [6].

Other reported [7,16] challenges faced by libraries when attempting to participate in the SW include:

- Cataloguing software can be inflexible in adapting to SW requirements.
- Many libraries use MARC21 format for generating bibliographic records, however the MARC data model is inadequate for direct use on the SW, and the processes and technologies used for transforming these records to a more suitable format are time-consuming and challenging.

The above indicates that, although librarians understand the benefits that the SW offers, they have difficulty engaging fully with it as LD technologies are not tailored to the library domain, or to the needs and expertise of librarians.

Librarians are an example of current, and potential, domain expert users of LD who may not have the technical expertise required to work with available LD technologies, but who have the potential to progress the development of the SW if given the opportunity to fully engage with it. Therefore it would be important to focus on exploring how librarians, who may not have a technical background, could engage more in the process of interlinking.

Interviews. Two informal semi-structured interviews were conducted with librarians who work as metadata catalogers, of both physical and digital assets respectively, in a large university library. These interviews were conducted in order to further investigate the challenges faced by librarians in engaging with LD. Both librarians had over a decade of experience working in bibliographic data management and both were familiar with the concepts of the SW and LD.

Common themes emerging from the interviews included:

- The librarians both noted that, although many libraries are exporting their catalogues in RDF, little further use is being made of the data.
- The above led to further discussion regarding the librarians' desire to be able to use published RDF datasets by interlinking them with data in their library catalogues.

- The librarians expressed a desire for a bespoke tool that would allow them to create links with LD datasets.
- The librarians highlighted a need for a tool that would allow them to create RDF records as part of their current cataloguing workflow. Both strongly stated that ideally such an interface would create RDF in the background of the cataloguing process.
- The above discussion led to both librarians expressing a certain level of frustration with current proprietary cataloguing software which does not facilitate the generation of RDF records, RDF ingestion, or interlinking.
- The interviewees highlighted how most current LD technologies do not target librarians or metadata cataloguers.
- The interviewees also expressed that libraries have a lot to offer the SW, both in terms of providing authority control and controlled vocabularies, and in providing "information about people, places and events... that would really bring value to the internet". It was also mentioned that the use of these controls provide a greater capacity for filtering searches.
- The above led to further discussion surrounding the librarians' concerns regarding authority control on the SW. The librarians felt that some current LD resources could be better controlled with increased validity if librarians were involved in the metadata creation process.
- Librarians need more use cases of LD being used effectively to enhance the visibility of resources and improve information searches in the library setting for them to allocate the necessary time and funding to LD creation.
- Difficulties in using MARC to create data in a format that is SW compatible were expressed.
- The interviewees both noted that it would be useful for librarians to have some basic training in coding, RDF and LD technologies so that they could better express their needs and so that they could interact with RDF datasets with greater ease.

From the interviews it was apparent that the librarians felt that libraries need more LD resources and tools targeting their specific needs in order for them to use the SW to its full potential. This was highlighted as a significant gap, with the interviewees feeling that most LD tools are not designed with librarians and their work processes in mind.

3 State of the Art

As prior sections of this paper indicate, librarians are currently unable to engage fully with the process of interlinking as a result of LD technologies being designed primarily for technical experts. The following section discusses some of the existing LD interlinking technologies as well as library projects that used LD interlinking in order to identify how interlinking is currently being achieved in the library domain.

OpenRefine [17] is an open source application that can be used for data cleanup and for transforming data to other formats. RDF Refine [18] is an extension of this tool which adds a GUI for exporting results in RDF and also allows for the reconciliation of collection specific vocabularies against controlled vocabularies expressed in RDF. This provides a way for users to interlink their LD resources to existing LD datasets by generating owl:sameAs links.

Silk [19] is a link discovery framework tool that can be used to generate links between related data items from different LD datasets. Like RDF Refine, the tool supports the generation of owl:sameAs links between resources as well as other types of RDF links. Silk also provides a Silk Workbench GUI for the creation of link specifications. LIMES [20] is another link discovery framework for the Web of Data and, like Silk, it can be used to discover links between LD resources using a GUI.

In 2013 the Library of the University of Nevada, Las Vegas embarked upon a focused LD project [21,22] where a collection of records from their Digital Library was uplifted to RDF and published as LD using OpenRefine. When creating their metadata records the library used controlled vocabularies and authorities from the Library of Congress [23], the FAST [24] subject heading schema, and the Getty Thesaurus of Geographic Names [25], among others. As these vocabularies are available as LD, the LD generated by the library was interconnected with these vocabularies. At the time of publication further interlinking between the library's dataset and other LD resources on the web was not completed, however it was reported that linking to DBpedia [26] and Europeana [27] was planned. In the case of OpenRefine and RDF Refine there are few examples of where the tool provided the user with a means of interlinking to datasets other than controlled vocabularies or large-scale general resources. There also appears to be little research exploring the usability of the tool from a librarian's or other domain expert user's perspective.

Research using Silk and LIMES has involved successfully creating links between large-scale LD resources such as DBpedia and GeoNames [28]. For instance, Swissbib, the meta-catalogue of Swiss University Libraries and the Swiss National Library, is currently being integrated into the SW with the generation of LD from their bibliographic metadata [29]. In this project the libraries' LD datasets have been successfully interlinked with DBpedia and The Virtual International Authority File (VIAF) [30] using both SILK and LIMES. However, as with OpenRefine, there are no apparent examples of the tools being used to successfully interlink with smaller LD resources exported from other libraries or cultural institutions. Additionally, little research with librarian or other domain expert users' of LD appears to have been completed.

Other methods of interlinking can be seen in the British National Biography LD project. In July 2011 the British National Library released the this resource as LD, converting the chosen records to RDF using XSLT [31]. Similar to the University of Nevada's Library project, the British National Library linked to library domain data sets such as VIAF and Library of Congress Subject Headings [32], by matching authorised headings in the records with the corresponding URI in the LD datasets available. In addition to this, the library also linked to other external data sets such as GeoNames and Lexvo [33] via Crosswalk Matching.

Although some interlinking was completed in the projects discussed above, the data sets were only linked with general resources or library domain data sets. Further interlinking could have been completed with smaller LD data sets emerging from other libraries or cultural heritage institutions.

4 Problem Statement and Contributions

As indicated above, librarians are currently not using LD interlinking to its full potential. Therefore the research question being investigated as part of this Early Stage PhD is:

– To what extent can librarians engage with LD interlinking tooling to manage, enhance and curate metadata records?

We hope to contribute to SW and library domain by:

1. Providing a means for librarians to engage in the process of interlinking in a more meaningful way.
2. Increase the potential for LD interlinking between institutions, and widen the possibilities of interlinking with smaller datasets.
3. Increase the potential for librarians' LD interlinking accuracy to equate to that of LD researchers.

The library domain will be used as a concrete environment in which this research question can be explored as librarians are considered experts of metadata management, enhancement and curation. With increasing numbers of libraries producing LD there is scope for them to interlink with LD datasets emerging from other cultural institutions, as well those produced by general resources, such as DBpedia or Library of Congress. Publishing LD datasets has the potential to open up and greatly enhance the metadata of digitised collections generated by libraries and other cultural heritage institutions [5]. Allowing the creators and curators of this data to become more involved in the LD process is likely to positively impact the use of LD and the development of LD tools [11].

It is hoped that this research will benefit librarians by providing them with a more automated way to engage in LD interlinking, and as such data curation, thus reducing the amount of time spent on the cataloguing process which could in turn allow librarians to use their skills and expertise on other areas.

5 Research Methodology and Approach

A Design Science approach will be followed for the purpose of this research. Design Science involves creating and evaluating artifacts designed to meet a particular need [34]. The process involves validating proposed artifacts, or products, through an iterative process of design and evaluation in order to design solutions to an identified problem [35].

We plan to initially investigate the current usage trends in respect to LD and LD tooling, and cataloguing tooling, within Ireland and further afield through the means of a state-of-the-art research as well as potential dissemination a survey. Based on these results a number of librarians will be interviewed in order to establish more specific needs and issues in working with LD. These interviews will have a particular focus on LD interlinking tools and the types of interfaces librarians are currently using when creating and managing LD.

A means of increasing librarians' engagement and interaction with LD, particularly at the interlinking stage, will be explored iteratively through a processes of problem identification, solution design, and user testing [36,37]. As we have worked closely with a number of librarians in the past, these librarians and their connections will be contacted to participate in the research.

6   Preliminary Results

A user interface for capturing bibliographic records using the Metadata Object Description Schema (MODS) RDF was developed (see Fig. 1).

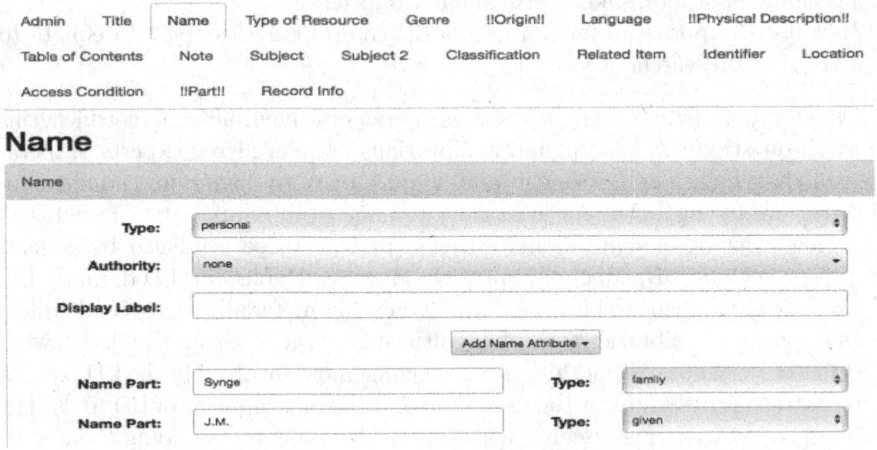

Fig. 1. Screenshot of part of the MODS cataloguing interface.

MODS is an XML schema for a bibliographic element set that can be used to catalogue library materials [38] and MODS-RDF is an OWL/RDF representation of the schema. The interface was developed in collaboration with metadata cataloguers from the Digital Resources and Imaging Services (DRIS) of the Library of Trinity College Dublin (TCD).

The interface was designed to constrain data entry options and to dynamically alter depending on the data entered. This was done to ensure that published MODS records met the requirements of DRIS and the minimum requirements for describing digital cultural-heritage and humanities-based scholarly resources

using MODS, as described by the Digital Library Federation Aquifer Initiative
[39]. Metadata entered into the interface is stored in a relational database. This
data can then be uplifted to MODS-RDF (see Fig. 2) using an R2RML mapping. Generating RDF records would allow for DRIS to publish and link their
data on the SW, the benefits of which have been discussed above. The resulting
MODS-RDF records can also be queried using SPARQL. SPARQL can be used
to integrate data from different resources through the use of federated queries
which can be conducted over multiple disparate datasets [40].

```
<http://data.library.tcd.ie/resource/typeOfResource/2>
        a       <http://www.w3.org/1999/02/22-rdf-syntax-ns#type> ;
        <http://www.w3.org/2000/01/rdf-schema#label>
                "Text" .

<http://data.library.tcd.ie/resource/titleinfo/2>
        a       <http://www.loc.gov/mads/rdf/v1#Title> ;
        <http://www.w3.org/2000/01/rdf-schema#label>
                "Transactions of the Institution of Civil Engineers of Ireland" .

<http://data.library.tcd.ie/resource/location/2>
        <http://www.w3.org/2000/01/rdf-schema#label>
                "TCD" , "http://digitalcollections.tcd.ie/home/#folder_id=1065&pidtopage=ICEI-026_101&entry_point=101" .

<http://data.library.tcd.ie/resource/subject/2>
        a       <http://www.loc.gov/mads/rdf/v1#Geographic> , <http://www.loc.gov/mads/rdf/v1#County> ,
        <http://www.loc.gov/mads/rdf/v1#Country> , <http://www.loc.gov/mads/rdf/v1#Topic> , <http://www.loc.gov/mads/rdf/v1#City> ,
        <http://www.loc.gov/mads/rdf/v1#Temporal> , <http://www.loc.gov/mads/rdf/v1#SimpleType> ,
        <http://www.w3.org/1999/02/22-rdf-syntax-ns#type> , <http://www.loc.gov/mads/rdf/v1#Region> ;
                <http://www.w3.org/2000/01/rdf-schema#label>
                        "Dublin" , "Transactions" , "Ireland" .

<http://data.library.tcd.ie/resource/genre/2>
        a       <http://www.loc.gov/mads/rdf/v1#GenreFormElement> ;
        <http://www.w3.org/2000/01/rdf-schema#label>
                "journals (periodicals)" .
```

Fig. 2. Sample MODS RDF output from the cataloguing interface.

The cataloguing interface was tested by observing the DRIS metadata cataloguer using the tool to create a bibliographic record for an item in the repository.
Results indicated that the metadata cataloguer was satisfied with the progress
of the interface and DRIS indicated a strong interest in the ongoing development of the tool. Some interface layout issues and additional requirements were
identified. Future iterations of the tool will focus on overcoming these issues.

The aim of developing the cataloging interface was to provide a pathway for
DRIS to move towards publishing the bibliographic metadata of their digital
collections as RDF. If DRIS begin publishing RDF records, it is hoped that
DRIS may be used as an environment in which to explore the usability of LD
interlinking tools and interfaces.

7 Evaluation Plan

In keeping with the Design Science approach, evaluation of potential solutions
to the problems identified will occur iteratively throughout the research process.

Evaluation will likely involve usability testing which involves recruiting representative users, in this case librarians, to evaluate the degree to which a product meets specific usability criteria [41]. As usability testing seeks holistic information about the product, the setting and the user, it typically requires both quantitative and qualitative data [36].

8 Conclusions

It is hoped that facilitating increased engagement of domain expert users with the process of LD interlinking will aid in the realisation of the full vision of the SW.

Acknowledgments. This study is supported by the Science Foundation Ireland (Grant 13/RC/2106) as part of the ADAPT Centre for Digital Content Platform Research (http://www.adaptcentre.ie/) at Trinity College Dublin.

This research is being completed under the supervision of Prof. Declan O'Sullivan and Dr. Christophe Debruyne, both of the ADAPT Centre, Trinity College Dublin.

References

1. Berners-Lee, T., Hendler, J., Lassila, O.: The semantic web. Sci. Am. **284**, 1–5 (2001)
2. W3C (2015). https://www.w3.org/standards/semanticweb/
3. Hastings, R.: Linked data in libraries: status and future direction. Comput. Libr. **35**, 12–16 (2015)
4. Gonzales, B.M.: Linking libraries to the web: linked data and the future of the bibliographic record. ITAL **33**, 10–22 (2014)
5. Ryan, C., Grant, R., Carragin, E., Collins, S., Decker, S., Lopes, N.: Linked data authority records for Irish place names. IJDL **15**, 73–85 (2015)
6. OCLC: Online Computer Library Center (2017). http://www.oclc.org/research/themes/data-science/linkeddata.html
7. Hallo, M., Lujan Mora, S., Trujillo Mondejar, J.C.: Transforming library catalogs into linked data. In: ICERI (2013)
8. Mitchell, E.T.: Library linked data: early activity and development. Libr. Technol. Rep. **52**, 5–33 (2016)
9. Neish, P.: Linked data: what is it and why should you care? Aust. Libr. J. **64**, 3–10 (2015)
10. Miller, E., Westfall, M.: Linked data and libraries. Serials Libr. **60**, 17–22 (2011)
11. Shvaiko, P., Euzenat, J.: Ontology matching: state of the art and future challenges. IEEE Trans. Knowl. Data Eng. **25**, 158–176 (2013)
12. Greenberg, J.: Advancing the semantic web via library functions. CCQ **43**(3–4), 203–225 (2006)
13. Harper, C.A., Tillett, B.B.: Library of congress controlled vocabularies and their application to the semantic web. CCQ **43**(3–4), 47–68 (2007)
14. Bergman, M.: A decade in the trenches of the Semantic Web (2014). http://www.mkbergman.com/1771/a-decade-in-the-trenches-of-the-semantic-web/
15. Bizer, C., Heath, T., Ayers, D., Raimond, Y.: Interlinking open data on the web. In: 4th European Semantic Web Conference, Austria (2007)

16. Cole, T.W., Han, M.J., Weathers, W.F., Joyner, E.: Library MARC records into linked open data: challenges and opportunities. J. Libr. Metadata **13**(2–3), 163–196 (2013)
17. OpenRefine (2017). http://openrefine.org
18. RDF Refine (2017). http://refine.deri.ie
19. Volz, J., Bizer, C., Gaedke, M., Kobilarov, G.: Silk - a link discovery framework for the web of data. In: LDOW 2009, Spain (2009)
20. Ngomo, A.-C.N., Auer, S.: LIMES - a time-efficient approach for large-scale link discovery on the web of data. In: Proceedings of IJCAI (2011)
21. Lampert, C.K., Southwick, S.B.: Leading to linking: introducing linked data to academic library digital collections. J. Libr. Metadata **13**, 230–253 (2013)
22. Southwick, S.B.: A guide for transforming digital collections metadata into linked data using open source technologies. J. Libr. Metadata **15**, 1–35 (2015)
23. Library of Congress Linked Data Service (2017). http://id.loc.gov
24. OCLC FAST (2016). http://fast.oclc.org/searchfast/
25. Getty Thesaurus of Geographic Names Online (2015). http://www.getty.edu/research/tools/vocabularies/tgn/
26. DBpedia (2017). http://wiki.dbpedia.org
27. Europeana Collections (2017). http://www.europeana.eu/portal/en
28. GeoNames (2017). http://www.geonames.org
29. Bensman, F., Prongu, N., Hellstern, M., Kuntschik, P.: Swissbib goes linked data. In: SWIB (2016)
30. VIAF: The Virtual International Authority File (2016). http://viaf.org
31. Deliot, C.: Publishing the British national bibliography as linked open data. Catalogue Index **174**, 13–18 (2014)
32. Library of Congress Subject Headings (2017). http://id.loc.gov/authorities/subjects.html
33. Lexvo (2016). http://www.lexvo.org
34. Hevner, A.R., March, S.T., Park, S.: Design science in information systems research. MIS Q. **28**, 75–105 (2004)
35. van Aken, J.E.: Management research as a design science: articulating the research products of mode 2 knowledge production in management. Br. J. Manage. **16**, 19–36 (2005)
36. Emanuel, J.: Usability testing in libraries: methods, limitations, and implications. IDLP **29**, 204–217 (2013)
37. Nielsen, J.: Usability 101: Introduction to Usability (2012). http://www.nngroup.com/articles/usability-101-introduction-to-usability/
38. Library of Congress (2017). http://www.loc.gov/standards/mods/
39. Digital Library Federation (2009). https://wiki.dlib.indiana.edu/download/attachments/24288/DLFMODS_ImplementationGuidelines.pdf
40. W3C (2013). https://www.w3.org/TR/sparql11-federated-query/
41. Rubin, J., Chisnell, D.: Handbook of Usability Testing: How to Plan, Design, and Conduct Effective Tests. Wiley Pub, Indianapolis (2008)

A Knowledge-Based Framework for Improving Accessibility of Web Sites

Jens Pelzetter[✉]

FB3 Informatik, Universität Bremen, Bremen, Germany
jens.pelzetter@uni-bremen.de

Abstract. Many sites in the World Wide Web are, unfortunately, not accessible or usable for people with impairments despite several existing guidelines. This paper describes an approach for improving the accessibility of web sites using ontologies as the foundation for several tools. The approach is investigated as a PhD thesis as part of other research that uses ontologies to provide disabled or elderly people with assistance for several everyday tasks.

Keywords: Accessibility · World Wide Web · Ontologies

1 Introduction

For many people the World Wide Web has become their primary source of information. The technologies developed for the Web are used in many other areas. Many organizations have intranets to share informations with their employees. Many applications are provided as Web Applications. These only require a web browser as client.

Developing good web sites or web applications is a quite challenging task. Applications must be usable with a variety of devices (smartphones, tablets and PCs). On top of this, the web sites or web applications should be *accessible*. Accessibility of web sites or web applications is a wide field. Many people only think about blind people when they hear the term *accessibility* in association with web sites or web applications. Accessibility for the web includes much more. Impairments that can affect the way how people may interact with web sites and web applications include problems with all senses and also motoric impairments.

However there are not many tools that support developers to create accessible web sites. Many sites currently available on the web are not accessible and will not become accessible very soon. In this paper, an approach is proposed to use ontologies as the foundation of several tools to improve web accessibility.

The terms *web site* and *web application* will be used interchangeably in this paper since the difference has become very small in the last years. The term *web site* will be used for a collection of web pages, the term *web page* refers to a single page inside a web site.

© Springer International Publishing AG 2017
E. Blomqvist et al. (Eds.): ESWC 2017, Part II, LNCS 10250, pp. 226–235, 2017.
DOI: 10.1007/978-3-319-58451-5_17

2   State of the Art

There are several guidelines for accessible web sites. The most current and widely adopted standard are the *Web Content Accessibility Guidelines 2.0* (WCAG 2.0) [5] created by the W3 Consortium. Most of the other standards for accessible web sites are based on it, for instance the German BITV [22].

Nevertheless many web sites either ignore these standards completely or do not implement them correctly. One reason might be that there are no good, simple to use tools to test a web site for accessibility. All test procedures for accessibility we are aware of either only check some of the very basic requirements or require manual testing.

For the WCAG 2.0, there are several studies which examine, how reliable the results of these tools are [1,3,4]. These studies discovered that the reliability of the results depends on the experience of the tester. Unexperienced testers either find very few or too many problems.

Garrido et al. propose the use of *client side refactorings* to make web pages accessible [8]. Such refactorings use small pieces of JavaScript altering a web pages to make it more accessible. However, the user has to know which refactorings must be applied to a certain web page to make the site accessible for him or her.

The Cloud4all project [18] was a broad attempt funded by the European Union to improve the accessibility of IT technology. In this project an infrastructure was developed that should allow users to store a profile with their preferred settings in the Cloud. Applications can retrieve that profile from the Cloud. The appropriate settings from these profiles are applied to a specific environment using so called *matchmakers*[24]. The primary focus of the project was on traditional, native applications and the usage of native accessibility functions of the operating systems and desktop environments [9]. Besides implementations for Windows and Gnome [2], a proof of concept implementation for web sites was created [19].

Other researchers have tried to use semantic web technologies to enhance the accessibility of web pages and applications. Kouroupetroglou et al. [15] developed a framework which uses annotations in the pages. These annotations can be used by a user agent to provide a better user experience for users of assistive technology. Their research was focused on visually impaired people.

A similar approach is described by Semaan et al. [21], but with a stronger focus on describing the relationship between the several blocks of information on a web page. Their approach was to transform a web page into an RDF document which could then be viewed in the special browser. This special browser uses the additional informations about the document structure to enhance the user experience for users with special needs.

In 2014 the W3C has published ARIA 1.0 [7] (**A**ccessible **R**ich **I**nternet **A**pplications). ARIA uses an approach similar to the approaches described in [15,21]. A web page is annotated with special attributes. The informations provided by these annotations are used by the browser to provide assistive technology with additional information about the elements of a web page.

In other areas, such as mobility assistance, formal modeling approaches have been used with some success [16,20]. Our research group at Universität Bremen and DFKI Bremen has already created a large ontology in OWL-DL [14] describing illnesses, impairments and how they effect abilities such as sight. This ontology describes what of mobility assistance a person with certain impairments requires.

3   Problem Statement and Contributions

Despite the various approaches described in Sect. 2 and the availability of standards like the WCAG 2.0 [5] many web sites are still not accessible. There is also a lack of good tools for checking the accessibility of web sites. All tools which do an automatic check of the accessibility of a web page only check a limited range of requirements. An example for such an tool is the WAVE tool[1]. Test procedures which check a larger range of accessibility requirements require extensive manual work. An example is the BITV-Test[2].

But even if we get good tools for evaluating the accessibility of web pages there will still be many non accessible pages. Therefore it is also necessary to provide tools for disabled users to provide them with a better user experience when accessing non accessible web pages.

The primary research question of this work is whether ontologies can be used to model the knowledge about accessible web pages in a formal way *and* whether they can be used to automatically infer knowledge about accessible web page. One of the possible use cases is a tool which analyses a web page and then uses the knowledge from an ontology about accessibility for web pages to automatically apply refactorings as described in [8].

A common accessibility problem on web pages is an insufficient contrast between the background color and the color of the text. In many cases, this problem could easily be fixed by a client side refactoring. The WCAG 2.0 [5] contains two Success Criteria for contrast. Success Criterion 1.4.3 specifies the minimal requirement for contrast, Success Criterion 1.4.6 specifies an enhanced requirement. Often the only thing necessary to match the requirements and make a web page better readable for people with sight problems is to make the darker color a bit darker and the lighter color a bit lighter.

Another common accessibility problem is that many web sites do not specify their primary language (WCAG 2.0 Success Criterion 3.1.1). This information is needed by screen readers to choose the right pronunciation. A screen reader is a program, which presents the informations normally perceived visually either as speech or as tactile output using a Braille output device. In HTML it is also possible to specify the language of parts of a document by using an attribute (WCAG 2.0 Success Criterion 3.1.2). This information is also useful for screen readers. If the language is provided for a word or part of a web page, which is

---

[1] WAVE Tool: http://wave.webaim.org/.
[2] BITV-Test http://www.bitvtest.eu/.

not in the primary language of the web page, the screen readers can pronounce this word or part correctly.

A third example is the provision of alternative texts for images. These texts are often missing or applied incorrectly. The alternative text for an image is provided by the `alt` attribute of the `img` Element. For decorative images it is necessary to specify an *empty* `alt` attribute. Otherwise the screen readers use the filename as an alternative text. More details about alternative texts for images on web pages can be found in the description of the HTML element in the HTML5 standard [12] and in the description of technique H67 in [6].

There are several challenges along the way to accessible web pages. The first one is to translate the Web Content Accessibility Guidelines 2.0, a semi-formal specification written in natural language, to a formal description (an ontology).

The WCAG 2.0 consists of several documents. The primary one is written in a technology neutral manner. This document has not been updated since 2008. It describes several Success Criteria for accessible web sites, grouped into guidelines and principles. How these Success Criteria can be implemented is described in separate documents. The document describing the possible techniques [6] to implement the Success Criteria is regularly updated (last updated in October 2016) to include new technologies and other developments.

The W3 provides a tool [23] to connect the Success Criteria and the techniques. The challenging part for modeling the ontology is the connection between the Success Criteria and the techniques (which also describe test procedures). For some Success Criteria, the applicable techniques depend on certain conditions (called situations). For others this is not the case. Some techniques are meta techniques, which can be implemented by several other techniques. For some Success Criteria, two technologies are combined into a new one in the descriptions provided by the tool.

A second challenge lies in the nature of web sites. Web sites are written in HTML. The HTML standard has seen many different versions in the last 25 years, the current one is HTML 5 [12]. For several reasons, some web sites have been written in a very sloppy manner. Even today many web sites are not completely valid when checked with a validation tool for HTML. Users usually don't notice this because the user agents (browsers) have been become very good in making sense of defect HTML documents. Thus there are many invalid HTML documents out there. Analyzing them using formal methods should be quite challenging.

4   Research Methodology and Approach

To achieve the goals described above, the first step was to identify the relevant standards for accessible web sites including literature research about current approaches for test tools and methods for making inaccessible web sites accessible. To learn how users with impairments use the web sites, several afflicted users have been interviewed.

The next step is to create an ontology describing the relevant standards and methods to represent the properties of the web site under test. This also involves combining the ontology describing the WCAG 2.0 with the existing ontology of impairments and abilities.

Using the ontologies, some tools will be created as "proof of concept" and tested with users and web developers. The tools for web developers will use the ontologies to guide web developers through an accessibility test of a web site. The accessibility test itself will be semi-automatic. Some requirements can be checked without human interaction. Some requirements can not be check automatically, for example if the alternative text for an image is sufficient. For these requirements the test tool will guide the tester through the test procedure using structured questions.

The tools for users will include a browser plugin using the ontologies and automatic test procedures to automatically apply refactorings to web sites, depending on the abilities of the user and the properties of a web site. An example of an accessibility problem that can thus be fixed is insufficient contrast between foreground and background colors (cf. Sect. 3).

There are several different groups of impairments that affect how users can or can not use web sites. The most well-known are of course blindness or the inability to use standard input devices. However, there are many other forms and degrees of impairment that are relevant for accessible web sites, for example color blindness or a reduced field of vision. Moreover, people with cognitive impairments (caused for example by a head injury) might have problems using web sites. A test setup will be developed to evaluate how helpful the tools developed are for users with different kind of impairments. These relationships between the impairments of person how they effect the abilities of person and how they can compensated will be modelled in several interlinked ontologies.

There are several standards, which can be used by web sites to provide a formal description about their content. These include Microformats [17], Microdata [13], and RDFa [11]. If a web site provides such information, it should be possible to use this information to provide some kind of navigation assistance for the web site. This could be very useful for users with cognitive impairments who have difficulty finding information in a complex web site.

The ontologies developed as the foundation of the tools for web accessibility will be used to verify and test the pattern-based ontology tools developed by our research group. One focus of the ontology tools is to support ontology designers with safe maintenance support for ontologies.

5   Preliminary Results and Current Work

It has been more difficult than expected to translate the WCAG 2.0 into an ontology. Several versions of an ontology describing the WCAG 2.0 had to be developed to test different modeling approaches. Some relations between the concepts of the WCAG 2.0 and their instances could not be expressed with OWL 2 DL alone. To express these relations some SWRL rules [10] are used. Current work is focused on the ontology representing the WCAG 2.0 and the supporting documents using OWL 2 as well as a first simple tool.

The ontology describing the WCAG 2.0 has been divided into three parts representing the concepts and relations in these documents. They do not contain any of the Success Criteria, Techniques etc. Due to the amount of data – the *Techniques for WCAG 2.0* document for example contains several hundred techniques – a web scrapping tool has been developed that extracts the data from the web site of the W3C using the jsoup library[3]. The extracted data is used to create the OWL objects representing the Success Criteria etc. using the OWL API[4]. Fortunately this was quite easy thanks to the well-structured HTML format of the relevant documents. In addition to the six ontology documents for the WCAG 2.0, an additional ontology has been created containing some SWRL rules used to infer whether a web page is satisfying a conformance level.

The WCAG 2.0 defines several conformance levels for the accessibility of web pages. To achieve a conformance level, a web page has to meet several success criteria. This relation is represented by the `requiresSuccessCriterion` object property and the inverse object property `requiredByConformanceLevel`.

For each success criterion, several test cases are provided in the supporting documents, grouped into two major categories: *Techniques* describe an approach for meeting a success criterion in a specific situation; *Failures* describe the conditions under which a success criterion cannot be met by a web page.

The ontology contains classes for situations, techniques and failures, and the web pages under evaluation. Success criteria and situations are related by the object properties/inverses `hasSituation/isSituationForSuccessCriterion`, a success criterion and a failure by `hasFailure/isFailureForSuccessCriterion`, and `hasSufficientTechnique/isSufficientTechniqueForSituation` relates, which techniques are sufficient for a specific situation.

To achieve a particular conformance level, a web page must meet all its success criteria. A success criterion is met by a web page, if the web page does not contain any of the failures and meets the requirements for all situations of the success criterion. To meet the requirements of a situation, the web page has to implement at least one of the sufficient techniques for the situation successfully. The requirements for a situation to be met, if the situation is not applicable for a webpage, are also considered. Whether a web page meets a success criterion, contains a failure, etc., is represented by several object properties in the ontology. The foundation are the object properties `successfullyImplementsTechnique`, `containsFailure` and `notContainsFailure`.

To infer that a web page does not match a success criterion, if one of the failures is present on the web page, the following rule is used:

```
WebPage(?p), SuccessCriterion(?sc), Failure(?f),
isFailureForSuccessCriterion(?f, ?sc), containsFailure(?p, ?f)
-> notMeetsSuccessCriterion(?p, ?sc)
```

Whether a web page matches the requirements for a specific situation is inferred using two rules:

---

[3] jsoup library: https://jsoup.org.
[4] OWLAPI: https://github.com/owlcs/owlapi.

```
WebPage(?p), Situation(?s), notAppliesToSituation(?p, ?s)
-> matchesRequirementsForSituation(?p, ?s)

WebPage(?p), Situation(?s), Technique(?t),
isSufficientTechniqueForSituation(?t, ?s),
successfullyImplementsTechnique(?p, ?t)
-> matchesRequirementsForSituation(?p, ?s)
```

The first rule simply states that a web page matches the requirements for a situation, if the situation is not applicable for the web page, even if the web page does not implement any of the sufficient techniques for the situation. If the web page implements at least one of the sufficient techniques for a situation successfully, the web page matches requirements for that situation.

Now we need to define an rule to infer that a web page meets a success criterion, if the web page matches the requirements for situations of the success criterion and does not contain any of the failures for the success criterion. But due to the Open World Assumption of OWL 2, we cannot simply assert this. Unless stated otherwise, there may be failures or situations that are not described in the ontology. Therefore it is necessary to explicitly assert that there are no other situations or techniques. For this purpose, two classes are added for each success criterion, which contain only the situations and failures for this success criterion.

An example for failures of Success Criterion 1.1.1 in OWL functional syntax is

```
Declaration(Class(wcag20tf:FailureForSuccessCriterion-1-1-1))
EquivalentClasses(wcag20tf:FailureForSuccessCriterion-1-1-1
    ObjectOneOf(<wcag20-techniques#F13> <wcag20-techniques#F20>
                <wcag20-techniques#F72>))
SubClassOf(wcag20tf:FailureForSuccessCriterion-1-1-1
          wcag20-techniques:Failure)
SubClassOf(wcag20tf:FailureForSuccessCriterion-1-1-1
          ObjectHasValue(wcag20tf:isFailureForSuccessCriterion
                <wcag20#successCriterion-1-1-1>))
```

An earlier version of the ontology used cardinality assertions to achieve the same effect, but these made reasoning extremely slow. Using these classes, the following rule states that a web page meets a specific success criterion:

```
WebPage(?p), (matchesRequirementsForSituation
    min 6 SituationForSuccessCriterion-1-1-1)(?p),
(containsFailure max 0 FailureForSuccessCriterion-1-1-1)(?p)
-> meetsSuccessCriterion(?p, successCriterion-1-1-1)
```

This rule uses a class expression with a cardinality requirement. It requires that the web page ?p have a associated to at least six instances of the class SituationForSucessCriterion-1-1-1 by the object property matchesRequirementsForSituation.

The same approach is used to infer that a web page achieves a conformance level. Which success criteria are required by a conformance level is asserted using a class to infer whether a web page achieves a particular conformance level:

```
WebPage(?p), (meetsSuccessCriterion min 25
   SuccessCriterionForConformanceLevel-A)(?p)
-> compliesToConformanceLevel(?p, ConformanceLevel-A)
```

The next step will be to develop ontologies for the test procedures for the techniques, the refactorings and the requirements of users with impairments.

6 Evaluation Plan

When the first tools are ready to use the tools will be tested with various users. The first group of testers will be users with impairments, who will test the tools that automatically apply refactorings to a web site. The sites used in these tests will be evaluated with the tools developed before for testing web sites for accessibility problems to find out, which accessibility problems they might have. The users will have to execute several tasks, such as finding a specific information, on each site. For these tests the users will be split into two groups. The first group will execute the tasks without the support of the tools developed. The second group will use the tools developed for automatically applying client side refactorings and execute the tasks with the support of these tools. The results of the two groups will be compared to find out whether the tools improve the usability of the web sites for these users. To ensure that both groups are balanced regarding their abilities we will do interviews with each participant before they execute the tasks.

The second group of testers will consist of several web developers, who will use the tools in their daily work. This group will include web developers in larger companies with a solid background in programming and web design, but also developers and designers from small companies, who only occasionally develop web sites and have little programming background. Before the testers will start to use the tools, they will be asked to fill in a questionnaire with some questions estimating their experience in the field of accessibility. After about four to eight weeks, the testers will interviewed about their experience with the tools. The web sites that have been created with the help of the tools developed will be analyzed to investigate whether they have less accessibility problems than average sites.

7 Conclusions

The primary goal of the PhD thesis outlined in this paper is to develop a solid foundation for tools to improve the accessibility of web applications and web sites. This will allow developers to provide tools and better web applications and web sites. Users, especially those who rely on assistive technologies, will get web sites and web applications that are hopefully more accessible and thus better usable.

Accessibility for web sites and web applications is a complex domain. The lessons learned while developing the ontologies describing the knowledge about this domain will be helpful for other developers, who create ontologies in other similarly complex domains.

Acknowledgments. I would like to thank my supervisor Bernd Krieg-Brückner and my mentor Vojtěch Svátek for their valuable feedback.

References

1. Alonso, F., Fuertes, J.L., González, Á.L., Martínez, L.: On the testability of WCAG 2.0 for beginners. In: Proceedings of the 2010 International Cross Disciplinary Conference on Web Accessibility (W4A). p. 9. ACM (2010)
2. Antúnez, J.H., Clark, C., Markus, K.: The GPII on desktops in PCs OSs: Windows and GNOME. In: Stephanidis, C., Antona, M. (eds.) UAHCI 2014. LNCS, vol. 8516, pp. 390–400. Springer, Cham (2014). doi:10.1007/978-3-319-07509-9_37
3. Brajnik, G., Yesilada, Y., Harper, S.: Testability and validity of WCAG 2.0: the expertise effect. In: Proceedings of the 12th International ACM SIGACCESS Conference on Computers and Accessibility, pp. 43–50. ACM (2010)
4. Brajnik, G., Yesilada, Y., Harper, S.: Is accessibility conformance an elusive property? A study of validity and reliability of WCAG 2.0. ACM Trans. Accessible Comput. (TACCESS) **4**(2), 8 (2012)
5. Caldwell, B., Cooper, M., Reid, L.G., Vanderheiden, G.: Web Content Accessibility Guidelines (WCAG) 2.0, December 2008. http://www.w3.org/TR/WCAG20/
6. Cooper, M., Kirkpatrick, A., Connor, J.O.: Techniques for WCAG 2.0 (2016). https://www.w3.org/TR/2016/NOTE-WCAG20-TECHS-20161007/
7. Craig, J., Cooper, M., Pappas, L., Schwerdtfeger, R., Seeman, L.: Accessible rich internet applications (WAI-ARIA) 1.0. Technical report, W3C, March 2014. https://www.w3.org/TR/wai-aria/
8. Garrido, A., Firmenich, S., Rossi, G., Grigera, J.: Personalized web accessibility using client-side refactoring. IEEE Internet Comput. **17**(4), 58–66 (2013)
9. Gemou, M., Bekiaris, E., Vanderheiden, G.: Auto-configuration through cloud: initial case studies for universal and personalised access for all. In: 2013 IST-Africa Conference and Exhibition (IST-Africa), pp. 1–8. IEEE (2013)
10. Harrocks, I., Patel-Schneider, P.F., Boley, H., Tabet, S., Grosof, B., Dean, M.: SWRL: A Semantic Web Rule Language Combining OWL and RuleML (2004). https://www.w3.org/Submission/SWRL/
11. Hermann, I., Adida, B., Sporny, M., Birbeck, M.: RDFa 1.1 Primer - Third edition. Rich Structured Data Markup for Web Documents (2015)
12. Hickson, I.: HTML5. A vocabulary and associated APIs for HTML and XHTML, May 2011. http://www.w3.org/TR/html5/
13. Hickson, I.: HTML microdata (2013). https://www.w3.org/TR/microdata/
14. Hitzler, P., Krötzsch, M., Parsia, B., Patel-Schneider, P.F., Rudolph, S.: OWL 2 Web Ontology Language Primer, December 2012. https://www.w3.org/TR/owl2-primer/

15. Kouroupetroglou, C., Salampasis, M., Manitsaris, A.: A semantic-web based framework for developing applications to improve accessibility in the WWW. In: Proceedings of the 2006 International Cross-disciplinary Workshop on Web Accessibility (W4A): Building the Mobile Web: Rediscovering Accessibility? W4A 2006, NY, USA, pp. 98–108 (2006). doi:10.1145/1133219.1133238

16. Krieg-Brückner, B.: Generic ontology design patterns: qualitatively graded configuration. In: Lehner, F., Fteimi, N. (eds.) KSEM 2016. LNCS (LNAI), vol. 9983, pp. 580–595. Springer, Cham (2016). doi:10.1007/978-3-319-47650-6_46

17. Microformats 2. http://microformats.org/wiki/microformats2

18. Ortega-Moral, M., Peinado, I., Vanderheiden, G.C.: Cloud4all: scope, evolution and challenges. In: Stephanidis, C., Antona, M. (eds.) UAHCI 2014. LNCS, vol. 8516, pp. 421–430. Springer, Cham (2014). doi:10.1007/978-3-319-07509-9_40

19. Peinado, I., Ortega-Moral, M.: Making web pages and applications accessible automatically using browser extensions and apps. In: Stephanidis, C., Antona, M. (eds.) UAHCI 2014. LNCS, vol. 8516, pp. 58–69. Springer, Cham (2014). doi:10.1007/978-3-319-07509-9_6

20. Rink, M., Krieg-Brückner, B.: Wissensbasierte Konfiguration von Mobilitäts-Assistenten. In: VDE e.V. (ed.) Zukunft Lebensräume Kongress 2016 (ZL 2016), pp. 201–206. VDE Verlag, April 2016

21. Semaan, B., Tekli, J., Issa, Y.B., Tekli, G., Chbeir, R.: Toward enhancing web accessibility for blind users through the semantic web. In: 2013 International Conference on Signal-Image Technology Internet-Based Systems, pp. 247–256, December 2013

22. Verordnung zur Schaffung barrierefreier Informationstechnik nach dem Behindertengleichstellungsgesetz, Barrierefreie Informationstechnik-Verordnung - BITV 2.0 (2002)

23. How to Meet WCAG 2.0. A customizable quick reference to Web Content Accessibility Guidelines (WCAG) 2.0 requirements (success criteria) and techniques. https://www.w3.org/WAI/WCAG20/quickref/

24. Zimmermann, G., Strobbe, C., Stiegler, A., Loitsch, C.: Global public inclusive infrastructure (GPII)-personalisierte benutzerschnittstellen/global public inclusive infrastructure (GPII)-towards personal user interfaces. i-com 13(3), 29–35 (2014)

Iterative Approach for Information Extraction and Ontology Learning from Textual Aviation Safety Reports

Lama Saeeda[✉]

Faculty of Electrical Engineering, Czech Technical University in Prague,
Prague, Czech Republic
saeeda.lama@fel.cvut.cz

Abstract. Textual aviation safety reports are one of the main resources that contain valuable information to understand incidents and accidents in a high-risk industry such as the aviation domain. The reporting process, hence, is essential to provide these reports. Most of the time, the reporting process is done manually, and typically, poorly structured data are provided by the reporters. Automated content analysis for these reports has attracted researchers to extract the required information to perform many tasks, and they used several techniques to achieve it. Ontologies provide formal and explicit specifications of conceptualizations and play a crucial role in the information extraction process. In this paper, we propose a novel iterative ontology-based approach of information extraction and semantic annotations for aviation safety reports and augmenting back the aviation safety ontology with new concepts and relations depending on the terms already annotated in the discovered report model.

Keywords: Information extraction · Domain ontology · Ontology learning · Safety reports

1 Introduction

Aviation safety reports play a crucial role in data-driven safety oversight in the aviation safety field. Although the content of the reports is typically highly informative, its transformation into structured form (e.g. by means of dedicated reporting forms) is lossy and imprecise which negatively influences their potential for proper safety analyses.

Initial incident and accident reports are the best source of information for extracting the most important knowledge to feed the preliminary[1] reports' building process [20]. One of the main tasks in the process of building the preliminary report is to detect the type of event described in the initial aviation safety report so that an appropriate form can be displayed to the user to capture additional required information regarding the specific accident or incident.

---

[1] A preliminary report is created by the safety department of an organization and sent to the authority.

© Springer International Publishing AG 2017
E. Blomqvist et al. (Eds.): ESWC 2017, Part II, LNCS 10250, pp. 236–245, 2017.
DOI: 10.1007/978-3-319-58451-5_18

Text processing for such reports is essential for simplification of the safety reporting process. Many techniques have been proposed for the purpose of classification, cause identification, and knowledge extraction from the aviation safety reports. Most of these techniques are mainly following linguistic or statistical methods. Semantic web technologies provide the advancement in information systems by assigning semantics to information by means of shared formal ontologies. These ontologies are also used as the main resource for semantic annotation authoring. Ontologies are especially useful because they support the exchange and sharing of information as well as reasoning tasks [14]. An ontology containing a class hierarchy relevant to the aviation safety domain can simplify the reporting process by enabling the reporter to process the data in a controlled way by the mean of the ontology, and hence, a more relevant and accurate data will be provided by the reporter. This will assure a better experience for the safety management of the statistics business intelligent (BI) user, who will benefit from the targeted, without noise, and less biased statistics, which will improve the quality of the data, the speed of the reporting process on the general level and provide more precise results of the BI. However, using such ontologies depends directly on the availability of this ontology in the target domain. The domain ontology construction process consumes a lot of time and resources and requires a lot of efforts by knowledge engineers and human experts. Researchers have been discussing the process of automatically as well as semi-automatically building ontologies. Ontology learning from the textual corpus is the set of methods and techniques used for building an ontology from scratch, enriching, or adapting an existing ontology in a semi-automatic fashion using several knowledge and information sources [16].

In this paper, we propose to use a knowledge-based approach for extracting useful information from the aviation safety reports, as well as we propose a methodology for semi-automated ontology learning for the aviation safety ontology (Fig. 1).

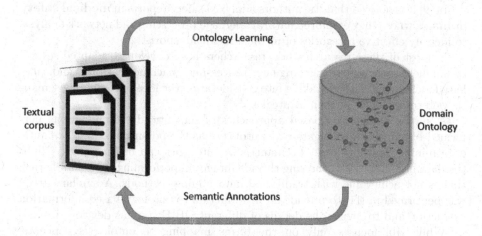

Fig. 1. The iterative approach

2 State of the Art

In the recent years, researchers have paid increasing attention to automatically analyzing textual safety reports in different domains.

For example, **transportation domain**, like analyzing maritime accident investigation reports [8], where text mining methods were applied to extract causal relations from maritime accident investigation reports collected from the Marine Accident Investigation Branch, and railroad accident investigation reports [3], where accident reports were analyzed using the text mining techniques of probabilistic topic modeling and k-means clustering to identify the recurring themes in major railroad accidents. Also, the performance of four machine learning paradigms applied to modeling the severity of injury that occurred during traffic accidents were performed in [17].

Also, safety reports in aviation domain were carefully studied. In [1] the authors presented a method of automatic Aviation Safety Reporting System (ASRS) shaping factor classification based on the most relevant words from a subjectivity lexicon, and [7] where the problem of cause identification from aviation safety reports was introduced to the NLP community as a multi-class, multi-label text classification task, and a bootstrapping algorithm was presented that automatically augments a training set by learning from a small amount of labeled data and a large amount of unlabeled data. Also, a survey of different NLP techniques designed and used to manage and analyze aviation incident reports was performed in [2].

Safety reports were also studied in other domains, such as **construction safety domain**, where in [5] the methodology consists of pre-processing the accident reports and weighting terms in order to apply a data-driven unsupervised K-Means-based clustering approach, in order to classify the collected reports in four clusters, each reporting a type of accident. and in [4], a natural language processing system based on hand-coded rules and dictionaries of keywords was proposed to extract precursors and outcomes from unstructured injury reports.

In [9] it is noticed that the authors analyzed safety reports in **medical safety domain**, where they employed natural language processing and network analysis to identify effective categories of medical incident reports.

In most of these research work, researchers have developed/adapted several techniques, such as natural language processing, machine learning, and rule-based techniques, without taking into consideration the idea of combining semantic web technologies in their strategies.

A survey of ontology-based approaches of semantic data mining was performed by [6], that investigate why ontology has the potential to help semantic data mining, and how formal semantics in ontologies can be incorporated into the data mining process, showing the advantages in performing data mining task that is not achievable with traditional data mining methods. A similar survey was performed in [11] to provide an introduction to ontology-based information extraction and to review the details of different OBIE systems developed.

While [10] focuses only on the terms mapping to ontology's concepts, performing simple NLP pre-processing and string matching to the labels, other studies made use of the hierarchy (structural nature) of the ontology, and the use of the relations between the extracted terms as in [22,23].

Using domain-ontologies relies on the availability of this ontology in the domain of study. However, the automatic ontology construction is not a trivial task and requires lots of human intervention in some stages of ontology construction. There are various approaches and tools available for automatic construction of ontology from a textual corpus. In [13] they Focus on presenting a method for learning axioms from text based on named entity recognition. [14] describes a new ontology learning approach that consists of a method for the acquisition of concepts and its corresponding taxonomic relations, where also axioms disjointWith and equivalentClass are learned from text without human intervention. [15] focuses on identifying the relationships between medical concepts as defined by the REMed (Relation Extraction from Medical documents) solution that aims at finding the patterns that lead to the classification of concept pairs into concept-to-concept relations.

Creating large annotated textual corpus for the training and evaluating tasks is, however, prohibitively expensive. To mitigate this problem, the notion of supplementing labeled data with features derived from large amounts of unlabeled data has been explored in [12].

3   Problem Statement and Contributions

Reporting process of aviation safety incidents and accidents needs to be done clearly and easily. To achieve that, a reporting tool has been built on the top of the aviation safety ontology in [18]. In order to make reporting process more user-friendly, as well as make it easy and logical, a smart form generation based on the event-type and other attributes is needed, in order to support the reporting process by reducing the list of attributes that have to be filled, only to those related and relevant for a specific event type. In order to detect the event-type in the initial safety report, a comprehensive textual analysis process has to be performed, taking into consideration the unstructured nature of the initial input report, which is usually full of jargon.

So far, our work has focused on text annotations. It has achieved high precision semantic annotations for aviation safety reports, detecting the main event and event-type in the safety report based on the aviation safety ontology as well as all the participants, temporal and spatial information, including all the information that can help to construct a dynamic form suited to the actual report. However, a large space of improvement can be done regarding the detection of the terms which have not been yet introduced in the ontology, based on the iterative method of detection and learning. The full scenario is explained in the Fig. 2.

To overcome these problems, we are planning to proceed to perform several tasks, including indicate a more accurate mapping to the corresponding concept in the ontology on the level of concepts, as well as to make use of the ontology hierarchy to detect the relations between the concepts for the better understanding of the aviation safety reports. Also, making use of the already-detected

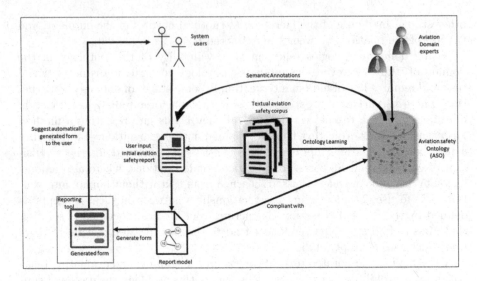

Fig. 2. Full scenario of the iterative approach of annotating and learning back the ontology

concepts for augmenting the ontology with new concepts and adding them in the proper position in the ontology hierarchy with a context-based approach.

Achieving these tasks will help to investigate several research questions as following:

- What is the impact of utilizing the domain-specific ontology on classification performance?
- How this iterative approach of recognition of the terms and augmenting back the ontology will improve the background knowledge and hence, the event type detection.

4   Research Methodology and Approach

We aim to improve the results of semantic annotations in our previous work [20]. This includes improving annotations of the terms, disambiguating them, and detecting the relations and the facts. This will be done by optimizing the pipeline described in Fig. 3, tuning the parameters of the tools and exploring new algorithms to be included in the pipeline.

In Fig. 2 we are showing the full scenario. Following, the most important relevant components are explained:

Semantic annotations: We are currently developing the Aviation Safety Text Analyzing Tool. Entity recognition tools that allow the usage of custom background knowledge inside were tested separately, then the most accurate tools that served our purpose the best were chosen. We combined these tools in one

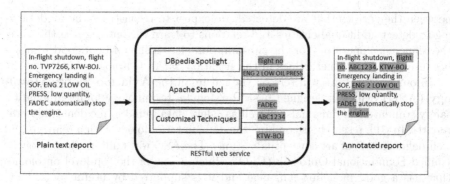

Fig. 3. Text annotation processing pipeline

pipeline to cooperate together. As shown in Fig. 3, Apache Stanbol[2] is one of the tools that provides the ability to work with custom vocabularies and creating custom indexes upon it. It also comes with a list of enhancement engines implementations, with the ability to build a specific one to get the most benefit out of the tool. This allowed us to build a chain of enhancement engines that fits perfectly to the aviation-safety concepts detection. DBpedia Spotlight[3] is another entity recognition tool that offers to create a spotlight model on the user's own server to model the occurrences of resources with the context in which they are mentioned. We are also taking into consideration the entities that are not possible for the current tools to detect, in spite of their ability to detect mentions from a specific terminology. For this complication that stems directly from the nature of the aviation domain, such as callsigns, registration marks, flight numbers, airport names abbreviations, etc., we are using special techniques for every case as discussed in [20]. The output of Apache Stanbol, DBpedia spotlight and the techniques for the special terms were parsed, merged, and optimized in a RESTful web service and output the mentions being detected with their proper mapping to the ontology hierarchy.

The clear line we are walking through now is the relationships recognition. Taking into consideration the ontology subclasses and subproperties, the detected entities that are mapped to the ontology, and the corpus as an input, then using the relations between the subclasses to detect the relationships and the facts between the entities.

Ontology learning: In [21], researchers proposed learning from structured knowledge (XML documents of card design). They used some ontological refinements. They designed an ontology on the top of the documents, then based on the data, the ontology was used to refine the learning task so they can learn according to the hierarchy. In this work, we have a more difficult task because of the unstructured nature of the input. Based on the entities and relations

---

[2] https://stanbol.apache.org.
[3] https://github.com/dbpedia-spotlight/dbpedia-spotlight.

between the entities that are detected, we expect to reconstruct the model trying to detect additional entities and relations to learn the ontology. In this case, the refinement operators for the subsumption classes should be a good solution because of the huge hierarchies in the Aviation Safety Ontology (ASO).

In our approach we will rely on the nature of the **Aviation Safety Ontology (ASO)**, which is designed based on domain terminology found in aviation safety standards and manuals, as well as safety data found in incident and audit reports. In [19], researchers combine the domain terminology with a relevant and well-defined state of art conceptualizations. The ASO was built on the top of the Unified Foundational Ontology (UFO), which is one of the top-level ontologies that has a good modeling language and it is supported by useful tools. UFO presents a level of abstraction that forms a perfect point to start with the annotation and learning process. Then, as mentioned in [19], during the analysis of the aviation safety domain, several modules and relationships among them could be identified. For example, the Aviation Safety Core (ASC) ontology which contains the Aviation ontology, that consists of the common aviation domain terms, such as Aircraft and Flight. and Safety ontology, which consists of the fundamental conceptualizations necessary for the management of safety information. This way of design can lead to a big potential of improvement in semantic annotations of the aviation safety reports, and potentially, makes the ontology learning process accurate and very specific task.

This iterative text annotating and ontology learning methodology should guarantee more reliable understanding for the new reports that need to be processed, as well as enriching the Aviation Safety Ontology with new terms and relations, and supporting the complicated and expensive process of building the domain ontology.

We will also take into consideration the nature of the targeted **domain corpus** and the characteristics and the structure of the aviation safety reports, (i.e. common abbreviations, registration marks, pieces of controlled language, event chains, etc.).

5   Preliminary Results

To this end, different Linked Data Knowledge Extraction tools with respect to a domain-specific vocabulary had been tested, then we chose the tools that allow the best results of entity recognition, combining them into one pipeline, and making them working together, as well as with other features that we added, taking into consideration some very specific terms and abbreviations used in the aviation field.

We built a tool that integrates several techniques inside in order to provide high precision reports' annotations in aviation safety domain in order to be used directly in practice. In our preliminary results, the precision scores high rates in most of the cases, it even reaches to 100% rate for some reports. On the other hand, the recall scores low rates. You can check the preliminary results of comparison of ontology-based entity recognition to the non-ontological approach of samples of reports in Table 1 and Sect. 4 in [20].

Currently, we are facing the problem of improving the achievement of reliable entity recognition, with the iterative methodology which will, potentially, improve the recall in a significant percentage.

Also, relations detection between the concepts, and the results for the new potential concepts for semi-automatic ontology learning. Our target now is to proceed with the iterative methodology that we discussed before, of ontology learning and augmenting the ontology with new terms, and relations between the terms, in a graph building process within a context-based approach, and to measure how much improvement will be gained with the semantic annotations.

6   Evaluation Plan

Due to a noticeable fact in the aviation safety domain, many public safety reports are available. However, most of these reports are unannotated or poorly automatically annotated, while a very few reports are actually well annotated. This makes the process of corpus construction a very hard task for the evaluation process, that requires extensive time and effort of the experts.

As mentioned in [20], we are creating a high quality, very precise gold standard corpus out of, mainly, initial aviation safety reports taken from different authorities' resources, for example, UZPLN[4], where they have their public aviation investigation incidents and accidents reports. Also, the corpus contains confidential data provided by the partners of the current project that this work is included in.

Experts in aviation domain manually annotated domain terms (entities) in each report with respect to the Aviation Safety Ontology (ASO) that mentioned earlier. Technically, they used the General Architecture for Text Engineering (GATE) tool[5].

To this end, this corpus consists of 80 high quality annotated documents that will hopefully grow fast through time. We need this kind of corpus for the evaluation process of the annotation pipeline using the well-known recall, precession and F1 score metrics. Regarding the ontology learning evaluation task, we will only be able to evaluate the algorithms and the methods of the generated concepts and relations but not the quality of the ontology. Creating ontologies automatically doesn't guarantee the quality of these ontologies. We want the ontology that we are creating to keep some reliable level of quality. In order to achieve that, aviation ontology is revised by experts who understand the problem. This way experts will not be forced to process or to read every single report. Instead, they will be offered some typical notions, relationships, and data patterns that they can generalize into a logical knowledge.

7   Conclusions

In this proposal, a novel iterative methodology of semantic annotations and ontology learning has been presented, in order to improve the understanding

---

[4] http://www.uzpln.cz/.
[5] https://gate.ac.uk/.

of the huge jargon unstructured incidents and accidents safety reports in the aviation domain, and supporting the smart form generation for easier and more efficient reporting process.

Acknowledgments. I want to thank Dr. Petr Křemen for his support in accomplishing this proposal. Furthermore, this work was partially supported by grants No. TA04030465 Research and development of progressive methods for measuring aviation organization's safety performance of the Technology Agency of the Czech Republic, No. SGS16/229/OHK3/3T/13 Supporting ontological data quality in information systems of the Czech Technical University in Prague.

References

1. Switzer, J., Khan, L., Bin Muhaya, F.: Subjectivity classification and analysis of the ASRS corpus. In: 2011 IEEE International Conference on Information Reuse & Integration (2011)
2. Tanguy, L., Tulechki, N., Urieli, A., Hermann, E., Raynal, C.: Natural language processing for aviation safety reports: from classification to interactive analysis. Comput. Ind. **78**, 80–95 (2016)
3. Williams, T., Betak, J., Findley, B.: Text mining analysis of railroad accident investigation reports. In: 2016 Joint Rail Conference (2016)
4. Tixier, A.J.-P., Hallowell, M.R., Rajagopalan, B., Bowman, D.: Automated content analysis for construction safety: a natural language processing system to extract precursors and outcomes from unstructured injury reports. Autom. Constr. **62**, 45–56 (2016)
5. Chokor, A., Naganathan, H., Chong, W.K., Asmar, M.E.: Analyzing Arizona OSHA injury reports using unsupervised machine learning. Procedia Eng. **145**, 1588–1593 (2016)
6. Dou, D., Wang, H., Liu, H.: Semantic data mining: a survey of ontology-based approaches. In: Proceedings of the 2015 IEEE 9th International Conference on Semantic Computing (IEEE ICSC 2015) (2015)
7. Persing, I., Ng, V.: Semi-supervised cause identification from aviation safety reports. In: Proceedings of the Joint Conference of the 47th Annual Meeting of the ACL and the 4th International Joint Conference on Natural Language Processing of the AFNLP, ACL-IJCNLP 2009, vol. 2 (2009)
8. Tirunagari, S.: Data mining of causal relations from text: analysing maritime accident investigation reports. arXiv preprint arXiv:1507.02447 (2015)
9. Fujita, K., Akiyama, M., Park, K., (Nakagami) Yamaguchi, E., Furukawa, H.: Linguistic analysis of large-scale medical incident reports for patient safety. In: MIE, pp. 250–254 (2012)
10. Sfakianaki, P., Koumakis, L., Sfakianakis, S., Iatraki, G., Zacharioudakis, G., Graf, N., Marias, K., Tsiknakis, M.: Semantic biomedical resource discovery: a natural language processing framework. BMC Med. Inform. Decis. Mak. **15** (2015)
11. Wimalasuriya, D.C., Dou, D.: Ontology-based information extraction: an introduction and a survey of current approaches. J. Inf. Sci. **36**, 306–323 (2010)
12. Henriksson, A., Kvist, M., Dalianis, H., Duneld, M.: Identifying adverse drug event information in clinical notes with distributional semantic representations of context. J. Biomed. Inf. **57**, 333–349 (2015)

13. Rios-Alvarado, A., Lopez-Arevalo, I.: Ontology learning from text: method for learning axioms. Technical report (2012)
14. Rios-Alvarado, A.B., Lopez-Arevalo, I., Tello-Leal, E., Sosa-Sosa, V.J.: An approach for learning expressive ontologies in medical domain. J. Med. Syst. **39** (2015)
15. Barbantan, I., Porumb, M., Lemnaru, C., Potolea, R.: Feature engineered relation extraction - medical documents setting. Int. J. Web Inf. Syst. **12**, 336–358 (2016)
16. David Sanchez, R.: Domain ontology learning from the web. Ph.D. thesis, Universitat Politecnicade Catalunya (2007)
17. Chong, M., Abraham, A., Paprzycki, M.: Traffic accident analysis using machine learning paradigms. Informatica **29**(1) (2005)
18. Vittek, P., Lališ, A., Stojic, S., Plos, V.: Challenges of implementation and practical deployment of aviation safety knowledge management software. In: Communications in Computer and Information Science Knowledge Engineering and Semantic Web, pp. 316–327 (2016)
19. Kostov, B., Ahmad, J., Křemen, P.: Towards ontology-based safety information management in the aviation industry. In: 13th International Conference, IESD (2016)
20. Saeeda, L., Křemen, P.: Text analyzing of aviation safety reports. WIKT & Data a Znalosti (2016)
21. Žáková, M., Železný, F., Garcia-Sedano, J.A., Tissot, C.M., Lavrač, N., Křemen, P., Molina, J.: Relational data mining applied to virtual engineering of product designs. In: Muggleton, S., Otero, R., Tamaddoni-Nezhad, A. (eds.) ILP 2006. LNCS (LNAI), vol. 4455, pp. 439–453. Springer, Heidelberg (2007). doi:10.1007/978-3-540-73847-3_39
22. Wong, M.K., et al.: A multi-phase correlation search framework for mining non-taxonomic relations from unstructured text. Knowl. Inf. Syst. **38**(3), 641–667 (2012)
23. Reyes, J.A., Montes, A.: Learning discourse relations from news reports: an event-driven approach. IEEE Lat. Am. Trans. **14**(1), 356–363 (2016)

Integrative Data Management for Reproducibility of Microscopy Experiments

Sheeba Samuel[(✉)]

Heinz-Nixdorf Chair for Distributed Information Systems,
Friedrich-Schiller University, Jena, Germany
sheeba.samuel@uni-jena.de

Abstract. Reproducibility is a fundamental factor in every domain of science since it allows scientists to trust data and results. The scientific community is interested in the results of experiments which are reproducible, reusable and understandable. In this paper, we present our work towards reproducibility of scientific experiments taking into account the use case of microscopy. We aim to analyze the components that are vital for reproducibility and to develop an integrative data management platform for scientific experiments. In this article, we show the use of Semantic Web technologies to conserve an experiment environment and its workflow. This allows scientists to ask queries related to an experiment and compare results. We present our approach for scientists to represent, search and share their experimental data and results to the scientific community for better data interoperability and reuse. Our overall goal is to extend data management and Semantic Web technologies to enable reproducibility.

Keywords: Reproducibility · Experiments · Ontology · Microscopy · Provenance

1 Introduction

Recent advancements in science and technology have brought a new range of challenges for scientists regarding the reproducibility of their research. Reproducibility has become a lot more difficult to achieve today because experiments and their setup have become much more complex. A sustainable, reliable and scalable data management platform is required for scientific experiments which generate a large volume of heterogeneous data. Apart from data management, it is required to ensure reproducibility of experimental data and results.

An experiment is said to be reproducible [18] when it can be repeated under different conditions to get the same results. This can occur when the experiment is carried out by another scientist in a different location using different devices and materials. To make an experiment reproducible, the provenance of the experiment and processing environment must be captured. Provenance is the source of information that is used to describe the entities and processes involved

© Springer International Publishing AG 2017
E. Blomqvist et al. (Eds.): ESWC 2017, Part II, LNCS 10250, pp. 246–255, 2017.
DOI: 10.1007/978-3-319-58451-5_19

in generating a resource. The provenance of an experiment describes who performed the study and when, the materials used and how they were produced, the last time it was modified, the devices used and their settings, the experimental procedures used etc. [17]. Semantic Web-based representations of provenance help in better data interoperability and reuse.

The main focus of our research is to extend data management and Semantic Web technologies in order to better support reproducibility. In the initial phase, we work towards developing an integrative data management platform which can enable reproducibility of scientific experiments with the help of Semantic Web technologies. This research focuses on the use case of microscopy.

2   State of the Art

Scientists need to store information about an experiment and its workflow so that they can share this with other collaborators in an understandable way. This leads us to focus on three important aspects of our research which can aid work on the reproducibility of scientific experiments: (1) Scientific Data Management, (2) Semantic Web technologies for capturing provenance and (3) Scientific Workflows.

(1) *Scientific Data Management*

Advancements in data storage solutions allow successful preservation, processing and analysis of large volume and variety of data. Scientific data management brings challenges like scalability, heterogeneity, sharing, transformation and quality of data. Many platforms are developed to support scientific data management for general or specific requirements of projects such as the European Bioinformatics Institute[1] (EBI) for genomic data, BacDive[2] for bacterial data, BExIS[3] for biodiversity data, myExperiment[4] for sharing bioinformatics workflows. Demchenko et al. [6] show how cloud-based services provide support for scientific data infrastructures.

A number of platforms aim at providing data management and analysis of microscopic images. OMERO [1] and BisQue [9] are two examples for storage solutions for microscopy images. OMERO, developed by the OME consortium, is an open source client-server platform for visualization, management and analysis of images generated from a microscopy experiment. Since it provides a rich set of different image file formats and flexibility to extend features, we selected OMERO as a suitable data management platform to support microscopic data infrastructure.

(2) *Semantic Web Technologies for Capturing Provenance*

Semantic Web technologies provide possibilities to represent experiment data

---

[1] http://www.ebi.ac.uk/.

[2] http://bacdive.dsmz.de/.

[3] https://www.bexis.uni-jena.de.

[4] http://www.myexperiment.org.

in a more understandable and reusable way to scientists and in particular in a machine-understandable way. Various ontologies are developed and used in various domains to help scientists annotate resources and support data interoperability. PROV-O [10] is a general purpose ontology developed by the W3C working group to model the entities and the activities that produced them. It provides high flexibility for extension so that it can be used for specific applications.

PROV-O has been extended in various works for specific purposes. For example, Ciccarese et al. [3] evaluate why PROV-O was selected for capturing provenance of web resources. They present the PAV ontology which helps to capture provenance, authorship and versioning of the resources on the web. Compton et al. [4] combines the Semantic Sensor Networks Incubator Group's ontology (SSNO) and PROV-O to describe sensor data.

Several authors have built ontologies for microscopy. The Cellular Microscopy Phenotype Ontology (CMPO) [7] provides phenotypic observations related to cellular components. Another work [8] describes the development of an Ontology for an Integrated Image Analysis Platform to enable Global Sharing of Microscopy Imaging Data. They present an ontology to describe data from microscopic images by converting the Open Microscopy Environment (OME) data model to the Resource Description Framework (RDF) schema. These works focus on using ontologies for describing biological structures and annotating microscopic images.

Moreau in his paper [13] describes reproducibility semantics for the Open Provenance Model (OPM)[5]. It is a specification of a reproducibility service and defines reproducibility formally with a mathematical explanation of OPM graphs. This semantics which takes the form of a denotational semantics, is the basis of a theory of provenance-based reproducibility.

(3) *Scientific Workflows*

Scientific workflow management systems [5] model the flow of data through a series of computational steps performed in an experiment. Several workflow management systems have been developed over the past years which are either generic or specific to a domain. Systems like Kepler [11] and Vistrails [2] provide facilities to design, execute and rerun the scientific workflows and also provide a visual interface for composing workflows. Santana-Perez et al. [15] present a semantic approach to attain reproducibility of computational environments in scientific workflows by documenting the scientific workflow and conserving the execution environment using semantic vocabularies.

In spite of all the advantages and the richness in features, scientists are hesitant to abandon the tools they are used to and try new ones instead, thus, in many scientific communities, the uptake of scientific workflow management systems has been slow. This motivates the work for YesWorkflow [12] and noWorkflow [14]. YesWorkflow extracts comments from the scripts and provides a graphical rendering of the workflow. The noWorkflow tool

---

[5] http://openprovenance.org/.

transparently captures provenance from Python scripts and supports different kind of analysis on them. Recent work on combining YesWorkflow and noWorkflow captures provenance of results generated by scripts written by the scientists in an experiment [16]. This work takes benefits of both the systems by capturing provenance which is collected from the structure of scripts, events occurred during script execution, annotations in the comments of scripts and the files generated by the scripts.

We focus our research on end-to-end reproducibility of scientific experiments by integrating scientific data management and Semantic Web technologies.

3 Problem Statement

The main goal of our research is to enable reproducibility by extending data management and Semantic Web technologies. Inorder to start with the research, we focus on the use case of microscopy experiments. Recent advances in microscopy techniques make the study of biological systems more promising. The need for the research on this area arises from the Collaborative Research Center ReceptorLight[6]. Scientists from this joint project work together to understand the function of membrane receptors with the help of high-performance microscopy methods. Membrane receptors are protein molecules which receive chemical signals from outside a cell and distribute the signals to other parts of the cell. Through this project, scientists want to understand the minute interactions happening in the biological structures and processes. Using high-resolution imaging of these receptors, scientists can gain new insights on neurological autoimmune diseases and other areas.

Discussions with scientists from various domains working in this project brought light to the challenges they face related to the reproducibility aspects. One of the challenges faced by the scientific community is that most of the information related to the experiment are not integrated to the digital systems. They still use lab notebooks (analog) to record their data as they perform experiments. The information in these notebooks is of great value as they contain the description of experiment procedures, resources used, data and results. The difficulty arises when the data and results have to be shared between scientists from different locations. Shared understanding of data is essential so that the data can be reused for new experiments and analysis.

. Samples and resources used in the experiment cannot be preserved for long in many biological and medical studies, unlike the computational science domain. It is required to digitally conserve the execution environment of an experiment which consists of different devices, software and materials. Data collected by the experimenter from different data sources using USB sticks are stored in personal hard disks. There are greater chances of losing data and the different modifications of the data if they are not versioned.

Another challenge faced by the scientists is the big data. Each measurement of a microscopy experiment can produce terabytes of data and images. Scientists

---

[6] http://www.receptorlight.uni-jena.de/.

have to make a choice to keep the data they are interested in and discard the rest of the data. So it is important to have a scalable and high-performance data management approach to handling the large volume of experimental data and results.

We have identified four main research questions based on the challenges faced by the scientists and ways to achieve the vision of reproducibility.

RQ-1 - How to capture the provenance of a scientific experiment through an integrative data management? Which are the vital components required to attain reproducibility?

RQ-2 - How to represent a scientific experiment and its execution environment with the help of Semantic Web technologies to enable reproducibility, data interoperability and reuse?

RQ-3 - How to provide a scalable and high-performance platform to handle the experimental data and results?

RQ-4 - How to enhance current Semantic Web languages with reproducibility qualifiers?

We focus our research based on the hypothesis that Semantic Web enabled scientific data management platform can be created that facilitates the reproducibility of scientific experiments.

4   Research Methodology and Approach

The research methodology and approach we present here are based on the research questions defined in Sect. 3. A high-level view of our research methodology in the first phase is illustrated in Fig. 1. The first step in our approach is to collect requirements and understand what type of information scientists need to reproduce the experiments. In this phase, it is required to identify the factors that can enable reproducibility. Continuous discussions with scientists help us to gain the domain knowledge and the challenges faced by them.

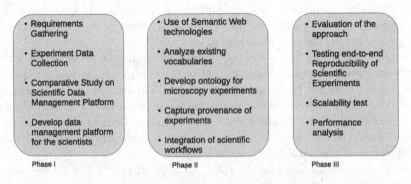

Fig. 1. Our research methodology and approach

For RQ-1, we explored the literature to find a suitable data management platform and selected a platform which can be extended to capture the provenance of microscopy experiments. We understood the deployment setup of the experiments and designed the system in such a way that scientists could conserve the experimental data along with the images in one place. In this phase, we consider different approaches to capture provenance and analyze whether those are sufficient to enable reproducibility. We are analyzing the various components required to enable reproducibility.

To answer RQ-2, we need to understand how data stored in computers in addition to their lab notebooks can benefit scientists. We realize the potential uses of semantic web technologies and how it can be used to describe the scientific workflow. Based on the literature review, we analyzed the different existing vocabularies in the scientific domain, particularly for microscopy. We comprehended the details of experiments we need to capture for our use-case and extended the existing ontology based on them. We will analyze whether all the questions related to reproducibility can be answered with this ontology or the ontologies need to be further refined. We will evaluate whether ontology-based representation of experimental data is enough to enable reproducibility.

For RQ-3, we will analyze the existing approaches for handling scalability. We will test the performance of the system and provide an optimal solution for scientists to handle their data.

For RQ-4, we will collect and analyze the questions that are asked by scientists concerning reproducibility. We will analyze how the questions and the approach be generalized for all the scientific domains. We will find out the type of qualifiers needed to extend Semantic Web languages. We will consider various methods to formalize the process needed to enable reproducibility.

5   Preliminary Results

Based on the literature review, we did a comparative study on two existing storage platform systems, OMERO [1] and BisQue [9]. We came to the conclusion to select OMERO for our requirements because of the richness in its features and higher flexibility to extend the platform. We collected data and requirements from the discussions we had with the scientists working in the CRC ReceptorLight project.

The initial goal of our approach is to document the description of an experiment and its execution environment conditions. The information includes experimental details, materials and devices used in the experiment and their settings, the time of each activity performed and the standard operating procedures used. The current prototype is developed [17] based on OMERO to achieve the initial goal.

We extended OMERO's server database to include the data schema provided by the scientists. OMERO's web client was extended to provide a facility to input the experimental data. It also helps the scientist to associate the experiment to the image results generated from the experiment. Users can view all the information of an experiment at one place.

Figure 2 presents the high-level system architecture developed for the data management of microscopy experiments. Scientists use different devices like a microscope, an electrophysiologic device for performing experiments. The workstations associated with them are installed with proprietary software of the device manufacturer. The description of the materials and the procedures used in the experiment and the execution environment parameters are noted down in the lab notebooks. The data and results obtained are stored in personal external devices. A desktop client was developed to deploy in the workstations associated with these devices as they are not connected to the internet due to security reasons. It allows the scientists to input the data as and when they perform experiments. Later they can upload this information to the server whenever the internet connection is available. Through the web client, they can share the experiment information to other scientists for their review. Based on the permissions and roles assigned to the scientists, they can view or edit the information related to the experiment. The web client also provides a facility to search experiment related information.

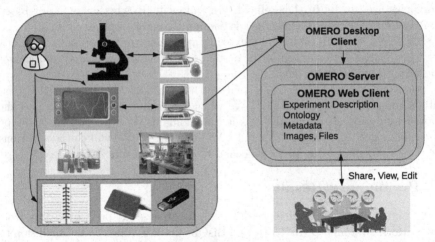

Fig. 2. The proposed approach

We have developed an ontology, REPRODUCE-ME (Reproduce Microscopy Experiments)[7] to describe an experiment and its execution environment. The ontology is built to capture the provenance of an experiment, the materials and devices used and their properties, standard operating procedures and the people who are responsible for an experiment. This is developed by extending the existing ontology, PROV-O. With the help of classes and properties of PROV-O and the new classes added in REPRODUCE-ME Ontology, it is possible to describe the entities, activities, agents and their role in a scientific experiment. The prefix "repr:" is used to indicate the namespace "http://fusion.cs.uni-jena.de/fusion/repr".

---

[7] http://fusion.cs.uni-jena.de/fusion/repr/.

Figure 3 shows the main concepts and properties of REPRODUCE-ME ontology. `prov:Entity`, `prov:Agents` and `prov:Activity` are the main concepts in PROV-O. The experiment materials and devices used in an experiment are extended from the concept `prov:Entity`. The people who are involved in producing materials and performing experiments are extended from `prov:Agents`. All the actions performed in an experiment are extended from the `prov:Activity` class. The standard operating procedures used in the experiment is extended from `prov:Plan`. Several objects and data properties are added to describe the experiment and its execution environment. Scientists can make semantic queries using SPARQL with the help of ontology.

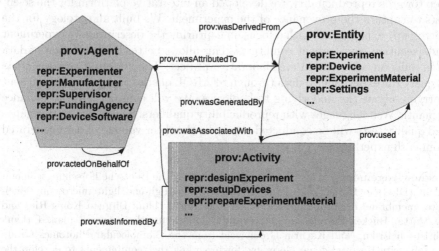

Fig. 3. A part of REPRODUCE-ME ontology

6 Evaluation Plan

To evaluate the research questions mentioned in Sect. 3, we will validate the approach using microscopy experiments. Later it will be validated by scientists from other domains. The validation of the approach will be achieved when scientists from different locations can reproduce the experiments based on the shared description provided by the REPRODUCE-ME ontology. The ontology developed for the microscopy experiments will be continuously validated, revised and corrected by the team of scientists. The detailed list of queries that a scientist would like to pose based on an experiment will be collected from the scientific community. The competency questions that the ontology can support will be clearly listed. We' will evaluate whether the provenance captured through the prototype is sufficient enough to attain end-to-end reproducibility of scientific experiments.

Scalability, performance and data quality will also be considered during the evaluation. The current prototype [17] will be manually tested by the domain scientists. Test cases will be formulated to validate the system and the results of queries. We are interested in scientists testing our system with a large amount of data and different experiment setups.

7   Conclusions

Reproducible research brings great value and benefits for the scientific community. The overall objective of our research is to extend data management and Semantic Web technologies to enable reproducibility. In this research, we aim to examine the suitable components required for reproducibility. As a first step towards reproducibility, we developed an integrative platform for the scientists to capture the provenance of the experiment. We built an ontology for the microscopy experiments with a focus on capturing the description of experiment and execution environment conditions. This allows better interpretation of data from different scientists in a collaborative project. The system allows scientists to query the experimental data through SPARQL queries and get results without worrying about the underlying technologies. We will consider ways to enhance Semantic Web languages with reproducibility qualifiers. Scalability, performance and quality tests will be conducted to handle the sheer volume of data generated from each experiment.

Acknowledgements. This research is supported by the Deutsche Forschungsgemeinschaft (DFG) in Project Z2 of the CRC/TRR 166 High-end light microscopy elucidates membrane receptor function - ReceptorLight. I thank Birgitta König-Ries and H. Martin Bücker for their guidance and feedback for the research plan. I thank Christoph Biskup and Kathrin Groeneveld from the Biomolecular Photonics Group at University Hospital Jena, Germany, for providing the requirements to develop the proposed approach and validating the system.

References

1. Allan, C., Burel, J.M., Moore, J., Blackburn, C., Linkert, M., Loynton, S., MacDonald, D., Moore, W.J., Neves, C., Patterson, A., et al.: OMERO: flexible, model-driven data management for experimental biology. Nat. Methods **9**(3), 245–253 (2012)
2. Callahan, S.P., Freire, J., Santos, E., Scheidegger, C.E., Silva, C.T., Vo, H.T.: VisTrails: visualization meets data management. In: Proceedings of the 2006 ACM SIGMOD International Conference on Management of Data, SIGMOD 2006, pp. 745–747. ACM, New York, NY, USA (2006). http://doi.acm.org/10.1145/1142473.1142574
3. Ciccarese, P., Soiland-Reyes, S., Belhajjame, K., Gray, A.J., Goble, C., Clark, T.: PAV ontology: provenance, authoring and versioning. J. Biomed. Seman. **4**(1), 37 (2013). http://dx.doi.org/10.1186/2041-1480-4-37
4. Compton, M., Corsar, D., Taylor, K.: Sensor data provenance: SSNO and PROV-O together at last. Terra Cognita and Semantic Sensor Networks, pp. 67–82 (2014)
5. Curcin, V., Ghanem, M.: Scientific workflow systems - can one size fit all? In: 2008 Cairo International Biomedical Engineering Conference, pp. 1–9 (2008)

6. Demchenko, Y., Zhao, Z., Grosso, P., Wibisono, A., de Laat, C.: Addressing big data challenges for scientific data infrastructure. In: Proceedings of the 4th IEEE International Conference on Cloud Computing Technology and Science, pp. 614–617 (2012)

7. Jupp, S., Malone, J., Burdett, T., Heriche, J.K., Williams, E., Ellenberg, J., Parkinson, H., Rustici, G.: The cellular microscopy phenotype ontology. J. Biomed. Seman. **7**(1), 28 (2016). http://dx.doi.org/10.1186/s13326-016-0074-0

8. Kume, S., Masuya, H., Kataoka, Y., Kobayashi, N.: Development of an ontology for an integrated image analysis platform to enable global sharing of microscopy imaging data. In: Proceedings of the ISWC 2016 Posters & Demonstrations Track co-located with 15th International Semantic Web Conference (ISWC 2016), Kobe, Japan, October 19, 2016 (2016). http://ceur-ws.org/Vol-1690/paper93.pdf

9. Kvilekval, K., Fedorov, D., Obara, B., Singh, A., Manjunath, B.: Bisque: a platform for bioimage analysis and management. Bioinformatics **26**(4), 544–552 (2010)

10. Lebo, T., Sahoo, S., McGuinness, D., Belhajjame, K., Cheney, J., Corsar, D., Garijo, D., Soiland-Reyes, S., Zednik, S., Zhao, J.: PROV-O: The PROV Ontology. W3C Recommendation 30 (2013)

11. Ludäscher, B., Altintas, I., Berkley, C., Higgins, D., Jaeger, E., Jones, M., Lee, E.A., Tao, J., Zhao, Y.: Scientific workflow management and the kepler system. Concurrency Comput. Pract. Experience **18**(10), 1039–1065 (2006). http://dx.doi.org/10.1002/cpe.994

12. McPhillips, T., Song, T., Kolisnik, T., Aulenbach, S., Belhajjame, K., Bocinsky, K., Cao, Y., Chirigati, F., Dey, S., Freire, J., et al.: YesWorkflow: a user-oriented, language-independent tool for recovering workflow information from scripts (2015). arXiv preprint arXiv:1502.02403

13. Moreau, L.: Provenance-based reproducibility in the semantic web. Web Seman. Sci. Serv. Agents World Wide Web **9**(2), 202–221 (2011)

14. Murta, L., Braganholo, V., Chirigati, F., Koop, D., Freire, J.: noWorkflow: capturing and analyzing provenance of scripts. In: Ludäscher, B., Plale, B. (eds.) IPAW 2014. LNCS, vol. 8628, pp. 71–83. Springer, Cham (2015). doi:10.1007/978-3-319-16462-5_6

15. Pérez, I.S., Pérez-Hernández, M.S.: Towards reproducibility in scientific workflows: An infrastructure-based approach. Sci. Program., 243180:1–243180:11 (2015). http://dx.doi.org/10.1155/2015/243180

16. Pimentel, J.F., Dey, S., McPhillips, T., Belhajjame, K., Koop, D., Murta, L., Braganholo, V., Ludäscher, B.: Yin & yang: demonstrating complementary provenance from noWorkflow & YesWorkflow. In: Mattoso, M., Glavic, B. (eds.) IPAW 2016. LNCS, vol. 9672, pp. 161–165. Springer, Cham (2016). doi:10.1007/978-3-319-40593-3_13

17. Samuel, S., Taubert, F., Walther, D., König-Ries, B., Bücker, H.M.: Towards reproducibility of microscopy experiments. D-Lib Mag. **23**(1/2) (2017)

18. Taylor, B.N., Kuyatt, C.E.: Guidelines for Evaluating and Expressing the Uncertainty of NIST Measurement Results. Tech. rep., NIST Technical Note 1297 (1994)

Towards an Open Extensible Framework for Empirical Benchmarking of Data Management Solutions: LITMUS

Harsh Thakkar[✉]

Enterprise Information Systems Lab, University of Bonn, Bonn, Germany
hthakkar@uni-bonn.de

Abstract. Developments in the context of Open, Big, and Linked Data have led to an enormous growth of structured data on the Web. To keep up with the pace of efficient consumption and management of the data at this rate, many Data Management Solutions There exists many efforts for benchmarking these domain specific DMSs, however, *(i)* reproducing these third party benchmarks is an extremely tedious task, and *(ii)* there is a lack of a common framework which enables and advocates the extensibility and re-usability of the benchmarks. We propose LITMUS, one such framework for benchmarking data management solutions. LITMUS will go beyond classical storage benchmarking frameworks by allowing for analysing the performance of DMSs across query languages. In this early stage doctoral work, we present the LITMUS concept as well as the considerations that led to its preliminary architecture, and progress reported so far in its realisation.

1 Introduction

Vast amounts of structured (following Linked Data principles) and un/semi-structured data is constantly being made available on the Web, often in an open manner[1], and within organisations. This rapid growth of data, available across organisations, has affected the data management layer of modern applications.

Consequently, organisations are increasingly facing the need to find data management tools suited for the specific tasks at the core of their information management. Choosing the best[2] data management solution is nonetheless challenging due to the limited comparability and compatibility of existing evaluation results and benchmarks. With regard to the limited domain expertise of the end user, the need for standardised frameworks to benchmark and analyse the existing diverse data management platforms is consequently of paramount importance.

Despite the growing interest and use in both research and the industry communities, currently the creators of benchmarks for Data Management Solutions (DMS) [1,4] do not offer a common suite/platform for performing cross-domain

---

[1] With *open* we follow the Open Data Definition (http://opendefinition.org/).
[2] We refer to best in terms of fitness for use.

© Springer International Publishing AG 2017
E. Blomqvist et al. (Eds.): ESWC 2017, Part II, LNCS 10250, pp. 256–266, 2017.
DOI: 10.1007/978-3-319-58451-5_20

benchmarks (i.e., one-to-one performance comparison of RDF, Graph, or Relational engines). In addition, there is no significant baseline to compare these cross-domain DMSs against each other. Moreover, reproducing benchmarks is a non-trivial problem owing to reasons such as non-standardised setup configurations, lack of publicly available resources (such as scripts, libraries or packages) and lack of transparent evaluation policies. Results in areas such as named entity recognition and linking [25] as well as question answering [23, 24] have, however, shown that the provision of standardised interfaces and measures can contribute to the improvement of the performance of software solutions.

In this early stage doctoral work we propose LITMUS, a generic approach for benchmarking DMSs. LITMUS aims to provide support to organisations aspiring to use Linked Data management technologies in a wide spectrum of applications and magnitudes. LITMUS will provide a realistic performance evaluation platform covering a plethora of heterogeneous technologies (see Sect. 4) for storage and query benchmarking. To put the reader into the context of this work, and to highlight the objectives of LITMUS, we present the following **user scenario**:

"The WDAqua research project[3] aims at building a data-driven question answering platform by using Web data, available in various formats, e.g., RDF, CSV, SQL, or Graph. Harsh, a researcher within the project, is responsible for ensuring efficient data management (storage and retrieval) for this project. There are a large number of DMSs, each deliberately tailored to handling specific formats of data and queries, which need to be benchmarked to select the best solution for the project's needs. However, benchmarking of DMSs is non-trivial: it takes large amounts of human effort in designing, administering, evaluating, and analysing the diverse systems involved. Additionally, for the research project, a large set of factors, e.g., query typology, indexing speed, index size, query response time, and dataset size, need to be considered to ensure reproducibility and generality of the observed experimental results. *Harsh wants to automate the whole benchmarking process, allowing easy integration, evaluation on custom stress loads, and fast analysis of the results.* He would also *expect the framework to be flexible to integrate new DMSs to the plethora of existing systems and benchmark them against a baseline*".

LITMUS will not only satisfy the requirement for automating the tedious benchmarking process, but will also offer: *(1)* an efficient way for replicating existing benchmarks (e.g., BSBM [4] or WAT-DIV [1]); *(2)* a wide set of performance evaluation metrics/indicators tailored specifically for the DMS being evaluated; and *(3)* quick analytical insights on performance comparison of benchmarked DMSs wrt various intrinsic factors (such as query length, query structure, etc.) employing visualisation via custom charts, graphs and tabular data.

The remainder of this article is organised as follows: Sect. 2 summarises the state of the art in benchmarking efforts, and their shortcomings, Sect. 3 sheds light on the foci, challenges, objectives and planned outcomes of LITMUS, Sect. 4 describes the conceptual architecture of LITMUS, its target audience, and Sect. 5 concludes with the work progress and future agenda.

---

[3] WDAqua ITN – (http://wdaqua.eu).

2 State of the Art

Benchmarking is widely used for evaluating data stores (DMSs). Benchmarks exist for a variety of levels of abstraction from simple data models to graphs and triple stores, to entire enterprise information systems. We describe the current state of the art in benchmarking, in particular for: (a) Relational databases, (b) Graph databases, (c) RDF stores, and (d) cross-domain benchmarking efforts. We identify the scope and shortcomings of existing benchmarking efforts, to determine the gaps that LITMUS needs to take into consideration.

1. In _Relational_ DMSs, the benchmarks of the Transaction Processing Performance Council (**TPC**) [14] are well established. TPC uses discrete metrics for measuring the performance of the relational DMS. The online transaction processing benchmarks **TPC-C** and **TPC-E** use a transactions per minute metric. The analytics **TPC-H** and decision support **TPC-DS** benchmarks use the queries per hour and cost per performance metrics, respectively.

2. For _Graph_ DMSs, there exist benchmarks, some of which are in their early stages (such as **HPC** Scalable Graph Analysis Benchmark [6], **Graph 500** [13], **XGDBench** [5]) that deal with graph suitability transformations and graph analysis. However, they do not succeed to define standards for graph modelling and query languages.

3. Benchmarking _RDF_ DMSs. The substantial increase in the number of applications that use _RDF_ data has encouraged the need for large-scale benchmarking efforts on all aspects of the Linked Data life cycle, mostly focusing on query processing [15]. RDF DMS benchmarks use real (i.e., DBpedia or Wikidata) and synthetic (i.e., Berlin SPARQL Benchmark or WAT-DIV) datasets to evaluate DMS performance over custom stress-loads and setup environments.[4] DBpedia SPARQL Benchmark (**DBPSB**) [12] assesses RDF DMSs performance over DBpedia by creating a query workload derived from the DBpedia query logs. The aim of the Lehigh University Benchmark (**LUBM** [8]) is to evaluate the performance of Semantic Web triple stores over a large synthetic dataset that complies to a university domain ontology. The Waterloo SPARQL Diversity TEST Suite (**WatDiv** [1]) provides data and query generators to enable benchmarking of RDF DMSs against a varying query structure (also complexity) to understand correlation of query typology with the variance in DMS performance. **SP2Bench** [21], one of the most commonly used synthetic data based benchmarks, uses the schema of the DBLP bibliographic dataset[5] to generate arbitrarily large datasets.

4. Benchmarking _Cross-domain_ DMSs. There are only a few efforts that benchmark cross-domain DMS so far. The Berlin SPARQL Benchmark (**BSBM** [4]) is a synthetic data benchmark, based on an e-commerce use cases built around a set of products offered by different vendors. It provides the dataset and queries for both RDF and Relational DMS benchmarking. **Pandora**[6],

---

[4] https://www.w3.org/wiki/RdfStoreBenchmarking.
[5] http://dblp.uni-trier.de/db/.
[6] http://pandora.ldc.usb.ve/.

uses the Berlin SPARQL Benchmark data to benchmark RDF stores against relational stores (Jena-TDB, MonetDB, GH-RDF-3X, PostgreSQL, 4Store). **Graphium** [7] is a similar study benchmarking RDF stores against Graph stores (Neo4J, Sparksee/DEX, HypergraphDB, RDF-3X) on graph datasets including a 10M triple graph data generated using the Berlin SPARQL Benchmark data generator. More recently, the **LDBC** [2] focused on combining industry-strength benchmarks for graph and RDF data management systems. The LDBC introduces a new choke-point analysis methodology for developing benchmark workloads, which tries to combine user input with feedback from system experts.

Efforts have so far been focused on benchmarking single-domain (RDF-vs-RDF stores, Graph-vs-Graph stores, etc.) DMSs, despite the need for integrating cross-domain DMSs and automating the benchmarking process. LITMUS aims at addressing these shortcomings and serve as an open, extensible platform allowing easy integration, benchmarking and performance comparison of diverse DMSs. To the best of our knowledge, no such extensible and reusable framework exists, which enables the exploration and analysis of a wide spectrum of DMSs.

3 Problem Statement and Contributions

The following generic research question acts as a guiding force to our efforts: *How can diverse cross-domain DMSs be benchmarked in an established standard environment[7]?* We hypothesise that: Devising a **generic** data and query translation mechanism together with a defined set of key performance indicators (KPIs) will enable the comparison of diverse cross-domain DMSs

3.1 Challenges to Be Addressed

The aim of the doctoral work is to validate the proposed hypothesis by developing such a benchmarking platform. In doing so we identify three key challenges (sub-research questions) which need to be addressed, namely:

- $\boxed{\text{C1}}$ **Data conversion**: This challenge mandates the development of a generic data conversion mechanism for converting the RDF data to a format interpretable by the corresponding DMSs (i.e., RDF, pure graphs, or SQL). The goal of this task is to efficiently represent RDF data in multiple formats, keeping the end user as secluded as possible from the underlying technicalities of the conversion. This leads us to our first research question: **RQ1: What are the methods to convert RDF into proprietary data formats?**
- $\boxed{\text{C2}}$ **Query translation**: Cross-domain benchmarking of DMSs demand that queries be represented in all languages and formats supported by the respective tools. Query languages differ in their structure and expressivity.

---

[7] By established standard environment we mean that all benchmarks will run under the same conditions and are not affected by external factors (e.g. different memory allocation by the OS).

For instance, complex path queries (in SPARQL, in particular Kleene stars) cannot be expressed in an equivalent SQL query [26]. Thus, there is a need to develop an intermediate mechanism to translate the queries from one form to the other (e.g., from SPARQL to Gremlin, SQL, etc.). This requires an exhaustive study of the query languages' specifications. The main challenge is to identify the correct mappings between different languages, preserving the semantics of the original query. Thus our second research question is: **RQ2: What are the semantic preserving methods/approaches for translating SPARQL queries to a graph query language[8] such as Gremlin?**.

- C3 **Performance indicators:** The performance of a DMS can be assessed with respect to a wide variety of indicators (referred to as performance metrics or key performance indicators (KPIs)). Dealing with the diverse characteristics of the DMSs, it is necessary to explore a range of performance indicators in contrast to traditional ones, namely precision, recall, index size, storage size, number of triples, number of unique instances, query response time, etc. The work by LDBC [2] presents a related study on this topic. We would like to dig deeper into this and other works, compare and analyse the strengths and limitations of the KPIs, ultimately select a set of KPIs to be considered for evaluation of these DMSs. Thus, **RQ3: What are the strengths and the limitations of the existing KPIs, and to what extent do they reflect the performance of a DMS.**

3.2 Focus of the LITMUS Framework

The focus of LITMUS is to bridge the gaps in adopting, deploying and scaling the consumption of Linked Data. LITMUS thrives on simplifying the use, assessment and the performance analysis of a wide spectrum of cross-domain DMSs. In particular, the LITMUS framework will:

- F1 enable a common platform for benchmarking and comparing a plethora of cross-domain DMSs, and reproducing existing third-party benchmarks;
- F2 create *(i)* interoperable machine-readable evaluation reports and *(ii)* scientific studies on the correlation of a variety of factors (such as query typology, data structures used for indexing, etc.) with respect to the performance of DMSs;
- F3 recommend particular DMSs and benchmarks based on a set of requirements predefined by the user.

3.3 Planned Outcomes

The planned artifacts resulting from the LITMUS project can be classified into two categories, namely **(A1)** scientific findings and **(A2)** software.

---

[8] We emphasise on graph query language in this question as there exists sufficient work addressing SPARQL-SQL (relational query language) translation problem.

A1 Scientific findings:

- An in-depth analysis of the (research challenges, ref. Sect. 3.1) (i) various RDF data representation formats and their conversion complexity, addressing challenge **(C1)**; (ii) query language expressivity and supported features striving to address the language barrier **(C2)**. These studies will provide us with deep insights about the functionality of various query languages, RDF data formats, their strengths and limitations.
- An exhaustive exploratory study on the selection of performance measures for evaluating cross-domain DMSs, addressing challenge **(C3)**

A2 Software *(i.e., algorithms, scripts, tools):*

- A novel data converter of RDF data to multiple data formats (such as CSV, JSON, SQL, etc.), providing compatible data as input to the cross-domain DMSs (i.e., the software implementation of outcome **A1.**(i), Sect. 3.3).
- A novel query translator for the automatic conversion of SPARQL to DMS-specific query language (e.g., Gremlinator, ref Sect. 4, etc.), enabling compatible query input for cross-domain DMSs (i.e., the software implementation of outcome **A1.**(ii), Sect. 3.3)
- An open, extensible benchmarking platform, for cross-domain DMS performance evaluation and easy replication of existing benchmarks.

4 Research Approach and Initial Results

Here, we present the conceptual architecture of LITMUS. It comprises of four major facets: Data Facet (F1), Query Facet (F2), System Facet (F3), and Benchmarking Core (F4) (ref. Fig. 1). The role of each facet is as follows:

Data Facet F1: The Data Facet consists of the (i) Dataset(s) and the (ii) Data Integration Module. Datasets chosen for benchmarking can be real datasets such as DBpedia[9], Wikidata[10], synthetic datasets such as the Berlin SPARQL Benchmarking (BSBM) [4], Waterloo SPARQL Diversity Test Suite (WatDiv) [1], or hybrid datasets comprising both real and synthetic data. The *Data Integration Module* is responsible for (a) making data available to the system in the requested formats (such as N-Triples, Graphs, CSV, SQL) by carrying out appropriate data conversion and mapping tasks (cf. Challenge **C1**), and (b) loading the desired format of data to the respective DMSs selected for the benchmark.

Query Facet F2: The Query Facet comprises of the (i) Queryset(s), and the (ii) Query Conversion Module. The *Queryset* refers to the set of query input files. The *Query Conversion Module* will be one of the key components addressing the language barrier (Challenge **C2**). It is responsible for converting the input SPARQL queries to the respective DMSs' query languages (such as Gremlin, SQL, etc.). The conversion will be performed by developing an intermediate

---

[9] http://wiki.dbpedia.org/.
[10] https://www.wikidata.org/.

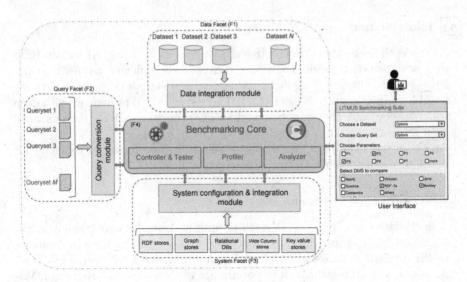

Fig. 1. The architectural overview of the LITMUS benchmarking framework [22].

language/logic representation of the input query. The aim of this module is to allow efficient conversion of a wide variety of SPARQL queries (such as path, star-shaped, and snowflake queries) to other query languages, ultimately breaking the language barrier.

System Facet F3: The System Facet consists of (i) DMSs and (ii) DMS Configuration and Integration module. The *DMS Configuration and Integration* module is responsible for (i) providing easy integration, via wrapper(s) or as a plug-in, of the DMSs, and (ii) monitoring and configuring the integrated DMSs for the benchmark. On top of this, this module makes use of *Docker containers*[11] to ensure a fair allocation of resources and to provide the necessary segregation required for conducting realistic benchmarks.

Benchmarking Core F4: The Benchmarking Core is the heart of the LITMUS framework, consisting of three modules: (i) Controller and Tester, (ii) Profiler, and (iii) Analyser. The *Controller and Tester* is responsible for executing the respective scripts for loading data, fetching the queries to their corresponding DMSs, validating the specified system configurations, and finally, executing the benchmark on the selected setting. The *Profiler* is responsible for: (a) generating and loading various profiles (stress loads, query variations, etc.) for conducting the benchmark tests and (b) storing the custom benchmark results. The *Analyser* is responsible for collecting the benchmark results from the *Profiler* and generates performance reports. It will perform correlation analysis between the parameters specified by the user. The final results (reports) will then presented to the end user in a suitable visualisation.

---

[11] https://www.docker.com/.

Initial results. We currently focus on curating the necessary benchmarking infrastructure for RDF and Graph DMSs. Thereafter, having achieved this milestone, we will cultivate the support for Relational DMSs. The preliminary results, can be clubbed together according to the planned outcomes (discussed in Sect. 3.3), addressing the research challenges and technological developments (ref. Sect. 3.1) of the framework, as follows:

1. **Research challenges**

    (i) Query translation: We are currently focused on addressing the query translation challenge [**RQ2**] (**C2**, Sect. 3.1) developing a novel SPARQL (defacto RDF query language) to Gremlin (graph traversal (query language)), **"Gremlinator"**. We choose Gremlin over other graph query languages (such as Cypher), owing to Gremlin's wide-spread popularity, coverage of graph DMSs and its strong support for both OLTP-based as well as OLAP-based graph processors. We are studying the underlying semantics and complexity of both the query languages for proposing a novel transformation function, mapping SPARQL algebra [3,17] to Gremlin traversals [18–20] ensuring soundness and completeness. Consequently devising a query engine for SPARQL queries to be able to exploit the benefits of existing graph database engines, e.g., neighbourhood indexes, transaction management, and built-in graph-based tasks.

    (ii) Data conversion: Our next milestone is to address the data conversion challenge [**RQ1**] (**C1** ref. Section 3.1). We start by first converting RDF to Graphs. Here, our goal is to propose a novel mechanism for generating Graphs from RDF data, theoretically transforming any RDF dataset to a pure Graph format. The related work in this topic includes efforts such as [9–11,16] who advocate the generation of property graphs using reification. We would like to study these and other works in detail and develop a generic RDF data converter as our ultimate goal.

2. **Implementation**

    This framework will be made available as open source software for encouraging research, open discussions and possible extensions to the idea. The source code, scripts, and other relevant modules are open-sourced at the Github organisation[12]. We are working on the query facet (**F2**) developing the query conversion module along with the continuous (incremental) development of the benchmarking core (**F4**) (ref. Fig. 1). We have developed bash-scripts and DMS dockers for the easy integration of DMSs, as a part of the System facet (**F3**). The overall development progress of the overall framework is around 25%.

5   Evaluation Plan and Conclusion

This doctoral work is dimensioned for three years, out of which the first year is dedicated for intense literature review. We identified the challenges and shortcomings of existing works summarised in Sect. 2 through 3. The literature review

---

[12] LITMUS Benchmark Suite: (https://github.com/LITMUS-Benchmark-Suite/).

confirms the absence of a cross-domain benchmarking platform. We first start by addressing the research challenges identified in Sect. 3.1, proposing the solutions (formally), implementing the solution (i.e., the components described in the architecture) and thereafter repeating this methodology for the planned architecture. We plan to devote a time period of six months for addressing each research challenge (i.e., **C1**, **C2** and **C3**) and the last six months for the integration, evaluation and testing of the overall framework. More than providing visualisation of DMS performance comparison and analysis scripts, what LITMUS will provide is a common open and extensible ground for independent evaluation and comparison of a given approach with respect of the state-of-the-art. This promotes and enhances not only reproducibility of the benchmarking results but also generality and experimental transparency.

Evaluation. We plan to evaluate our hypothesis by validating each research challenge/question defined in Sect. 3.1. The evaluation of the challenges **C1** and **C2** will be done by formally proving that the conversion/translation process is sound, complete and preserves the semantics (of the data and query). Furthermore, we will also evaluate the time complexity of the implemented converter and translator ensuring that a scalable solutions is possible for both **C1** and **C2**. We will evaluate challenge **C3**, by the means of empirical study. In this we will analyse and compare various KPIs using a wide variety of DMSs and datasets. Finally, for the whole platform, we plan to do an evaluation taking in consideration all the three components and define user scenarios (similar the one described in Sect. 1). These scenarios will be validated keeping in mind the existing benchmarks, thus proving its validity and strengths.

Acknowledgements. Supervised by Prof. Dr. Sören Auer and Prof. Dr. Maria-Esther Vidal. I would like to express gratitude to Prof. Dr. Jens Lehman, Dr. Christoph Lange, and Dr. Andreas Both for their quality insights on LITMUS. This work is supported by the H2020 WDAqua ITN (GA: 642795).

References

1. Aluç, G., Hartig, O., Özsu, M.T., Daudjee, K.: Diversified stress testing of RDF data management systems. In: Mika, P., Tudorache, T., Bernstein, A., Welty, C., Knoblock, C., Vrandečić, D., Groth, P., Noy, N., Janowicz, K., Goble, C. (eds.) ISWC 2014. LNCS, vol. 8796, pp. 197–212. Springer, Cham (2014). doi:10.1007/978-3-319-11964-9_13

2. Angles, R., Boncz, P.A., Larriba-Pey, J., et al.: The linked data benchmark council: A graph and RDF industry benchmarking effort. SIGMOD Rec. **43**(1), 27–31 (2014)

3. Angles, R., Gutierrez, C.: The expressive power of SPARQL. In: Sheth, A., Staab, S., Dean, M., Paolucci, M., Maynard, D., Finin, T., Thirunarayan, K. (eds.) ISWC 2008. LNCS, vol. 5318, pp. 114–129. Springer, Heidelberg (2008). doi:10.1007/978-3-540-88564-1_8

4. Bizer, C., Schultz, A.: The berlin SPARQL benchmark. Int. J. Semant. Web Inf. Syst. **5**(2), 1–24 (2009)

5. Dayarathna, M., Suzumura, T.: XGDBench: A benchmarking platform for graph stores in exascale clouds. In: CloudCom. IEEE Computer Society (2012)
6. Dominguez-Sal, D., Urbón-Bayes, P., Giménez-Vañó, A., Gómez-Villamor, S., Martínez-Bazán, N., Larriba-Pey, J.L.: Survey of graph database performance on the HPC scalable graph analysis benchmark. In: Shen, H.T., Pei, J., Özsu, M.T., Zou, L., Lu, J., Ling, T.-W., Yu, G., Zhuang, Y., Shao, J. (eds.) WAIM 2010. LNCS, vol. 6185, pp. 37–48. Springer, Heidelberg (2010). doi:10.1007/978-3-642-16720-1_4
7. Flores, A., Palma, G., Vidal, M.-E., et al.: GRAPHIUM: Visualizing performance of graph and RDF engines on linked data. In: Proceedings of the 2013th International Conference on Posters & Demonstrations Track-Volume, vol. 1035 (2013). CEUR-WS.org
8. Guo, Y., Pan, Z., Heflin, J.: LUBM: A benchmark for OWL knowledge base systems. Web Semant. 3(2–3), 158–182 (2005)
9. Hartig, O.: Reconciliation of RDF* and property graphs. CoRR, abs/1409.3288 (2014)
10. Hernández, D., Hogan, A., Krötzsch, M.: Reifying RDF: What works well with wikidata? In: Proceedings of the 11th International Workshop on Scalable Semantic Web Knowledge Base Systems co-located (ISWC 2015), Bethlehem, PA, USA (2015)
11. Hernández, D., Hogan, A., Riveros, C., Rojas, C., Zerega, E.: Querying wikidata: Comparing SPARQL, relational and graph databases. In: Groth, P., Simperl, E., Gray, A., Sabou, M., Krötzsch, M., Lecue, F., Flöck, F., Gil, Y. (eds.) ISWC 2016. LNCS, vol. 9982, pp. 88–103. Springer, Cham (2016). doi:10.1007/978-3-319-46547-0_10
12. Morsey, M., Lehmann, J., Auer, S., Ngonga Ngomo, A.-C.: DBpedia SPARQL benchmark – Performance assessment with real queries on real data. In: Aroyo, L., Welty, C., Alani, H., Taylor, J., Bernstein, A., Kagal, L., Noy, N., Blomqvist, E. (eds.) ISWC 2011. LNCS, vol. 7031, pp. 454–469. Springer, Heidelberg (2011). doi:10.1007/978-3-642-25073-6_29
13. Murphy, R.C., Wheeler, K.B., Barrett, B.W., Ang, J.A.: Introducing the GRAPH 500. Cray User's Group (CUG) (2010)
14. Nambiar, R., Wakou, N., Carman, F., Majdalany, M.: Transaction processing performance council (TPC): State of the council 2010. In: Nambiar, R., Poess, M. (eds.) TPCTC 2010. LNCS, vol. 6417, pp. 1–9. Springer, Heidelberg (2011). doi:10.1007/978-3-642-18206-8_1
15. Ngomo, A.-C.N., Röder, M.: HOBBIT: Holistic benchmarking for big linked data. ERCIM News 2016 (2016)
16. Nguyen, V., Leeka, J., Bodenreider, O., et al.: A formal graph model for RDF and its implementation. CoRR, abs/1606.00480 (2016)
17. Pérez, J., Arenas, M., Gutierrez, C.: Semantics and complexity of SPARQL. In: Cruz, I., Decker, S., Allemang, D., Preist, C., Schwabe, D., Mika, P., Uschold, M., Aroyo, L.M. (eds.) ISWC 2006. LNCS, vol. 4273, pp. 30–43. Springer, Heidelberg (2006). doi:10.1007/11926078_3
18. Rodriguez, M.A.: The gremlin graph traversal machine and language (invited talk). In: Proceedings of the 15th Symposium on Database Programming Languages, Pittsburgh, PA, USA, 25–30 October 2015 (2015)
19. Rodriguez, M.A., Neubauer, P.: The graph traversal pattern. In: Graph Data Management: Techniques and Applications (2011)
20. Rodriguez, M.A., Neubauer, P.: A path algebra for multi-relational graphs. In: Proceedings of the 27th International Conference on Data Engineering Workshops, ICDE 2011 (2011)

21. Schmidt, M., Hornung, T., Meier, M., et al.: SP^{2}Bench: A SPARQL performance benchmark. In: de Virgilio, R., Giunchiglia, F., Tanca, L. (eds.) Semantic Web Information Management, pp. 371–393. Springer, Heidelberg (2009)
22. Thakkar, H., Dubey, M., Sejdiu, G., et al.: LITMUS: An open extensible framework for benchmarking RDF data management solutions. CoRR, abs/1608.02800 (2016)
23. Tsatsaronis, G., Balikas, G., Malakasiotis, P., et al.: An overview of the BIOASQ large-scale biomedical semantic indexing and question answering competition. BMC Bioinform. **16**, 138 (2015)
24. Unger, C., Forascu, C., Lopez, V., et al.: Question answering over linked data (QALD-5). In: Working Notes of CLEF 2015, Toulouse, France (2015)
25. Usbeck, R., Röder, M., Ngomo, A.N., et al.: GERBIL: General entity annotator benchmarking framework. In: Proceedings of the 24th International Conference on World Wide Web, WWW 2015 (2015)
26. Zhang, X., Van den Bussche, J.: On the power of SPARQL in expressing navigational queries. Comput. J. **58**(11), 2841–2851 (2015)

Enhancing White-Box Machine Learning Processes by Incorporating Semantic Background Knowledge

Gilles Vandewiele[✉]

Department of Information Technology,
Ghent University - imec, IDLab, Ghent, Belgium
gilles.vandewiele@ugent.be

Abstract. Currently, most of white-box machine learning techniques are purely data-driven and ignore prior background and expert knowledge. A lot of this knowledge has already been captured in domain models, i.e. ontologies, using Semantic Web technologies. The goal of this research proposal is to enhance the predictive performance and required training time of white-box models by incorporating the vast amount of available knowledge in the pre-processing, feature extraction and selection phase of a machine learning process.

Keywords: White-box machine learning · Knowledge incorporation · Semantic knowledge bases

1 Introduction

Most machine learning techniques are data-driven and thus ignore most of the vast amounts of existing available knowledge already captured in domain models [1], such as SNOMED [2], SSN [3] and UMLS [4]. The advantage of applying a purely data-driven approach is that the model is robust to outliers and noise. The disadvantage is that a computationally expensive training phase needs to be executed and that valuable prior knowledge is not taken into account. In many critical domains such as electronic health care and law enforcement, wherein wrong decisions made can have significant repercussions, knowledge-based systems such as expert systems were long preferred [5] as they can easily give a comprehensible corresponding explanation with their predictions. Moreover, they can be deployed without requiring a lot of data, which was rather hard to collect prior to the big data era. The main disadvantage of a purely knowledge-based approach is that the performance is completely biased to the content of the knowledge base [6], which can take a lot of time to construct and maintain, and that it is not able to learn new patterns or insights. Moreover, this approach is often not robust, e.g. in the case of conflicting rules or samples that do not comply to any of the defined rules.

© Springer International Publishing AG 2017
E. Blomqvist et al. (Eds.): ESWC 2017, Part II, LNCS 10250, pp. 267–278, 2017.
DOI: 10.1007/978-3-319-58451-5_21

Within the data-driven approaches, two large families of techniques can be distinguished. First, there are black-box techniques, such as artificial neural networks, which are often able to learn features automatically, thus not requiring a feature extraction and selection phase, and tend to achieve high predictive performances [7]. However, they cannot provide an explanation for their predictions, making them impractical in applications where decision support, instead of decision making, is crucial. Secondly, white-box techniques, such as decision tree induction and classification rule mining, construct an easily comprehensible predictive model from the data. While the predictive performance of these technique tends to be lower than their counterpart, they are able to give a corresponding explanation, therefore being ideally suited to provide decision support for experts within critical domains.

Given the advantages of both data-driven and knowledge-based approaches, advancements within the machine learning domain, the growth of data within all domains [8] and the vast amount of prior knowledge already available on the Semantic Web, a hybrid approach seems to be ideal. In such an approach, a white-box predictive model, such as a decision tree or an ordered rule list, is constructed from the given data with incorporation of prior knowledge in each of its steps. Ideally, the advantages of both approaches would be retained, i.e. robustness to outliers and noise, ability to give a corresponding explanation, a less expensive and more performant training phase and the ability to deduce new insights and knowledge.

The remainder of this paper is as follows. A use case which will be used as a running example throughout the rest of this paper is presented in Sect. 2, followed by a discussion of the related work in Sect. 3. A problem statement with corresponding hypotheses and research questions are presented in Sect. 4. A methodology to provide an answer on these research questions in proposed in Sect. 5. Then, we discuss how our future research will be evaluated in Sect. 6 and finally, a conclusion is given in Sect. 7.

2   Use Case: Primary Headache Diagnosis

Primary headaches [9] are an increasingly common health issue in modern society, having a large prejudicial impact. In Europe, it has a prevalence of more than 50% and according to the World Health Organization (WHO), severe headache attacks are one of the top 10 most disabling conditions [10]. Currently, it costs a lot of time to diagnose a patient correctly because a lot of different aspects need to be taken into account and because many different types of primary headache exist. Furthermore, a lot of research by medical experts has already been done in the headache domain, resulting in a vast amount of available prior domain and expert knowledge [11]. Therefore, the automatic diagnosis of primary headaches seems an ideal use case to combine both the data-driven and knowledge-driven approach which can have a very positive impact. For my master dissertation, a mobile headache journal[1] was developed that allows headache patients to

---

[1] https://play.google.com/store/apps/details?id=be.ugent.chronicals&hl=en.

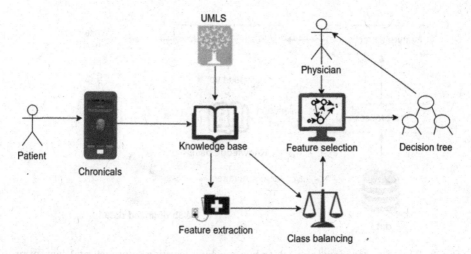

Fig. 1. Schematic overview of the machine learning work-flow, with incorporation of prior knowledge into the different phases, to diagnose a patient with primary headache. A patient enters all required information concerning his medicine consumption and headache attacks in a mobile application. This data is fed together with background knowledge to a machine learning process in order to discover new features, balance the distribution of classes and automatically select features. The feature selection process is based on a graph created by the physician.

register their headache attacks and medicine consumptions. The semantically annotated data generated by this mobile application, in combination with background knowledge [4], can be used to generate a decision tree in order to support an expert in making a correct diagnosis. An overview of this work-flow can be found in Fig. 1. This use case will be used as a running example throughout this paper and, in addition to well-known benchmark datasets, to evaluate the different proposed techniques.

3   Related Work

Combining the advantages of knowledge-driven and data-driven approaches, sometimes referred to as semantic data mining, has been investigated before. Two very thorough and recent surveys can be found in [12,13]. A traditional white-box data-driven approach consists of several main steps, which can be identified in Fig. 2. In a first step, numerical features that have a high discriminative power are extracted from the raw data, which is optionally pre-processed first. Pre-processing examples include applying transformations to the data or generating and removing samples to balance the dataset. When all features are extracted, a selection phase is applied in order to discard the uninformative features, which allows for better generalization. Finally, a white-box model is constructed from the selected features. In the following subsections, related work for each of these phases is presented.

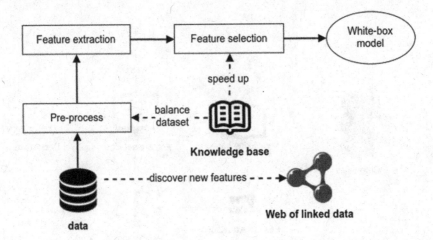

Fig. 2. The different steps of a white-box machine learning approach and how prior knowledge can be incorporated.

3.1 Automatic Feature Discovery

In a typical machine learning work-flow, a very large amount of time is spent on data cleaning and feature extraction. Generic features, which can be applied in a large number of problems, are available, but often, the most efficient features require some prior knowledge about the task to solve. Facilitating this feature extraction process by exploiting the concept of linked data to automatically discover new informative features could therefore significantly reduce the time required to create a predictive model. In order to do this, entities in the training set are mapped to a URI which corresponds to a node in the graph of linked data. From here, we can traverse edges to discover new features [14–16]. While this is a very interesting approach, there are many possible optimizations left, such as automatic measurements of feature importance, heuristics to decide when to stop traversing the immensely large graph and pruning parts of the graph in order to reduce the gigantic search space.

3.2 Class Balancing

In the classification domain, a dataset is called imbalanced when the distribution of the classes in the training set is skewed. An imbalanced dataset is very common in the financial and medical domain, e.g. fraud and epilepsy detection respectively. Class imbalance gives rise to a few potential problems. First, the classifier will be biased towards the largest populated class as this has the highest impact on the objective function it is trying to optimize, while this is often the class of least importance to the expert. Second, general metrics, such as accuracy, to evaluate the model give a wrong representation of the predictive performance [17,18]. Two large approaches to tackle with data imbalance can be identified. On the one hand sampling techniques can remove or create new samples in order to make the distribution of the classes more uniform [19]. On the other hand, the classification algorithm can be modified (e.g. adapting the objective function) to pay more attention to samples in the minor class [20, 21]. Sampling techniques are very interesting, as they can be applied as a pre-processing step of the machine learning work-flow, and can therefore be seen as model-agnostic. Sampling techniques can be divided in either oversampling, where the number of samples in the minor class in increased, or undersampling, where the number of samples in the major class is decreased. In current state-of-the-art oversampling algorithms, such as SMOTE [22] and ADASYN [23], virtual samples of the minority class are generated by using the small amount of data available and thus no prior knowledge is used. On the other hand, researchers have already attempted to generate 'virtual' samples solely based on the prior knowledge available [24–28]. While the latter research attempts were not done in the context of imbalanced dataset but more in the context of data augmentation, a hybrid approach, which combines the positive characteristics of both approaches, can be very interesting.

3.3 Feature Selection

When all of the possible features are extracted from the raw data, a selection phase can optionally be applied in order to remove uninformative features. This can mitigate the curse of dimensionality and thus possibly increases the generalization capability of the model while reducing the amount of training time required. The research field dealing with incorporating prior knowledge into the selection phase is still very young and pre-mature. In [29], the Semantic Sensor Network (SSN) ontology [3] is adapted to allow for automatic feature selection. Here, features are selected based on dependency relations defined by an expert between predictor variables or between a predictor variable and the target variable. This technique has a lower computational complexity than current feature selection techniques, as it is dependent only on the number of features and not on the number of data samples, which can become very large in many cases. Moreover, in contrast to dimensionality reduction techniques such as t-SNE [30] and PCA [31], interpretability of the features is maintained and the selection phase only has to be re-applied when new features are added to the model, instead of when a certain amount of new samples is added. Unfortunately, this technique is still rather simplistic and is equivalent to manual feature selection.

4   Problem Statement

By analysis of the state of the art, one open problem can be identified:

P1. Current white-box machine learning techniques learn from scratch and often only use a limited amount of information (i.e. the training set) as they do not make full use of the vast amount of prior background and expert knowledge available in ontologies and on the web of linked data [32].

To solve this problem, different research questions need to be resolved first:

Q1. Can we improve existing or develop new techniques that map the entities in the dataset to a URI identifiable on the web of linked data in order to traverse the graph of data to extract new relevant, discriminant features for the task to solve?

Q2. Can we develop a hybrid technique that uses both the limited amount of samples in the minority class and the knowledge about the minority class in order to generate new samples to balance the dataset? Moreover, how does this hybrid technique compare to the techniques where only one of the two is used?

Q3. Is it possible to improve the feature selection phase by creating a new algorithm that ranks the different features based on their relations defined by an expert?

Finally, the following hypotheses can be deduced:

H1. The automatic discovery of new features by exploiting the concept of linked data can lead to a reduction in the labor needed for feature extraction while resulting in an increase in the predictive performance of the model.

H2. Balancing the dataset using both knowledge and the limited amount of samples in the minority class will result in a better predictive performance for the minority class than sampling methods that are based only on this limited amount of samples.

H3. Applying feature selection based on a ranked list of features, generated by applying a ranking algorithm on a graph of features defined by an expert, will require less time than current feature selection techniques and result in a better generalization capability. Moreover, it allows for experts to have more control of the algorithm, which can increase their will to adopt such a system.

5   Methodology

5.1   Automatic Feature Discovery

In order to augment the data with information from the web of linked data, a mapping phase must first be applied. Here, the entities in the initial dataset are mapped to a URI identifiable on the web of linked data or on a semantically annotated electronic health record in the medical domain. This mapping has to occur with minimal user interaction. When each of the samples are mapped on a URI, we can try to find new features by doing a breadth-first search in the graph of linked data. The reason for a breadth-first strategy is because of the almost infinite depth of the graph. In order for a new candidate feature to be informative, not too many missing values may occur and there must be correlation with the target variable (or must improve the cluster quality in the unsupervised case). Since counting the number of missing values and calculating correlations between a new candidate predictor and the target variable for a large dataset can take a significant amount of time, a subset of the initial dataset can be used to provide an approximation. Moreover, to decide heuristically which feature-threshold combination results in the most optimal split of data from all possible candidates, the Hoeffding bound [33,34] can be applied. Since the graph we are traversing has an immense size, we need to define conditions when to stop the search, e.g. stop when we traversed k levels deeper in the graph without finding a new usable feature. Finally, pruning of the graph can optionally be applied by calculating semantic concept relatedness [35,36] between a new subject and the target concept. When there is almost no semantic relation between a new concept and the target concept, that part of the graph can already be pruned. Many different metrics exist to calculate this relatedness [37,38]. I will perform a clear evaluation of different metrics in order to find the most suited one for this task.

For the headache use case, a user profile in the mobile journal needs to be mapped to the patient's semantically annotated electronic health record. This can be done by joining on unique identifiable information such as the combination of name and email. As the electronic health record is semantically annotated, it can be seen as a graph, which can be traversed to discover new informative features that help in formulating a correct diagnosis for a primary headache patient. Moreover, datasets ideally suited for evaluation of this technique exist. Examples include the zoo dataset from UCI [39] and the datasets curated by the University of Mannheim [40]. The property of these datasets is that they contain a limited amount of information about rich concepts (such as cities or animals), and therefore rely on automatic feature discovery to obtain reasonable results.

5.2   Class Balancing

In order to balance the classes, oversampling as a pre-processing step will be investigated, enabling a model-agnostic approach. I will create a hybrid approach that combines the positive characteristics of data-based sampling algorithms, such as SMOTE [22] and ADASYN [23] and knowledge-based sampling algorithms, where samples are generated that comply to a pre-defined knowledge base. First, consistency of the knowledge base or the given data needs to be checked by evaluating whether the small amount of samples in the minority class complies to this knowledge. If this is not the case, there is either an anomaly in the data or a inconsistency/fault in the knowledge base that needs to be resolved. When we find that a certain fraction of the samples in the minority class do not comply to one specific rule in the knowledge base, chance are high that the rule is inconsistent with the ground truth and we can remove the rule. Else, the sample is probably an anomaly and can therefore be removed. Alternatively, both the rule and the sample can be removed. An evaluation is required to determine which technique (and threshold on the fractions of samples) is most suited for a dataset with certain properties. After this phase, data can be generated based on the knowledge base and on the small amount of samples in our dataset. For each dimension (i.e. feature) for which knowledge is available, these dimensions of a new virtual sample are set to values that comply to this defined knowledge (e.g. the value must be in a certain range). Of course, it is infeasible to have complete information about each dimension. For these dimensions, the values of samples in our dataset can be used as follows: we find the two nearest neighbors to our new virtual sample in the feature space defined by the features of which knowledge is available; then we can generate a random point on the link between these two neighbors.

One of the most severe primary headache types is cluster headache. It has been discovered quite recently and is a rather rare condition, with a prevalence of 1 out of 1000 [41] as opposed to 1 out of 7 for migraine [42], making it very hard to diagnose. This, in combination with the fact that a lot of domain knowledge is available [11], makes it an ideal use case to evaluate the new technique on.

5.3 Feature Selection

I will design a method that allows to represent the knowledge base as a graph, where each feature defined in the knowledge base or dataset corresponds to a node, and each relation between two features (such as dependsOn or independentOf) corresponds to an edge between their two corresponding nodes. I will then rehone a ranking algorithm, similar to e.g. Google PageRank [43], to calculate a weight for each of the nodes (or features) in the graph [44–46]. Finally, we can sort the features on their rank and return the top k features [47]. An example is given in Fig. 3.

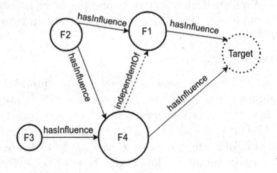

Fig. 3. Feature selection by applying a technique similar to PageRank to the knowledge graph of features.

For the headache use case, the newly discovered features (see Subsect. 5.1) and their corresponding descriptions, in combination with the features obtained from the semantically annotated information produced by the mobile headache journal, can be visualized for a neurologist in a GUI. The neurologist can then define relations between these features, analogue to Fig. 3. Finally, the ranking algorithm can be applied to create a list of features, ordered by their importance. This technique can easily be compared to other feature ranking techniques by taking the k top ranked features of both approaches and measuring the predictive performance of the model, trained on these features.

6   Evaluation

To evaluate the impact of prior knowledge incorporation in each of the phases, a comparison will be done between the process with and without incorporation regarding the following criteria (sorted by decreasing priority):

- predictive performance of the model: by calculating the accuracy, balanced accuracy, precision, recall, AUC, F-measure, etc.
- predictive model complexity: by visual inspection and counting the maximal depth, number of nodes or leaves in the resulting decision tree
- computational time: by timing the execution of each of the phases in the machine learning process

The evaluation will be done for both incorporation in each phase separately and incorporation in all (possible subsets) of the phases. To take the no-free-lunch theorem [48] into account, the evaluation will be done on multiple benchmark datasets with varying characteristics.

7   Conclusion

In this research proposal, related work and methodologies are presented to incorporate prior background and expert knowledge, represented using Semantic Web technologies, into the first phases of a white-box machine learning approach: data balancing and feature extraction & selection. We are convinced that the incorporation of prior knowledge into these phases will allow for higher predictive performances and reduced training times. An evaluation regarding computational time, model complexity and predictive performance will be done by comparing the process with and without incorporation on multiple benchmark datasets and a real-world use cases.

Acknowledgements. I would like to thank my promoters prof. Filip De Turck & dr. Femke Ongenae from Ghent University and my mentor, prof. Agnieszka Ławrynowicz from Poznan University, for their support and valuable input in the realization of this work. This research is funded by a PhD SB fellow scholarship of FWO (1S31417N).

References

1. Jan, T., Debenham, J.: Incorporating prior domain knowledge into inductive machine learning. J. Mach. Learn., 1–42 (2007)
2. Schulz, S., et al.: Snomed reaching its adolescence: ontologists and logicians health check. Int. J. Med. Inform. **78**, S86–S94 (2009)
3. Compton, M., et al.: The SSN ontology of the W3C semantic sensor network incubator group. Web Seman. Sci. Serv. Agents WWW **17**, 25–32 (2012)
4. Bodenreider, O.: The unified medical language system (UMLS): integrating biomedical terminology. Nucleic Acids Res. **32**(Suppl 1), D267–D270 (2004)
5. Kattan, M.W.: Expert systems in medicine. Elsevier Ltd. (2001)

6. Tresp, V., Bundschus, M., Rettinger, A., Huang, Y.: Towards machine learning on the semantic web. In: Costa, P.C.G., d'Amato, C., Fanizzi, N., Laskey, K.B., Laskey, K.J., Lukasiewicz, T., Nickles, M., Pool, M. (eds.) URSW 2005-2007. LNCS (LNAI), vol. 5327, pp. 282–314. Springer, Heidelberg (2008). doi:10.1007/978-3-540-89765-1_17

7. Lim, T.S., et al.: Comparison of prediction accuracy, complexity, and training time of thirty-three classification algorithms. Mach. Learn. **40**, 203–228 (2000)

8. Wu, X., Zhu, X., Wu, G.-Q., Ding, W.: Data mining with big data. IEEE Trans. Knowl. Data Eng. **26**(1), 97–107 (2014)

9. Caemaert, J., Baert, E.J.A.: Neurologie. Springer (2003)

10. Stovner, L.J., Zwart, J.-A., Hagen, K., Terwindt, G.M., Pascual, J.: Epidemiology of headache in Europe. Eur. J. Neurol. **13**(4), 333–345 (2006)

11. Levin, M.: The international classification of headache disorders. Headache J. Head Face Pain **53**(8), 1383–1395 (2013)

12. Dou, D., Wang, H., Liu, H.: Semantic data mining: a survey of ontology-based approaches. In: 2015 IEEE 9th International Conference on Semantic Computing (ICSC), pp. 244–251 (2015)

13. Ristoski, P., Paulheim, H.: Semantic web in data mining and knowledge discovery: a comprehensive survey. Web Seman. Sci. Serv. Agents World Wide Web **36**, 1–22 (2016)

14. Nickel, M., et al.: A review of relational machine learning for knowledge graphs from multi-relational link prediction to automated knowledge graph construction. Proc. IEEE, 1–18 (2015)

15. Paulheim, H., Ristoski, P., Mitichkin, E., Bizer, C.: Data mining with background knowledge from the web. In: RapidMiner World (2014)

16. Ristoski, P.: Towards linked open data enabled data mining. In: Gandon, F., Sabou, M., Sack, H., d'Amato, C., Cudré-Mauroux, P., Zimmermann, A. (eds.) ESWC 2015. LNCS, vol. 9088, pp. 772–782. Springer, Cham (2015). doi:10.1007/978-3-319-18818-8_50

17. Longadge, R., Dongre, S.: Class imbalance problem in data mining review. arXiv preprint arXiv:1305.1707 (2013)

18. He, H., Garcia, E.A.: Learning from imbalanced data. IEEE Trans. Knowl. Data Eng. **21**(9), 1263–1284 (2009)

19. Chawla, N.V.: Data mining for imbalanced datasets: an overview. In: Maimon, O., Rokach, L. (eds.) Data Mining and Knowledge Discovery Handbook, pp. 853–867. Springer, New York (2005)

20. Ganganwar, V.: An overview of classification algorithms for imbalanced datasets. IJETAE **2**(4), 42–47 (2012)

21. Tang, Y., Zhang, Y.-Q., Chawla, N.V., Krasser, S.: Svms modeling for highly imbalanced classification. IEEE Trans. Syst. Man Cybern. Part B (Cybern.) **39**(1), 281–288 (2009)

22. Chawla, N.V., Bowyer, K.W., Hall, L.O., Philip Kegelmeyer, W.: Smote: synthetic minority over-sampling technique. J. Artif. Intell. Res. **16**, 321–357 (2002)

23. He, H., et al.: Adasyn: adaptive synthetic sampling approach for imbalanced learning. In: IJCNN, pp. 1322–1328. IEEE (2008)

24. Niyogi, P., Girosi, F., Poggio, T.: Incorporating prior information in machine learning by creating virtual examples. Proc. IEEE **86**(11), 2196–2209 (1998)

25. Iqbal, R.A.: A generalized method for integrating rule-based knowledge into inductive methods through virtual sample creation. arXiv:1101.4924 (2011)

26. Yang, J., et al.: A novel virtual sample generation method based on Gaussian distribution. Know.-Based Syst. **24**(6), 740–748 (2011)

27. Lin, L.-S., et al.: Improving virtual sample generation for small sample learning with dependent attributes. In: 2016 5th IIAI International Congress on Advanced Applied Informatics (IIAI-AAI), pp. 715–718 (2016)

28. Li, D.-C., Wen, I.-H.: A genetic algorithm-based virtual sample generation technique to improve small data set learning. Neurocomputing **143**, 222–230 (2014)

29. Ringsquandl, M., Lamparter, S., Brandt, S., Hubauer, T., Lepratti, R.: Semantic-guided feature selection for industrial automation systems. In: Arenas, M., et al. (eds.) ISWC 2015. LNCS, vol. 9367, pp. 225–240. Springer, Cham (2015). doi:10.1007/978-3-319-25010-6_13

30. van der Maaten, L., Hinton, G.: Visualizing data using t-SNE. J. Mach. Learn. Res. **9**, 2579–2605 (2008)

31. Wold, S., Esbensen, K., Geladi, P.: Principal component analysis. Chemometr. Intell. Lab. Syst. **2**(1–3), 37–52 (1987)

32. Gülçehre, Ç., Bengio, Y.: Knowledge matters: importance of prior information for optimization. J. Mach. Learn. Res. **17**(8), 1–32 (2016)

33. Domingos, P., Hulten, G.: Mining high-speed data streams. In: Proceedings of the Sixth ACM SIGKDD International Conference on Knowledge Discovery and Data Mining, pp. 71–80. ACM (2000)

34. Terziev, Y.: Feature generation using ontologies during induction of decision trees on linked data. In: ISWC PhD Symposium (2016)

35. Gabrilovich, E., Markovitch, S.: Computing semantic relatedness using wikipedia-based explicit semantic analysis. In: IJCAI, vol. 7, pp. 1606–1611 (2007)

36. Bonte, P., Ongenae, F., De Turck, F.: Learning semantic rules for intelligent transport scheduling in hospitals. In: CEUR Workshop Proceedings, vol. 1586, pp. 1–6 (2016)

37. Hassan, S., Mihalcea, R.: Semantic relatedness using salient semantic analysis. In: AAAI (2011)

38. Gurevych, I.: Using the structure of a conceptual network in computing semantic relatedness. In: Dale, R., Wong, K.-F., Su, J., Kwong, O.Y. (eds.) IJCNLP 2005. LNCS (LNAI), vol. 3651, pp. 767–778. Springer, Heidelberg (2005). doi:10.1007/11562214_67

39. Lichman, M.: UCI machine learning repository (2013)

40. Ristoski, P., de Vries, G.K.D., Paulheim, H.: A collection of benchmark datasets for systematic evaluations of machine learning on the semantic web. In: Groth, P., Simperl, E., Gray, A., Sabou, M., Krötzsch, M., Lecue, F., Flöck, F., Gil, Y. (eds.) ISWC 2016. LNCS, vol. 9982, pp. 186–194. Springer, Cham (2016). doi:10.1007/978-3-319-46547-0_20

41. Fischera, M., et al.: The incidence and prevalence of cluster headache: a meta-analysis of population-based studies. Cephalalgia **28**(6), 614–618 (2008)

42. Burch, R.C., Loder, S., Loder, E., Smitherman, T.A.: The prevalence and burden of migraine and severe headache in the united states: updated statistics from government health surveillance studies. Headache J. Head Face Pain **55**(1), 21–34 (2015)

43. Page, L., Brin, S., Motwani, R., Winograd, T.: The pagerank citation ranking: bringing order to the web. Technical report, Google (1999)

44. Thalhammer, A., Rettinger, A.: PageRank on wikipedia: towards general importance scores for entities. In: Sack, H., Rizzo, G., Steinmetz, N., Mladenić, D., Auer, S., Lange, C. (eds.) ESWC 2016. LNCS, vol. 9989, pp. 227–240. Springer, Cham (2016). doi:10.1007/978-3-319-47602-5_41

45. Wade, A.D., et al.: Wsdm cup 2016: entity ranking challenge. In: Proceedings of the Ninth ACM International Conference on Web Search and Data Mining, pp. 593–594. ACM (2016)

46. Lee, S., et al.: Random walk based entity ranking on graph for multidimensional recommendation. In: Proceedings of the Fifth ACM Conference on Recommender Systems, RecSys 2011, pp. 93–100. ACM, New York (2011)

47. Ienco, D., Meo, R., Botta, M.: Using pagerank in feature selection. In: SEBD, pp. 93–100 (2008)

48. Wolpert, D.H., Macready, W.G.: No free lunch theorems for optimization. IEEE Trans. Evol. Comput. $1(1)$, 67–82 (1997)

Author Index

Printed in the United States
By Bookmasters